Territorial Changes and Territorial Restructurings in the Himalayas

Territorial Changes and Territorial Restructurings in the Himalayas

Edited by Joëlle Smadja

ADROIT

The authors of this book are grateful to Bernadette Sellers (UPR 299, Centre for Himalayan Studies) for proofreading the texts.

Published by
ADROIT PUBLISHERS
4675/21, Ganpati Bhawan, Ansari Road,
Daryaganj, New Delhi-110 002
Phones : 011-23266030, 23263506

Published in association with
Centre for Himalayan Studies, France (CNRS)

Distributors
AKHIL BOOK DISTRIBUTORS
e-mail: akhilbooks@yahoo.com
akhilbooks@hotmail.com

ISBN : 978-81-87393-01-6

Printed in India on behalf of M/s Adroit Publishers by
Arpit Printographers E-mail : arpitprinto@yahoo.com

List of Contributors

Joëlle Smadja is a geographer, director of research at the CNRS in France, and head (2005-2012) of the research unit 'Centre for Himalayan Studies'. She has been carrying out research in Nepal since the 1980s in the fields of geomorphology, land use, resource management, climate change,perceptions and representations of the environment, environmental policies and their repercussion on rural societies, and territorial restructurings. Since 2005, she has extended her fieldwork to North-East India, in particular to Arunachal Pradesh and the Brahmaputra plain in Assam, where she focuses on the mobility of land and of populations, and on protected area policies. She has edited *Reading Himalayan Landscapes over Time: Environmental Perception, Knowledge and Practice in Nepal and Ladakh*, InstitutFrançais de Pondichéry, 2009, (first published in French in 2003). E-mail: jsmadja@vjf.cnrs.fr

Hermann Kreutzmann is Chair of Human Geography at the Institute of Geographic Sciences and Director of the Centre for Development Research at FreieUniversität Berlin. His research focuses on issues such as migration, conflict, development, water utilisation, pastoralism and political geography in Central and South Asian contexts (Afghanistan, PR China, India, Kyrgyzstan, Nepal, Pakistan, Tajikistan). His recent publications include: *Pastoral practices in High Asia. Agency of 'development' effected by modernisation, resettlement and transformation*, Kreutzmann H. (ed.), Dordrecht, Springer 2012; "Boundary-making as a strategy for risk reduction in conflict-prone spaces", in Müller-Mahn, D. (ed.), *The spatial dimension of risk. How geography shapes the emergence of riskscapes*, London, Routledge 2013, pp. 154-171. E-mail:h.kreutzmann@fu-berlin.de

SanjibBaruah is Professor of Political Studies at Bard College, Annandale-on-Hudson, New York. His publications include *India against Itself: Assam and the Politics of Nationality* (University of Pennsylvania Press and Oxford University Press 1999), *Durable Disorder: Understanding the Politics of Northeast India* (Oxford University Press, 2005),*Postfrontier Blues: Towards a New Policy*

Framework for Northeast India (East-West Center) and the edited volumes *Beyond Counterinsurgency: Breaking the Impasse in Northeast India* (Oxford University Press, 2009) and*Ethnonationalism in India: A Reader* (Oxford University Press, 2010). E-mail: sanjibbaruah@yahoo.com

Philippe Ramirez is an anthropologist and senior research fellow at the Centre for Himalayan Studies (CNRS). His first site for fieldwork was Nepalon which he centred his anthropological investigations over a 15-year period. His main study concerned the political anthropology of a Hindu village community of Central Nepal. His publications on Nepalinclude: *De la disparition des chefs. Uneanthropologiepolitiquenépalaise*(2000); and, as an editor, *Resunga, The Mountain of the Horned Sage: Two districts in Central Nepal* (2000). Since 2002, he has been trying to decipher the anthropological complexity of North-East India, with particular focus of the relationships between kinship and ethnicity on the Assam-Meghalaya borderlands. E-mail: pramirez@vjf.cnrs.fr

Narendra Raj Khanal, professor at the Central Department of Geography, Tribhuvan University, Kathmandu, Nepal, has over 34 years' experience in teaching and research. Trained at the Department of Geography, University of Colorado, Boulder, USA and at the Institute for Geoecology, Technical University Brauschweigh, Germany, he completed his PhD in 2003 and did his post doctoral research at the Laboratory of Geoecology, Graduate School of Environmental Earth Science, Hokkaido University, Sapporo, Japan. He has actively contributed to many research projects at international, regional and national level. He has prepared more than 100 research reports and published several books and articles focusing on regionalization, environmental issues, vulnerability and risk assessment, and adaptation strategies. His main area of interest is mountain environment. E-mail: nrkhanal.geog@gmail.com

Krishna Hachhethu is Professor of political science and a faculty member of the Central Department of Political Science, Tribhuvan University, Kathmandu, Nepal. He is the author of a dozen books, including *Party Building in Nepal* (2002), *Nepal in Transition* (2008), *State Building in Nepal* (2009), and has published many articles in books and journals in Nepal and abroad. He was a visiting scholar at Oxford University in 2005. He is a former member of the High Level Commission of State Restructuring of Nepal. Prof.Hachhethu is country coordinator in a South Asia Democracy Study network. He has a solid experience of working with international organizations i.e. International IDEA, DEFID, UNDP, UNDP, NORAD on subjects such as democracy, political party, governance, election, state restructuring etc. E-mail: krihac@gmail.com

Dilli Ram Dahal obtained his PhD in Anthropology from the University of Hawaii in 1983. He is a retired Professor of Anthropology from Tribuvan University, Kirtipur, Nepal. He has published several books and articles, including: *National Dalit Strategy Report* (Volume 1, 2 and 3, 2002, co-author, NPC, Kathmandu); *Forestry User Groups: A Case Study of Forestry User Groups in Dhankuta,*

Sankhuwasabha, and Ilam Districts of the Eastern Hill Region of Nepal, Kathmandu, International Centre for Mountain Research and Development, 1994. His recent articles include: "Social Exclusion/Inclusion among the Madhesis: A Case Study of Yadavs of Central Nepal Tarai",*Contributions to Nepalese Studies,* Volume 39, Special Issue, 2012, pp 1-34. E-mail: dahal.dilliram485@gmail.com

Anne de Sales is an anthropologist and senior research fellow at the 'Laboratoired'Ethnologieet de Sociologie Comparative' (CNRS and University of Nanterre-Paris X). Her research interests include shamanic practices and ritual language in Nepal, the anthropology of social and cultural transformations and the history of anthropology. She has published extensively on a tribal population in Western Nepal, the Kham-Magars, including the monograph, *Je suis né de vosjeux de tambours. La religion chamanique des Magar du Nord* (1991) and on the impacts of the Maoist insurrection on rural communities. In collaboration with Robert Parkin, she edited a collection of articles on ethnographic theory in French anthropology (2010, *Out of the Study, into the Field,Ethnographic Theory and Practice in French Anthropology,* Oxford, Berghahn).E.mail: anne.de-sales@mae.u-paris10.fr

Martin Gaenszle is Professor in Cultural and Intellectual History of Modern South Asia at the Department of South Asian, Tibetan and Buddhist Studies, University of Vienna, Austria. His research interests include ethnicity, local history, oral traditions and religious pluralism in North India and Nepal. He is the author of *Origins and Migrations: Kinship, Mythology and Ethnic Identity Among the MewahangRai* (Mandala Book Point, 2000) and *Ancestral Voices: Oral Ritual Texts and their Social Contexts among the MewahangRai of East Nepal* (LIT Verlag, 2002) and has co-edited various books on rituals, discourses and space in South Asia. E-mail: martin.gaenszle@univie.ac.at

Chetan Singh is Professor of History at Himachal Pradesh University, Shimla. He is the author of *Natural Premises: Ecology and Peasant Life in the Western Himalaya, 1800-1950* (Oxford University Press, Delhi, 1998);*Region and Empire: Panjab in the Seventeenth Century* (Oxford University Press, Delhi, 1991) among other books. He has also edited *Recognizing Diversity: Society and Culture in the Himalaya* (Oxford University Press, New Delhi and Indian Institute of Advanced Study, Shimla, 2011). Besides articles on medieval Indian history, Chetan Singh has written numerous articles in peer-reviewed journals and edited volumes on the environment-society relationship, religion and culture in Himachal Pradesh. E-mail: profcsingh@gmail.com

Brigitte Steinmann is Professor of Anthropology at the University of Lille I (France). She obtained her first PhD at Paris-Nanterre University in 1985 and qualified to supervise higher degree research candidates (habilitation) in 2001. She is member of the scientific council of her Faculty and of the EcoleFrançaised'Extrême Orient.She coordinates several University exchange projects, including one between Tribhuvan University and Lille I (International

Erasmus program), and one between Beijing University and Lille I (CaiYuanpei project). Her research focuses on the socio-religious history of Buddhism among populations of central Nepal (Tamang), and SikkimeseLepcha. She has carried out extensive inquiries into the Maoist war in Nepal. She is extending her research to the sociology of migration and labour (Nepal/Europe). She is the author of five books and a number of articles (seehttp://www.digitalhimalaya.comandhttp://clerse.univ-lille1.fr). E-mail: brigitte.steinmann1@gmail.com

Ben Campbell is a lecturer in anthropology at Durham University. He has worked in Rasuwa District with Tamang-speaking communities since 1989. His research focuses on issues about ethnicity and development, agro-pastoral livelihoods, environmental knowledge, and the politics of nature conservation. His recent research interests include renewable energy technologies for sustainable mountain development. His book *Living Between Juniper and Palm: Nature, Culture and Power in the Himalayas* is published by Oxford University Press, 2012. In 2009 his film *The Way of the Road* was released, made with cameraman and brother Cosmo Campbell. E-mail: ben.campbell@durham.ac.uk

BlandineRipert is a geographer and anthropologist. She is a research fellow and currently (2010-2013) head of the Centre for South Asian Studies (CNRS-EHESS, Paris). Her research concerns contemporary dynamics in Nepal, and more recently in India. She has studied the process of Christianization among a Buddhist population in central Nepal, the Tamangs, and has worked on the impact of globalization on them. Her work also includes research on environmental politics and changes in the landscape in Nepal, in North-East India and in the Alps, and the spread and impact of information and communications technology in India, in particular in Tamil Nadu. More recently, she began researching the ethnicisation of politics in Nepal. E-mail: ripert@ehess.fr

Gérard Toffin is a social anthropologist and director of research at the Centre for Himalayan Studies (CNRS). He has done extensive fieldwork among the Newars of the Kathmandu Valley and among the Tamangs of the Ganesh Himal range. His research currently focuses on a Krishnaite sect between Nepal and India (Gujarat and North Bengal), the Hindu city model in South Asia, the construction of democracy in Nepal and the anthropology of theatre. He is the author of: *Sociéteet religion chez les Néwar du Népal*, CNRS 1984; *Newar Society. City, Village and Periphery*, Himal Books 2007. His edited volumes include: *Man and His House in the Himalayas*, Sterling 1991; *Nepal, Past and Present*, CNRS/Sterling 1993; (with Joanna-Pfaff-Czarnecka) *The Politics of Belonging in the Himalayas*, Sage, Delhi, 2011. E-mail: gtoffin@vjf.cnrs.fr

Pascale Dollfus is an anthropologist and senior research fellow at the Centre for Himalayan Studies (CNRS). She has been working since 1980 on Tibetan-speaking populations in the western Indian Himalayas. She focuses her research on material culture, territorial and social organisation and the religion of sedentary and nomadic communities in Ladakh and Spiti. She recently broadened her fieldwork to neighbouring Kinnaur and Lahaul. Since 2010 she has also been working in North-East India where her interests lie in the Sherdukpens and their relationship with their neighbours. She is the author of *Lieu de neigeet de genévriers*(CNRS 1989, 2005); *Ghurras. From the Farm to the Museum, the Farmer to the Collector* (le Toit du Monde, 2007); *Les Bergers du Fort Noir- Nomades du Ladakh*(Sociétéd'ethnologie, 2012). E-mail: pdollfus@gmail.com

Tristan Truslé is a geographer, he is a research fellow at the Centre for Himalayan Studies (CNRS). His interests lie in international labour migration and diasporas. While building on the case of Nepalese migrants, he addresses topics such as migrants' integration in the host society, transnationalism, relationships to places, the effects of migration on communities of origin and digital diasporas. He has been studying Nepalese migrants in India and in Qatar. He edited a special double issue of the *European Bulletin of Himalayan Research* related to "Nepalese migrations" (2009-2010, 35-36). E.mail: Tristan.brusel@vjf.crs.fr

Contents

Introduction. Legacies and Current Trends: Characteristics of Territory-Making in the Himalayan Region

Joëlle Smadja

"A little territoriality creates sociality and solidarity, a lot kills them". (Théry and Brunet 1992a: 436)[1]

Like identities that are not fixed but permanently reworked, territories are permanently being rebuilt. They are remodelled within their limits, their content, their organisation and their functionality; they undergo a permanent requalification and resemantization process. This, of course, has always occurred worldwide, but what do recent changes mean? Geographers who study the process of territorial restructurings in Latin America,[2] in South Africa and in parts of India (Antheaume et al. 2003, Gervais Lambony et al. 2003) highlight the fact that the fragmentation, separation patterns and specialisation of space are becoming more marked and that many different languages of territorialisation coexist, and often clash, in the same space. It is not therefore a question of the "end of territories" as announced by Badie (Badie 1995), but of a territorial complexity, of a growing number of territorial entities that lead some geographers, such as Jean Renard, to evoke the image of a "territorial jam" since their different functions often overlap (Renard 2000: 255). What about the Himalayan region? What are the characteristics and consequences of territorial restructurings there, and do they contribute to making this region a specific entity? These are the questions addressed in this book that takes up various challenges.[3] It discusses territory-making although the notion of territory itself is not shared alike by scholars worldwide and is a controversial issue with regard to the Indian world.[4] It sets out to study these territorial restructurings on different scales, from the nation-state scale to the village scale. It brings together Indian, Nepalese, and European scholars who carry out research both in various parts of the Himalayan region—from the far western to the far eastern part—and in various disciplines, mainly anthropology, geography and history. This is an ambitious undertaking and our only hope here is to open a discussion for future debate.

First of all, to avoid any misunderstanding, let us rapidly define what territory means here. Generally speaking, this word is not used in the same way by French scholars as it is by Anglo-Saxon and most of the English-speaking ones who, depending on the country they come from, attribute many different meanings to it.[5] In this book, its acception is the one adopted by French authors for whom the word is polysemic: in a French geographical dictionary there are no fewer than nine definitions[6].

The words "*territoire*" in French and "territory" in English have the same Latin etymology, they both come from "*territorium*" which means "area in which a human group lives", which itself comes from "*terra*" meaning "land". In French, the Latin word "*terra*" has been declined: it has given "*terre*" for "land", and "*terroir*", "*territoire*" (territory), "*territorialité*" (territoriality), etc.

The first meaning of the word is shared by and large by everyone. It refers to a space that is demarcated and controlled and which falls under an authority and jurisdiction. It has a political, juridical and administrative meaning and is used mainly to speak about a state, a nation, a country and about their administrative and political units. It mainly refers to delimited, bounded areas. This meaning is the most common and the oldest, it dates back to the eighteenth century. It has been used in political science and international relations and adopted by geographers.

The word has subsequently been used in biology, in works on ethology, to talk about an area appropriated by an animal or by a group of animals. Territory is thus an exclusive, controlled space: a definition extended to population groups that protect their own territory. The word is also used with this meaning more or less worldwide, depending on the countries (see Sack 1983).

Since the 1980s, the word has been taken up again by social scientists in France and is used in anthropology, sociology and geography. In this respect, "researchers all insist on the double dimension of territory, both material and ideational, ecological and symbolic"[7] (Debarbieux 1999: 36).[8] Among French-speaking scholars, the notion of "territory" has become the perfect conceptual tool to combine, on the one hand, material realities and spatial processes and, on the other hand, socio-symbolic constructions. It has also become an interdisciplinary tool in the same way as the notion of "landscape". At the same time concepts such as "lived space" have been introduced into geography, and there has been renewed interest in the field of "cultural geography". Since then, in France, territory has also been understood as a space used and appropriated by a population and which bears an identity reference.[9] Thus, a territory can be both a political, juridical, administrative space and a space appropriated by a community that transforms it in a physical and symbolic way, depending on its own conceptions of the world and which, in return, informs the community of its own identity. Moreover, this notion is of interest because it "allows a temporal decomposition of a spatial process"[10] (Ozouf-Marignier 2009: 35). Indeed, a territory is produced over time (Di Méo 2004) thanks to regular use of the space, of the milieu and of its resources. A collective memory built up over time thereby leads to defining a specific "Us" and a feeling of belonging to a territory (see Jolivet and Léna 2000: 8, Pfaff-Czarnecka and Toffin 2011), a process which is a gradual social construction. Under these conditions, belonging to a territory means feeling pride in it and this may contribute to valorising it. The space used by a population therefore becomes a space of identification and of memory where people can build a project (Deffontaines 2006, Deffontaines and Prod'homme 2000). The population is no longer only an "economically active population"; it is an organised civil society with its own territory, capable of making choices to influence its development (see Jolivet and Léna 2000: 6). Here territory is thus regarded as the expression of the relationship between man and his milieu, and it is an integral part of his culture and identity (Smadja 2011). Consequently one also uses the term "territoriality", which is a process of building space to express the individual or collective relationship with a territory that has been appropriated. Yet we are well aware that the notion of territory refers to a production, a process, it is not a given reified entity which would correspond to a unified society free of any conflict (see Banos 2009).

One way of appropriating space and of making a territory out of it is to name it, thus people give a name to a territory and to the different places within it which are a memory of their practices and in which knowledge is recorded: "The Earth becomes men's Earth when, ceasing to be anonymous, it is named by them" (Pinchemel and Pinchemel 1988),[11] and "to name is to territorialise"[12] says Debarbieux (1989). Furthermore, toponymy is one of the features of the process of legitimizing a new territorial basis, hence political control of a territory starts with its linguistic marking (Akin 1999). Several chapters of this book reveal the extent to which the "naming game", as Hermann Kreutzmann puts it—through which some conflicts are expressed—is an integral part of the territorial restructuring process. In India and in Nepal, in several cases it is symptomatic of names related to reliefs or to nature that may be used to avoid ethnic or religious connotations.

As already mentioned, not everybody shares the same idea of territory, and this is the case in the Himalayan region itself as underlined by Brigitte Steinmann in this book. The Lepchas from Sikkim, for example, do not assert that they "own a territory" and they do not consider their territory to be "a space inhabited by men" or "a political space". They look upon their living space as consubstantial to their group: an infinite diversity of animals and plants as well as a non-human world of spirits inhabit the territory alongside human beings. This conception refers us to the debate about place and environment animated by Descola (2005), Ingold (2000), Viveiros de Castro (2009) and Latour (1991) for example.

In fact, the definition that French scholars give to 'territory' is close to what the Anglo-Saxons attribute to the word 'place'. The same debate the French have had about 'territory'[13] has occupied Anglo-Saxon scholars regarding the word 'place' (see among others, Casey 1996, Duncan 1989, Entrinkin 2002, Hirsh 1995, etc.). Hence, in many instances in this book, the word 'place' might have been used instead of the word 'territory', yet the argument here is that we need both. For many of us from the French School, a place is a component of a territory. A territory is made up of places and links, and a landscape is a visual expression of a territory (of a space appropriated by a population). By keeping the word territory with all its meanings, which imply notions of citizenship, belonging, appropriation, identity, memory, heritage and the like, the authors also affirm that for people inhabiting it, a territory is never simply political, administrative or economic but also cultural and symbolic. It implies both a sense of belonging (I am from there) and, often, of appropriation (this is mine, this is my land, my space…) (Théry and Brunet 1992b: 436).

If we accept this definition, forms of territory vary greatly, which enables us to study them on very different scales. They may correspond to an adjacent stretch, a surface, whatever the scale, or they may be reticular and correspond to an arrangement of disjointed sites or places linked together by elements of networks whether materialised or not in space. Thus, nomadic societies are also territorialised; no antinomy exists between nomadism and territory, no more so than between mobility and territory as shown by the texts in the second part of this book.

In this work, we focus on two processes which do not exclude each other. The first one is that each community tends to use its identity to push forward claims that are often spatial and territorialised and to redefine itself according to new territories. The other one concerns territories that result from modernity, development programmes, mobility and exclusion. These are dealt with in sixteen texts providing data on the characteristics of territorial changes and territorial restructurings in the Himalayan area, pinpointing the ambiguities in the various claims and the difficulties in implementing them. The book addresses the question of internal borders rather than any international borders since, whether federal state territories or autonomous territories, or new environmental borders, they all create new limits: in the Himalayas, the number of internal borders is on the increase.

In this region, if we are to consider nation-states and political and administrative units, we simply have to look at maps from the last three centuries to observe the many changes that have occurred (see for example, the Schwartzberg atlas, 2006 [1978]): states and districts are forever being redefined revealing for the latter an increasing parcelling off. In the first part of this book, all the texts, whether about Pakistan, India or Nepal, demonstrate the fact that a large part of current territorial restructurings within states are linked to federal and autonomous territorial claims, and are—setting aside any political and economic criteria—based on religious, ethnic, language, identity criteria, and on ancestral realms. These criteria are used by groups to claim sovereignty over a territory which overlaps traditional administrative units. In atlases, maps representing the distribution of Himalayan population groups often have indistinct frontiers: as a result of current territorial restructurings they are set to become maps which represent territories with clear-cut frontiers, that is to say new political and administrative maps of countries. As Kidwai says about the cases of Awadh and of Rohilkhand:

> "Even when they have carried significant historical weight, politico-administrative units on a regional scale produce weak territorial identities because they barely mark the public consciousness. They acquire sentimental value only if a politico-administrative identity and a 'social' identity coincide over a long and durable time period […]." (Kidwai 2003: 182).[14]

As evidenced by some authors of this book, territorial dynamics in part of the Himalayas found its roots in policies implemented at the time of British rule that considered the Himalayas as a specific area and gave special status to their populations (mainly to tribal groups) and territories. The Indian government has pursued these policies since Independence. However, Nepal, which never came under British rule but where many tribal groups live, has followed the same tendency. According to these authors, throughout the Himalayas, from East to West, we have witnessed an "ethnicization of territories" (Philippe Ramirez) or an "ethno-territoriality process" (Sanjib Baruah), which started at the time of British colonisers. Their aim, as Hermann Kreutzmann explains, was to "protect the Jewel in the Crown", that is to say to protect the wealth of the Indo-Gangetic plain and the 'civilized' world—which is governed by general law—, from the 'tribal' people—who are casteless and governed by customary law. They subsequently drew borders and demarcated territories separating the 'hill tribes', in their supposedly natural habitat, from communities living in the plains. This gave rise to the North-West Frontier or to the North-East Frontier Agency, which are frontiers that still exist today. And, as Sanjib Baruah points out, people from the North-East are always classed as belonging either to the hills or to the plains: a classification that continues to apply in the Indian Census. Based on this colonial classification, a number of tribal groups in the Himalayas are now demanding greater democracy and more rights by making claims over ethnic territories, often alluding to ancient kingdoms to justify this. This is also the case in Nepal where claims to territories are likewise based on ethnicity and on indigeneity. However, these claims are often ambiguous and, according to Sanjib Baruah, the exclusionary territorial imaginaries that shape territorial projects in North-East India are grounded in the colonial ethno-territory framework, not on reality. Hermann Kreutzmann makes the same statement about tribal zones in Pakistan. Philippe Ramirez highlights this statement explaining that the Balkanisation of North-East India does not correspond to the recognition of old sovereignties. He emphasizes: "the strict correlation between culture and identity is far from being firmly established" [...] "In several areas [...], local communities are not the subset of a particular ethnicity, but an association of lineages asserting various ethnicities or none at all." Moreover, anthropological territories rarely correspond to current political territories, and people of a given group do not always represent the majority in the territory they claim as their own, all the more so now that people are very mobile, which is furthermore demonstrated in the three texts about Nepal and about the making of federalism there (see the texts by N. Khanal, D.R. Dahal and K. Hachhethu). Thus, to delimit territories based on ethnicity and culture means reifying these features. As Dilli Ram Dahal puts it "The basic assumption that people will remain static or stay in one place or that culture does not change is a short-sighted way of viewing the future of Nepal and of the Nepalese people". Indeed, these territorial claims are often contradictory because globalisation, modernity, development and accelerated migration and urbanisation, lead to considerable ethnic diversity everywhere. Lastly, not all people of a given group conceive of their claimed autonomous territory in the same way, as Krishna

Hachhethu demonstrates in a text about Nepal.

Another ambiguous aspect of claims over ethnic territory is well illustrated in North-East India where particularly virulent autonomist movements and many Autonomous District Councils founded on ethnicity have emerged.[15] These have been set up to limit injustices and inequalities. Here, people claim land rights, the right to a bilingual education, to self-management of local resources and of the environment, but in many cases it is forbidden to transfer land rights to anybody who is not from the territory's ethnic group: these territories are also used to counter the arrival of migrants or to exclude foreigners (see S. Baruah). This process has also been observed in South Africa (Guyot 2006) or in Central America (Martinez Mauri 2007) for example; in India, it may, in certain cases, be endorsed by Hindu fundamentalists such as the RSS[16] (see Jaffrelot 2007), or other movements close to it (see J. Smadja in this book). Whatever the case, as in other places in the world, the political instrumentalization of territorialized ethnic entities constitutes one of the main factors of instability and of violence in the Himalayas.

On a village-community scale, a significant feature related to territory in the Himalayas has been historically the legitimization of community territories by deities, with in return the offering up of rituals to these divinities that have reaffirmed the sense of belonging to a territory[17]. This historical and religious link to territory is illustrated in the texts by Chetan Singh, Martin Gaenzle and Anne de Sales. The legacies, and the way people use this link to territory (C. Singh) or reinvent it (M. Gaenzle) or the way it is instrumentalized by others to serve their cause (A. De Sales) are also identified. Thus, we can see the process used by people or groups to legitimize their claims over ethnic territories, and how for example "territories that used to be a blank space on the map have turned into the heart of a nation", as the Maoists have done with the Kham Magar territory in Nepal (A. de Sales). Other groups in Nepal, such as the Rais, have strengthened their ethnic identity within a nation-state by reinventing their rituals, giving them a new meaning and unifying them in order to display a unity even though major differences have always existed from one village to another among the very same group (M. Gaenzle). In areas such as Himachal Pradesh, as shown by Chetan Singh, the power of the community—reaffirmed through current rituals performed for the divinities of the territory—is still very strong. This is expressed in modern disputes (over ski resorts, dams), in which the power of village communities, their territorial gods and the ancient kings of these territories play a significant role. This process explains the saliency of these legacies in current policies and decisions.

Development is another feature that contributes to current territorial restructurings. In implementing development projects, new spaces are created whenever administrative or political entities overlap. Hermann Kreutzmann demonstrates this using the example of Gilgit, Baltistan and Chitral which NGOs regard as a single entity. Not only do development programmes create new spaces, but they also provide insight into the welfare of the population and of the territories, which may be very different from the population's expectations. They may produce the opposite effect to the one desired, as explained in texts by Ben Campbell, Brigitte

Steinmann and Joëlle Smadja. These authors also evidence the way dams, protected areas, and roads restructure space and the way the management of natural resources has led to significant territorial restructurings. This has been done on a village scale, modifying distances to resources and networks (B. Campbell and J. Smadja), as well as on a state scale. Referring to the concept of nature as an objective ontological field has led to delimiting new environmental borders and to creating "natural spaces" such as national parks which are both new territories and which at the same time restructure national territories. These new territories, set up in accordance with international rules and with the intervention of international agencies, are emblematic of globalisation. The development, or sustainable development programmes dams and protected areas result in often go hand in hand with population displacements, that is to say—in the Himalayan region—of tribal people who live off the natural resources available in these spaces. With the hundreds of protected areas and dams, and the numerous tribes living there, the Himalayas are paradigmatic of the policies linked to these kinds of projects.

The examples given in this book on a local scale provide food for thought regarding spatial justice, particularly those denouncing the role of large NGOs in these development processes (see for example Amelot, Moreau and Carrière 2011). They highlight the difference between the ambitions expressed on a global scale and the reality on a local scale. This refers us to J. Theys' statement:

> "If there is somewhere a democratic link to be found between the three constitutive dimensions of sustainable development—the social, the ecological, and the economic—, it is with no doubt on a local level that it can be built more effectively, given that it is also on this scale that their contradictions or their synergies appear with the most force and evidence [...]" (Theys 2010: 10).[18]

And, as O. Godard puts it:

> "Sustainable development is not fractal" [...] "[…] the ecological constraint has an absolute meaning only at world level. At other levels, exchanges, substitutions and imbalances are possible and ecological constraints are relative" (Godard 1996: 33, 34 in Pecqueur and Zuindeau 2010: 51).[19]

Other than the forced displacement of people in the case of development projects, migrants and new settlers, for the most part in cities, illustrate cases of displacement and mobility that create or reshape territories. Let us recall here that in Nepal for example, since 1991 the Terai plain has been more densely populated than the mountains, and that the population in this plain is mostly concentrated in cities. Whether national or international, mobility structures space and contributes to restructuring territories both of origin and of arrival. For displaced populations, the new space needs to be 'experienced' over time for a non-place, an "outplace" (extra-territorial space), to be turned into a territory, a space bearing a meaning to which they can feel they belong and where they can undertake projects. "Migration and population displacements have produced new peripheral spaces, zones of uncertainty, poverty and unemployment, 'outplaces' such as slums, which neither belong to the urban territory nor to its outside space" as Gérard Toffin highlights in his study based in the Kathmandu Valley. Similarly, as described by Pascale Dollfus, most of the Ladakhi mobile community who have sold their livestock and have migrated to the outskirts of Leh, "live as though they are in a waiting room". In this process of building new territories, migrants, displaced peoples, new settlers create new links to the space where they have settled, "affective ties with the material environment" or *topophilia* (Bachelard 1961 [1957], Tuan 1990), as explained by Tristan Bruslé. At the same time, they create new bonds with their territory of origin (B. Ripert), their village, a native village which is often idealized (T. Bruslé).

Content of the book

The book is made up of two parts. The first part focuses on territory-making in the light of the historical role of legacies. It addresses administrative and political restructurings referring to a somewhat classical notion of territory and which is expressed characteristically in the Himalayan regions by autonomist and federalist claims based mainly on ethnic and/or religious criteria. They are sometimes the source of violent conflicts. This is examined from West to East in the Himalayan area and on different scales, from state to village.

In the first text, by discussing the status of Gilgit-Baltistan within the scope of the Kashmir conflict, Hermann Kreutzmann explores the historical developments that led to boundary-making during the British Empire and he links these findings to the prevailing crisis. He shows how today's borders and territories are now experiencing the repercussions of former policies. Claims to federalism are linked to the recognition of elementary rights and to more democracy; in this process, development projects implemented by NGOs add another layer to territorial restructurings. All these territorial changes are associated with a "naming game" reflecting "the power struggles and regional tension that are prevalent in the Northern Areas".

In a text about the politics of territoriality in North-East India, Sanjib Baruah explains how these politics are influenced by colonial ethnography—"the colonial ethno-territorial framework" that has made a clear-cut distinction between hill tribes and plain tribes— which is the reason for the numerous territorial claims and conflicts in this area. Since Independence these claims have relied on the Sixth Schedule of the Indian Constitution and, as he points out, for the most part they are based on imaginary realms. He warns of the dangers of politics of territoriality shaped by the institutional legacy of colonial knowledge, arguing for multi-level (Indian and state) citizenship as an alternative to the ethnic homeland model.

Although claims over autonomous territories rely on the inseparability of identity, culture and ancestral realms, by taking the case of a hilly region straddling Assam and Meghalaya in North-East India, Philippe Ramirez emphasizes that, "the strict correlation between culture and identity is far from being firmly established". He adds: "the very question of ethnic affiliation may not arise at all", and suggests that ancient social identities attributed less importance to ethnicity than is generally believed. He demonstrates how the Karbis, Khasis and Tiwas' "anthropological territories" do not coincide with current political territories and how, contrary to the continuous spaces that are being imposed as territories on the nation-state model, these groups have been organized according to interwoven village networks.

In the series of three texts that follow, related to federalism in Nepal, the authors explain how difficult it is to delimit territories founded on ethnicity.

Narendra Khanal presents the different models proposed for restructuring Nepal and their respective maps. Most of the proposals include ethnicity, language and cultural identity as the major criteria in delineating federal units. He compares the advantages and disadvantages of some models: those taking into consideration the complementary features of different ecological zones by including the plain, the hills and mountains in each federal unit; and those based only on ethnicity, language and cultural identity as proposed by the Maoists, for example. Other than the drawbacks of the latter models, he demonstrates that for most groups, the majority of the population of the ethnic group concerned resides outside the claimed federal units. He also points out the importance of naming the different units within the various models.

Based on accurate fieldwork data, Krishna Hachhethu exposes, on the one hand, the revival of Gurung ethnicity—this group's search for a better representation associated with their claim over an ethnic-based province, i.e. Tamuwan—and, on the other hand, the difficult task in demarcating such a territory in which the Gurungs would be the majority. His study reveals that not all Gurungs conceive of Tamuwan in the same way, whether they are leaders of political parties, activists, members of civil society or otherwise. Gurung activists claim that the western hills, the Gandaki zone in particular, is their traditional homeland. Yet K. Hacchethu wonders to what extent it would be practical to create a province in the name of and for the benefit of one particular ethnic group (Gurung) in an area where another ethnic group (Bahun/Chhetri) represents more than twice the population of the group in question (Gurung).

In a text on cultural territoriality taking Adivasi/Janajati and Madeshi groups as examples, Dilli Ram Dahal shows that the ethnic/caste-based model of a separate cultural territory or of an autonomous region is not viable for Nepal. This model loses some of its edge if we take into account modernization, globalization, changing demographic factors and, above all, the broader process of migration and urbanization that has taken place in Nepal over the last 50 years, as well as the spread of various groups over different regions and over Nepal as a whole. Given these data, he considers that Nepal has to be fully united within diversity and pluralism, and thereby needs to take the path from segmentation to integration.

The next three texts focus more on the village-community roots of territory-making, on the link populations forge with their territories through their divinities, on the history of these territories and on the way the power acquired via the link to territories is used or instrumentalized.

By considering the historical and linguistic territory of the Kham Magar groups, as well as the territory delimited by shamans, Anne de Sales sheds light on the different territorial frames defined by the group. She depicts how historical features associated with their territory have led to a lack of integration and visibility of

Kham-Magar country at national level, a situation that has been used by the Maoists to their own advantage. We therefore understand how a peripheral, marginal, impoverished territory has been made the emblem of the nation, how Kham speakers who have always been considered the Magars' "poor relatives" have, through the Maoist insurrection, been given the opportunity to occupy an eminent place in national politics and vis-à-vis their more privileged Magar cousins, and how, in a word, "the Maoists turned a blank space on the map into the heart of the new nation".

Among Rai groups in eastern Nepal, the major fertility rituals offered up to a form of the "divinity of the soil" are associated with a notion of ancestral territory, the settlement, its lands, and its 'original' inhabitants. Martin Gaenzle analyses how these rituals—that vary to a considerable extent among the numerous Rai "subtribes"—have been reinvented and unified in order to strengthen Rai ethnic identity within a nation-state, given that the place of the Rais and of the eastern hills in a federal state has become a burning issue since the 1990 democratic movement. He argues that there has been a shift in the intended efficacy of these rituals: in villages they correspond to a renewal of the link with ancestral forces as well as with the space and fertility of the territory; whereas in the urban context such rituals of a newly defined ethnic unity make a significant statement that comes with a political message, that is they express claims related both to ethnic kinship and territorial belonging.

Taking the example of communities in Himachal Pradesh, Chetan Singh shows how one goes from building communities to building territories in pre-Independence India: he explains the way deities legitimize a community's territory on different scales and the way, in return, by giving them the names of the places, populations consecrate these territorial deities. He underlines that the latter often reigned over their followers as would a king over his subjects. Surprisingly, despite the numerous changes that accompanied the Independence of India, and among them the creation of Himachal Pradesh, this situation is still of major importance today since legacies of the past are expressed through rituals that reaffirm the sense of belonging to a territory. This may be witnessed in modern conflicts over projects such as ski resorts or dams where village communities, their territorial divinities and the ancient kings of these territories are still powerful. C. Singh therefore demonstrates how legacies are highly salient in current policies and decision-making and how local communities dominate politics in Himachal Pradesh today.

The second part of the book deals with territories of modernity, territories of exclusion. On the one hand, it addresses the emergence of new territories associated with development programmes and displays the difference between various conceptions of territory and of resources. On the other hand, it is devoted to the way migrants, displaced people, and new settlers make new territories and forge bonds with them. Unlike the first part of the book that discusses structured societies that have projects and often wish to strengthen their territory by making autonomy and federalism claims, the cases examined here focus more on destructured societies that have lost their original territory and are now experiencing new projects that they themselves are no active part of.

After an overview of the territorial restructurings which heighten the split between nature and society on a village scale, Joëlle Smadja focuses on protected areas. Using the example of Kaziranga National Park in Assam, she demonstrates how, as enclaves within states, and with new environmental borders, these protected areas constitute new environmental territories at the heart of debates that are not merely limited to nature protection issues. She argues that internal boundaries related to the environment are of relevance and importance in the structuring of space and societies at different levels, and can generate issues that fuel the numerous conflicts already raging in the Himalayan area.

In a text about the Lepchas in Sikkim, Brigitte Steinmann explains why they are ready to die for their motherland. Through examples taken from hunting tales, she illustrates the close links they have established between human artefacts and nature, and their conception of their territory: a territory that, save for humans, is made up of spirits, animals and plants and that they consider as consubstantial to their groups. Unlike the current trend, the Lepchas do not assert that they "own a territory", a conception that has been used by the state to displace them. Their territory is now threatened with destruction due to the building of hydroelectric dams.

Through a film he shot in a Tamang village in the remote district of Nuwakot in Nepal, Ben Campbell analyses the passage of a new road in this village and the territorial restructurings linked to it. He indicates that the territoriality expressed by villagers differs from what is subsumed by development programmes such as those of the Asian Development Bank for example, which funds the building of the road and whose aim is to relieve poverty in the area. It appears that the latter has neglected both the history of this society—that was linked to Tibet via an important trade route crossing their village—and current conditions of trans-national labour migration as well as the country's political situation. He concludes that "the road intended to relieve poverty will do the opposite of the stated goal and make them poorer".

The next four texts are devoted to migrants' territories and to the bonds they have with them.

While also studying Tamang people from Nuwakot district in Nepal, Blandine Ripert pays particular attention to the migrations of individuals over the course of their lifetime. She focuses on the effects mobility has had on bonds with territory and on how these bonds relate to the group of origin. Through three individual itineraries that prove that migration strategies are not always accommodated in a broader family context, she shows how bonds with territory, and consequently the sense of belonging to a territory, differ both with regard to the type of migration—temporary, definitive, forced, successful or unsuccessful—and as to whether or not a return to the territory of origin is envisaged.

By considering slum settlements along a riverbank in the Kathmandu Valley, Gérard Toffin points out that migrations and population displacements have created new peripheral spaces, *outplaces*, which neither belong to the urban territory nor to its outside space. Contrary to these migrants' previous territories of origin which were based mainly on kinship, ethnic group and caste hierarchy, these spaces throw together people from different geographical origins and are principally multicaste, multiethnic and multilingual. People settling there are regarded as pariahs, are implicated in national causes of conflict, and are easily manipulated by political leaders and organisations. A new social fabric is emerging from these settlements, and the author argues that the study of these urban fringes sheds light on the broader political context and is a valuable key to understanding contemporary urban entities.

As for the Ladakhis from Karnak, who after selling their livestock have migrated to the outskirts of Leh—to what they call Kharnakling—, Pascale Dollfus explains the extent to which territory occupies a central place in building their identity and in the way they refer to a group. Indeed, Ladakhis usually refer to their native land to define their identity, and ethnonyms have been built on place names: they are "those from". For them, it will take time to build a new territory in Kharnakling. While older people have not yet done so and "live as though in a waiting room", for the younger generation, their territoriality corresponds to a network made up of far more localities than simply their permanent urban settlement.

To round off this book,Tristan Bruslé presents a text related to the "migration territory" of Nepalese migrants, both in India and in Qatar, arguing that it throws light on Nepal's current socio-spatial developments. His text based on the concept of *topophilia* takes into account the affective ties migrants have with the material environment when building a territory. He shows how they appropriate it in a different way from the appropriation of their place of origin to which they consider they belong. He ultimately demonstrates how migration strengthens one's love for one's own country and one's own village.

***

To sum up, this overview about territorial changes and territorial restructurings in the Himalayas highlights at least three points.

Firstly, two factors seem to play an increasingly important role in the territorial restructurings at work in the Himalayan region: ethnicity and/or religion on the one hand and nature conservation or more broadly development programmes on the other hand which, in both cases, lead to a differentiation and to a specialisation of space.

Secondly, it appears that the administrative structures of nation-states have a real challenge to face in establishing a decentralised and federalist system between, on the one hand, the various development programmes—leading in some cases to new territories such as protected areas—that depend for a large part on international bodies and, on the other hand, autonomous regions which are, by definition, managed by indigenous actors. The tension between global and local scales is one of the particularities of these territorial restructurings. This is a highly sensitive issue in the Himalayas where there are both numerous ethnic groups and development programmes.

Thirdly, territories may be restructured when there is a project, a project targeting society in a given space. However, when the structure of society itself collapses, when there is a loss of identity, people abandon land or pastures for the slums. Time is therefore needed to reinvest meaning in a living space and to rebuild a territory.

This book does not of course claim to address all the features related to territorial changes and restructurings in the Himalayan region. Yet those presented here highlight the particular nature of the area. In a way, it appears that federalist and autonomous claims could be considered as a direct consequence of the population's situation in the area at least up until the 1950s, as expressed in the notion of Zomia defined by Willem Van Schendel (2002) and James Scott (2009): an area encompassing the mountains from the western part of the Northern Himalayas to those of the lower end of the Southeast Asian peninsular which is characterized by populations living (up until 1950 at least) far from the centres of power and which is therefore not administratively, economically and politically characterised by states or the centre of power, but by their periphery. Nevertheless, if we consider new territories which derive from development projects as well as migrants territories, it appears that the remote areas described within the Zomia are now, more than others, linked to the world web of international programmes, INGOs, diaspora networks, etc. These features in no way diminish the particular nature of the area, but they do now have to be taken into consideration to characterize it.

REFERENCES

Akin, S. (ed.). 1999. *Noms et re-noms : la dénomination des personnes, des populations, des langues et des territoires*, Rouen, Publications de l'Université de Rouen.

Alphandéry, P. and Bergues, M. 2004. "Territoires en questions", *Ethnologie française*, January-March 2004, vol. 34, no. 1.

Amelot, X., Moreau, S. and Carrière, S. M. 2011. "Des justiciers de la biodiversité aux injustices spatiales. L'exemple de l'extension du réseau d'aires protégées à Madagascar" in Blanchon D, J. Gardin and S. Moreau (Eds), *Justice et injustices environnementales*, Paris, Presses Universitaires de Paris Ouest, pp 193-216.

Antheaume, B., Giraut, F. (eds.). 2005. *Le territoire est mort, vive les territoires !*, Paris, IRD Editions.

Antheaume, B., Giraut, F. and B. Maharaj (eds.). 2003. *Territorial restructurings, comparisons and innovations*, proceedings of the French-South African meeting on territorial innovation, http://iga.ujf-grenoble.fr/territoires/publications.

Bachelard, G. 1961[1957]. *La poétique de l'espace*, Paris, Les Presses universitaires de France.

Badie, B. 1995. *La fin des territoires. Essai sur le désordre international et sur l'utilité sociale du respect*, Paris, Fayard.

Banos, V. 2009. "Repenser le couple "territoire-lieu" : pour une géographie de la démocratie ?" in M. Vanier (ed.), *Territoires, territorialité, territorialisation. Controverses et perspectives*. Presses universitaires de Rennes, pp. 157-165.

Berti, D. 2006. "Ritual Kingship, Divine Bureaucracy, and Electoral Politics in Kullu", *European Bulletin of Himalayan Research*, no 29-30, Summer 2006: 39-61

Berti, D. 2009. "Divine Jurisdictions and Forms of Government in Himachal Pradesh" in Berti D. and Tarabout G. (eds), *Territory, Soil and Society in South Asia*, Delhi, Manohar, pp. 311-339.

Berti, D. and Tarabout, G. (eds). 2009. *Territory, Soil and Society in South Asia*, Delhi, Manohar.

Blanc Pamard, C. 1999. "Les savoirs du territoire en Imerina (Hautes terres centrales de Madagascar)" in J. Bonnemaison, L. Cambrezy and L. Quinty-Bourgeois (eds.), *Les territoires de l'identité. Le territoire, lien ou frontière ?*, Paris, L'Harmattan, vol. 1.

Bonnemaison, J. 1981. "Voyage autour du territoire", *L'espace Géographique*, no. 4, 1981: 249-262.

– 1996. *Gens de pirogue et gens de la terre. Les fondements géographiques d'une identité : L'Archipel du Vanuatu*, Paris ORSTOM, vol. 1.

– 1997. *Les gens des lieux. Histoire et géosymboles d'une société enracinée : Tanna. Les fondements géographiques d'une identité. L'archipel du Vanuatu, Essai de géographie culturelle*, Paris, ORSTOM, vol. 2.

Bonnemaison, J. and Cambrezy, L. 1996. "Le lien territorial entre frontières et identités", *Géographie et Cultures*, no. 20: 7-18.

Bonnemaison, J., Cambrezy, L. and L. Quinty-Bourgeois (eds.). 1999. Les territoires de l'identité. Le territoire, lien ou frontière ?, Paris-Montréal, l'Harmattan, 2 vols.

Casey, E. 1996. "How to get from space to place in a fairly short stretch of time: phenomenological prolegomena" in *Senses of place*, Feld S. and K. Basso (ed.), Santa Fe: School of American Research Press, pp 13-52.

Claval, P. 1996. "Le territoire dans la transition de la postmodernité", *Géographies et Cultures*, no. 20: 93-110.

Collignon, B. 1996. *Les Inuit. Ce qu'ils savent du territoire*, Paris, L'Harmattan, Géographie et Cultures.

Debarbieux, B. 1989. "Les statuts implicites de la montagne en géographie", *Revue de géographie Alpine*, 1: 279-292.

– 1999. "Le territoire : histoires en deux langues. A bilingual (his-)story of territory", in C. Chivallon, Ragouet P. et M. Samers (eds.), *Discours scientifiques et contexts culturels. Géographies françaises et britanniques à l'épreuve postmoderne*, Talence, Maison des sciences de l'homme d'Aquitaine, pp 33-46.

– 2006. "Prendre position : réflexions sur les ressources et les limites de la notion d'identité en géographie", *L'Espace géographique*, no. 4: 340-354.

Deffontaines, J.P. 2006. Rôle du territoire dans l'ajustement des discordances entre dynamiques sociales et biophysiques ? in C. Beck, Y., Luginbühl and T. Muxart (eds.), *Temps et espaces des crises de l'environnement*, pp. 333-340. Paris, Editions Quae, collection Indisciplines.

Deffontaines, J.P., and J.P. Prod'homme (eds.). 2000.*Territoires et acteurs du développement local, de nouveaux lieux de démocratie*, Paris, Editions de l'Aube.

Descola, P. 2005. *Par-delà nature et culture*, Paris, Editions Gallimard.

Di Méo, G. 1998. *Géographie sociale et territoire*, Paris, Nathan Université, Fac Géographie.

– 2000. "Que voulons-nous dire quand nous parlons d'espace ?" in J. Lévy and M. Lussault (eds.), *Logiques de l'espace, esprit des lieux. Géographies à Cerisy*, Paris, Belin, pp. 37-48.

– 2004. "Composantes spatiales, formes et processus géographiques des identités", *Annales de géographie*, no. 638-639: 339-362.

Duncan, J. 1989. "The Power of Place: Sri Lanka: 1780-1980", in Agnew J and Duncan J. (dir), *The Power of Place: Bringing together Geographical and Sociological Imagination*, Boston/London, Unwin Hyman, pp. 185-201.

Entrikin, J.N. 2002. "Democratic place-making and multiculturalism" *Geografiska Annaler B*, vol. 84, no. 1: 19-25.

Ferrier, J.P. 1984. *Antée 1 : la géographie, ça sert d'abord à parler du territoire*, Aix-en-Provence, Edisud.

Gervais Lambony, P., F. Landy, and S. Oldfield (eds.). 2003. *Espaces arc-en-ciel. Identités et territoires en Afrique du Sud et en Inde*, Nanterre, Paris, Johannesburg, Géotropiques, KARTHALA, IFAS.

Grandjean, P. (ed.). 2009. *Construction identitaire et espace*, L'Harmattan.

Guyot, S. 2006. *Rivages Zoulous*, Paris, Karthala, Collection Tropiques.

Hirsh, E.,1995. "Landscape: between space and place" in Hirsh E. and M. O'Hanlon (ed.) *The anthropology of landscape. Perspectives on place and space*, Oxford: Clarendon Press, pp. 1-30.

Ingold, T. 2000. *The Perception of the Environment: Essays on Livelihood, Dwelling & Skill*, London & New York, Routledge.

Jaffrelot, C. 2007. *Hindu Nationalism: A Reader*, Princeton and Oxford, Princeton University Press.

Jolivet, M.J. (ed.). 2000. Logiques identitaires, logiques territoriales, *Autrepart, Cahier des sciences humaines*, Editions de l'Aube, IRD, n.s., 14.

Jolivet, M.J., and P. Léna. 2000. "Des territoires aux identités", *Autrepart, Cahier des sciences humaines*, Editions de l'Aube, IRD, n.s.,14: 5-16.

Kidwai, A.H. 2003. "Contextualisation des identities territoriales et sociales. Deux régions historiques et une minorité en Inde du Nord", *in* P. Gervais-Lambony, F. Landy and S. Oldfield (eds), *Espaces arc-en-ciel. Identités et territoires en Afrique du Sud et en Inde*, Paris, Editions Karthala, pp 167-185.

Latour, B. 1991. *Nous n'avons jamais été modernes*, Paris, La Découverte.

Lévy, J. 2003. "Territoire" in Lévy J. and Lussault M. (eds.), 2003, *Dictionnaire de la géographie et de l'espace des sociétés*, Paris, Belin.

Lévy, J. and Lussault, M. (eds.). 2003. Dictionnaire de la géographie et de l'espace des sociétés, Paris, Belin.

Luchesi, B. 2006. "Fighting enemies and protecting territory : deities as local rulers in Kullu, Himachal Pradesh", *European Bulletin of Himalayan Research*, no 29-30, Summer 2006: 62-81

Martinez Mauri, M. 2007. "Une réserve de la biosphère non achevée. Les Kunas et la conservation de la nature", *Cahiers d'anthropologie sociale*, no. 3: 97-107.

Ozouf-Marignier, M.-V. 2009. "Le territoire, la géographie et les sciences sociales : aperçus historiques et épistémologiques", in M. Vanier (ed.), *Territoires, territorialité, territorialisation. Controverses et perspectives*. Presses universitaires de Rennes, pp. 31-35.

Pecqueur, B. and Zuindeau, B. 2010. "Espace, territoire, développement durable", *in* B. Zuindeau (ed.), *Développement durable et territoire*, Villeneuve d'Ascq, Presses Universitaires du Septentrion, pp 49-58.

Pfaff-Czarnecka, J. and Toffin, G. (eds.). 2011, *The Politics of Belonging in the Himalayas, Local attachments and Boundary dynamics*, Delhi, Sage Publications.

Pinchemel, P. and Pinchemel, G. 1988. *La face de la Terre*, Paris, Armand Colin.

Piolle, X. 1991. "Proximité géographique et lien social, de nouvelles formes de territorialité?", *Espace géographique*, no. 4, Dec. 1991: 349-358.

Raffestin, C. 1986. "Ecogénèse territoriale et territorialité", in Auriac F. and R. Brunet (eds.), *Espaces, jeux et enjeux*, Paris, Fayard-Fondation Diderot.

Renard, J. 2000. "Les espaces flous entre agglomérations et pays : l'exemple de la Loire-Atlantique", *Territoires vécus, territoires prescrits : au coeur des recompositions des espaces ruraux*, Abstracts of the 24th ARF Workshop, Toulouse, 25-27 Oct: 253-263.

Revel-Mouroz, J. (eds.). 1989. *Pouvoir local — Régionalismes — Décentralisation — Enjeux territoriaux et territorialité en Amérique latine*, Collection Travaux et mémoires de l'IHEAL, vol. 3, no. 47.

Sack, R. D. 1983. *Human territoriality. Its theory and history*, Cambridge, Cambridge University Press.

Sax, W. 2003. "Divine Kingdoms in the Central Himalayas" in Gutschow N., Michaels A., Ramble C., Steinkellner E., (eds.), *Sacred landscape of the Himalaya*, Vienna, Austrian Academy of Sciences Press, pp 177-194.

Sax, W. 2006. "Rituals of the Warrior *khund*", *European Bulletin of Himalayan Research*, no 29-30, Summer 2006: 120-134

Schwartzberg, J.E. (ed.). 2006[1978]. *A Historical Atlas of South Asia, Digital South Asia Library*, http://dsal.uchicago.edu/reference/schwartzberg/.

Scott, J. C. 2009. *The Art of Not Being Governed. An Anarchist History of Upland Southeast Asia*, New Haven and London, Yale University Press.

Smadja, J. 2011. "Belonging, Protected Areas and Participatory Management. The Case of Kaziranga National Park (Assam) and of the Misings' Shifting Territory", *in* Pfaff-Czarnecka J. and Toffin G. (eds.), *The Politics of Belonging in the Himalayas, Local attachments and Boundary dynamics*, Delhi, Sage Publications, pp 246-271.

Sutherland, P. 2006. "T(r)opologies of Rule (*Raj*): Ritual Sovereignty and Teistic Subjection", *European Bulletin of Himalayan Research*, no 29-30, Summer 2006: 82-119

Théry, H, and R. Brunet. 1992a. "Territorialité", in Brunet R., R. Ferras and H. Théry (eds.), *Les mots de la géographie. Dictionnaire critique*, Paris, Reclus—La Documentation Française, p. 436.

– 1992b. "Territoire", in Brunet R., R. Ferras and H. Théry (eds.), *Les mots de la géographie. Dictionnaire critique*, Paris, Reclus – La Documentation Française, p. 436.

Theys, J. 2010. "Préface" *in* B. Zuindeau (Ed.), *Développement durable et territoire*, Villeneuve d'Ascq, Presses Universitaires du Septentrion, pp 9-11.

Tuan, Y.F. 1990[1974].*Topophilia. A study of environmental perception, attitudes, and values*, New York, Columbia University Press.

Van Schendel, W. 2002. "Geographies of knowing, geographies of ignorance: jumping scale in Southeast Asia" *Environment and Planning D: Society and Space*, vol. 20: 647-668.

Vanier, M. (ed.). 2009. *Territoires, territorialité, territorialisation. Controverses et perspectives*, Rennes, Presses universitaires de Rennes.

Viveiros de Castro, E. 2009. *Métaphysiques cannibales*, Paris, Presses Universitaires de France.

Notes

1. Translated from the French: "Un peu de territorialité crée de la socialité et de la solidarité, beaucoup de territorialité les assassine." (Les mots de la géographie, dictionnaire critique: The words of Geography, critical dictionary).
2. See for example the conference: "Frontières, territoires et pouvoirs en Amérique latine", organised by Marie-France Prévot-Schapira, CREDAL, Paris, 2005 and Revel-Mouroz, 1989.
3. This book is based on the proceedings of a workshop (December 2007) funded by the CNRS (National Centre for Scientific Research, France), by the French-Indian exchange programme run by the MSH (*Maison des Sciences de l'Homme*) and by the French National Research Agency via the North-East India Programme. The final texts making up this book were updated at the end of 2010 and therefore show the state of the political situations and of territorial restructurings at this date.
4. On this topic, see Berti and Tarabout, 2009.
5. On this topic, see Bernard Debarbieux, 1999.
6. See Lévy 2003, Théry and Brunet 1992b.
7. Translated from the French: "les chercheurs insistent tous sur la double dimension, matérielle et idéelle, écologique et symbolique du territoire".
8. Who also refers to Bonnemaison 1996, Bonnemaison and Cambrezy 1996, Ferrier 1984, Piolle 1991, Raffestin1986.
9. The identity component is put to the fore by many French authors, among them: Alphandéry P. and Bergues M., 2004, Blanc-Pamard 1999, Bonnemaison 1981, Bonnemaison 1997, Bonnemaison and Cambrezy 1996, Claval 1996, Collignon 1996, Debarbieux 2006, Di Méo 1998, Grandjean 2009, Jolivet 2000, etc.
10. Translated from the French: "elle permet la décomposition temporelle d'un processus spatial"
11. Translated from the French: "La Terre devient Terre des hommes quand, cessant d'être anonyme, elle est nommée par eux.
12. Translated from the French: "nommer c'est territorialiser".
13. See for example the texts from Lévy, Debarbieux, Ferrier at the entry "Territoire" in the dictionary edited by Lévy and Lussault (2003) pp 907-917, as well as all the texts in Vanier (2009).
14. Translated from the French: "Les unités politico-administratives d'échelle régionale, même lorsqu 'elles ont eu un poids historique important, produisent des identités territoriales faibles car elles marquent peu la conscience populaire. Elles n'acquièrent de valeur sentimentale qu'en cas de correspondance longue et durable entre une identité politico-administrative et une identité 'sociale ' […]."
15. The process began, after India's Independence, with the Sixth Schedule of the Indian Constitution and the establishment of Autonomous District Councils, the aim of which was to protect hill and other tribal communities from the power of the groups from the plains and to preserve their distinct identity and autonomy. This has led to retracing administrative boundaries. Several groups that obtained these Autonomous Councils have achieved territorial autonomy since the 1970s. This is the case of the Khasis and Garos who carved out Meghalaya from Assam in 1972 and who, with the Jaintia, now have their own District Councils in this state. Then the North Cachar Hills and Karbi Anglong also obtained their District Councils. The Bodos acquired their autonomy with the Bodoland Territorial Autonomous District in 2003. Other groups that already have Autonomous councils, such as the Misings, Tiwas, Thengal-

Kacharis, and Deuris are now demanding Territorial Autonomous Districts. Many other communities wish to be included in the Sixth Schedule of the Constitution. In Arunachal Pradesh, demands to create Autonomous District Councils have been made.

16. RSS: Rashtriya Swayamsevak Sangh.
17. On this topic see, among others, texts from Berti, Luchesi, Sax, Sutherland, in the European Bulletin for HimalayanResearch (29-30 Summer 2006), as well as Sax 2003, Berti 2009, etc.
18. Translated from the French: "S'il y a quelque part une articulation démocratique à trouver entre les trois dimensions constitutives du développement durable — le social, l'écologique et l'économique —, c 'est sans aucun doute au niveau local qu'elle pourra le plus concrètement être construite — dans la mesure où c 'est aussi à cette échelle que leurs contradictions ou leurs synergies apparaissent avec le plus de force et d'évidence [...]."
19. Translated from the French: "Le développement durable n'est pas 'fractal'" [...] "[...] la contrainte écologique n'a de sens absolu qu'au niveau planétaire. Aux autres niveaux, des échanges, des substitutions et des déséquilibres sont possibles et les contraintes écologiques sont relatives".

Part I: Claims to Territories and Role of Legacies in Territory-making

1.1. CLAIMS TO TERRITORIES IN THE WESTERN AND EASTERN HIMALAYAS

1.1.1.

Boundary-Making and Border Practices in Northern Pakistan[1]

Hermann Kreutzmann

Introduction

Boundaries are linked to division and convey a message of separation. Territories of dominance are carved out of other structures, integrated into empires or states[2] and defended against outsiders. Milder terms address them as spheres of influence and control areas. In all respects, past and present borders can be perceived as an expression of attribution, interest and power. South Asia is a perfect laboratory for studying border issues and separate entities. The present-day nation states are the result of division. The populations on the subcontinent have experienced tragedies and traumas linked to boundary-making.[3] The legacies are felt in everyday matters when disputed areas, undefined international boundaries and unsolved territorial claims govern the political agenda shared with neighbours. But even within nation states, the division of provinces and the provision of regional autonomy are boundary-making practices. In Pakistan, frequent suggestions have been made to solve issues of regionalism by dividing the existing spatial entities and by creating new provinces. Several models have been discussed ranging from a one-unit solution to an approach comprising 14 provinces.[4] The vast majority of these proposals do not offer any solution to the Kashmir dispute and/or to the status of Gilgit-Baltistan.[5] Most suggestions avoid alluding to the 'burning issue' of contested territories. Only one author has suggested merging the Northern Areas with the neighbouring region to create a province called Hazara (Fig. 1). No serious debate has emerged from these proposals to settle the ongoing dispute. As these recent proposals regularly omit historical references, a different path is followed here attempting to retrieve some aspects of the historical developments which led to boundary-making during the British Empire and to relate these findings to the prevailing crises. The disputed boundaries provide ample space for debating history and interpreting memory, for building territorial entities and highlighting their importance for actors and stakeholders. The status of Gigit-Baltistan within the arena of the Kashmir conflict is explored in the following sections.

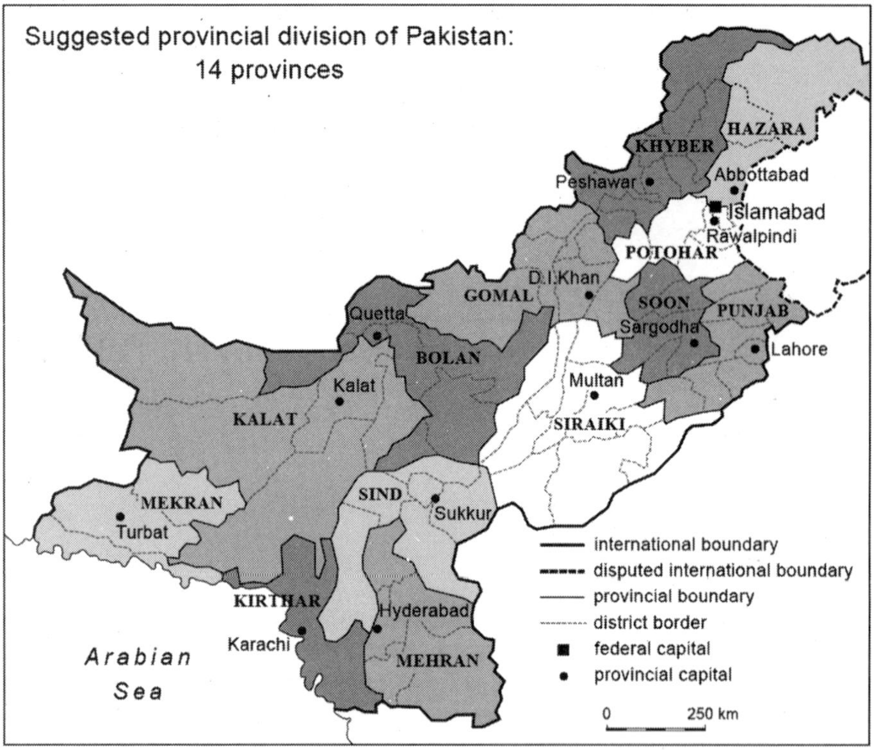

Source: own design based on Tanveer Kayani & Khan 1998, p. 265

Figure 1: Suggested provincial division of Pakistan

Borders

Borders are often visible and identifiable divisions between different entities. Political boundaries separating nation states are the result of developments and the 'making of a frontier'.[6] The period running from the second half of the nineteenth century until the end of the First World War can be perceived as the climax of imperial intervention, the division of the world, and concomitant boundary-making. At the time, the debate held by geographers, historians, politicians and diplomats focussed on the philosophical interpretation of terms such as border, boundary and frontier.[7] Their implementation was discussed using terms such as 'stromstrich' and 'thalweg', which helped boundary commissions to negotiate on the ground what their superiors had discussed as 'natural' and/or 'scientific borders' in the 'Indian Borderland'.[8] The debate revolved around the search for obvious solutions to enable colonial administrations to work on the ground. The British explorer Ney Elias was in favour of the 'hill frontier' as opposed to rivers separating spheres of influence. For him, border practice implies finding "a possibility of coming to an arrangement with Russia on the subject, under which each party should keep the territory he now possesses. […] the Afghans should consolidate the territory they now hold in these regions. For this purpose the three chief steps required are (1) to define the boundaries in every direction; (2) either to conciliate or thoroughly overawe the discontented inhabitants; (3) to make no embarrassing claims for more territory, but rather abandon old claims if more desirable boundaries can thereby be secured. They have a large tract of poor mountainous country divided into a number of petty provinces, the borders of which are still open to dispute. These provinces are inhabited by people who have little or nothing in common with the Afghans, and who hate them with the two kinds of which, taken together, make up perhaps the most intense form of enmity. They hate them with race hatred—both Tajik and Turk; and they hate them as conquerors."[9] Ney Elias's statement clearly shows that the prime interest was to define and separate in order to secure territory. In a second step the affordable cost for colonial power needs to be assessed, either in establishing practices of indirect rule or in controlling the people. Both steps are necessary to find the "desirable boundaries", which he identifies as lying in the high mountain areas.

Information reported by Elias in Badakhshan and East Turkestan, the assumption confirmed by the Lockhart and Woodthorpe mission (Gilgit, Hunza, Chitral and Wakhan) that the northern mountains were impassable for a large Russian army, and enquiries led by Algernon Durand about the weakness of the Kashmiri administration on the Northern Frontier caused Mortimer Durand, foreign secretary to the British Indian government, to formulate an active British boundary policy for the entire tribal belt between Dera Ghazi Khan in the south and Gilgit in the north.[10] Major elements of his plan were diplomatic agreements about the conditions of power at the periphery of the Indian Empire, a definition of the boundaries with Afghanistan as a buffer state, and the avoidance of direct confrontation between Russia and British India. The cost factor is the guiding principle in finding the appropriate measure to tackle the problems of boundary-making.

Mainstream opinion about dealing with sought after territories has been contested by some contemporary critics. In numerous articles and commentaries published in journals and the *London Times*, Gottlieb Wilhelm Leitner challenged the need for territorial expansion and for extending Britain's borders. He influenced the debates and became the advocate of the mountain dwellers concerned. Discussions in political and diplomatic circles about a solution to the boundary issue were commented upon in a fierce critique of Britain's colonial policy in the Pamirs and Dardistan. Gottlieb Wilhelm Leitner had the local people's interests in mind and called for the de-militarization and autonomy of the ethnic groups in this region: "The neutralization of the Pamirs is the only solution of a difficulty created by the conjectural treaties of diplomatists and the ambition of military emissaries. Left as a huge happy hunting-ground for sportsmen, or as pasturage for nomads from whatever quarters, the Pamirs form the most perfect 'neutral zone' conceivable. That the wanderings of these nomads should be accompanied by territorial or political claims, whether by Russia, China, Afghanistan, Kashmir, or ourselves, is the height of absurdity. As for Hunza-Nagyr, the sooner they are left to themselves the better for us, who are not bound to help Kashmir in encroaching on them." (Leitner 1891: 73)

Nevertheless, Leitner's adversaries as the advocates of a 'forward policy' succeeded in their attempts to secure as much as possible for posterity's sake during the 'Great Game'. Boundary-making became a major subject in diplomatic training sessions and colonial administrations.

Formation and Legacies of South Asian Boundaries

South Asia is a prime example and laboratory for experiments and practices in this respect. From the North-West Frontier to the North-East Frontier Agency, the creation of so-called 'buffer zones' including tribal areas, neutralized bulwark states and semi-dependent or fully dependent weak neighbouring countries supported the goal to safeguard the 'jewel in the crown', the rich potential of the Indo-Gangetic Plains. Therefore wealthy princely states such as Kashmir and comparatively poor mountain communities—e.g. Hunza and Nager—became the focus of treaties and agreements, of leaseholds and indirect rule, with the aim of marking out boundaries in less-populated and strategically acceptable locations. Boundaries that served the purpose of the imperial rule and colonial administrations are not necessarily conducive to the aspirations of newly independent states. These boundaries are of central importance in areas of competition especially in cases where Partition is a vital element of the independence process and of post-colonial disputes—Bangladesh, India, Pakistan, Burma. Consequently, boundaries become sources of conflict, interrupting the flow of goods, ideas and people, as well as creating peripheries within nation states. In the case of Kashmir and the Northern Areas there seem to be no mutually accepted maps of border demarcation available. National statistics do not include contested territories and political representation is readily executed from the capital. The centre of power determines the fate of the periphery and reacts sensitively to all cases of internal unrest and border skirmishes.

In order to perceive the persistence of boundaries and their effects on local populations it is necessary to establish the context in which the creation of the Northern Areas of Pakistan or Gilgit-Baltistan is embedded. The debate on the Northern Areas—which formerly seemed to be synonymous with the colonial Gilgit Agency—is a long story, at least much longer than the time the term has been in use. The Maharaja of Kashmir tried to expand his influence there at the same time as British India and Russia became interested in the Himalayas-Karakoram. Consequently, the case of Gilgit-Baltistan is strongly linked to the Kashmir issue. In the Treaty of Amritsar of 1846 the expansionist policies by Maharaja Gulab Singh had been formalized when British India ceded the "hilly or mountainous country with its dependencies situated eastward of the river Indus and westward of the river Ravee" to the Dogra rulers. Although the description of the territory is rather vague, it leaves ample scope for competition and expansion. The immediate crisis occurred a century later at the time of Partition.

The Kashmir stalemate situation: the roots of an ongoing dispute

India, Pakistan and the people of Kashmir disagree about the constitutional and territorial status of the formerly largest princely state of the Indian Empire (Fig. 2). The continuing disputes originate from two viewpoints.

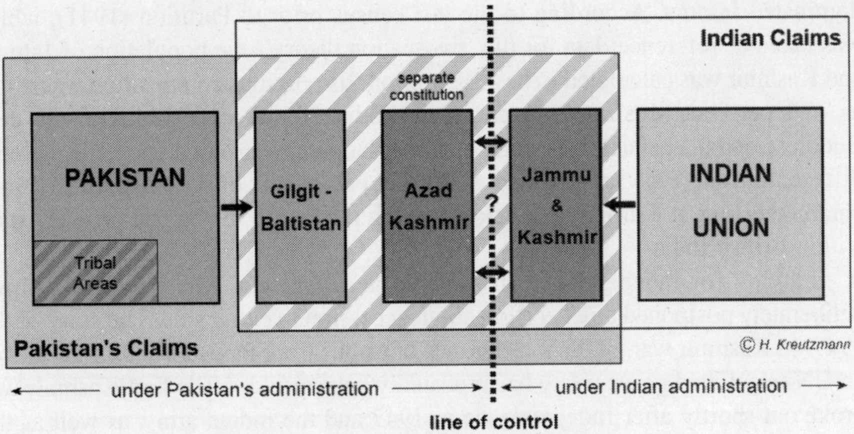

Figure 2: The Kashmir stalemate situation

First, the implementation of the so-called 'two-nation theory' has failed in Kashmir. The Maharaja of Kashmir, Hari Singh, who belonged to the Hindu Dogra dynasty, ruled over a population the majority of which embraced the Muslim faith. Exceptions to this rule occurred in Buddhist-dominated Ladakh/Zanskar and Hindu-

Language groups 1941

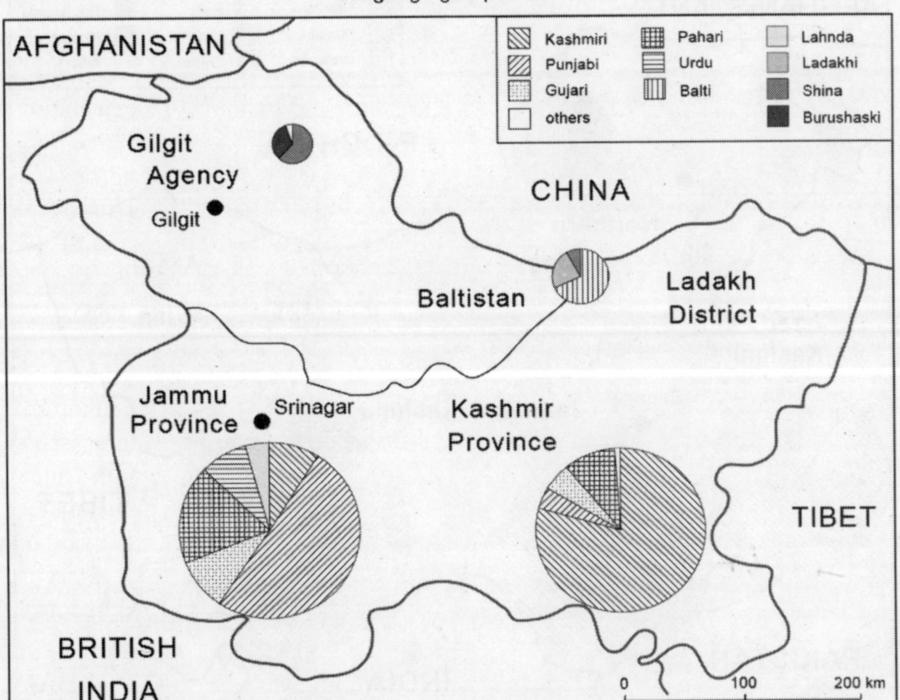

Source: own data processing on the basis of Census of India 1941 © H. Kreutzmann

Figure 3a: Language groups 1941

dominated Jammu. According to the last census prior to Partition (1941), which was taken as reference data for the 'two-nation theory', the population of Jammu and Kashmir was calculated to be 4.02 million. Its religious composition was given as 77.1 per cent Muslim, 20.1 per cent Hindu, 1.7 per cent Sikh, 1.0 per cent Buddhist, and 0.1 per cent Christian. As far as languages are concerned, the regional differentiation (Fig. 3a) showed a similar patchy, inconsistent picture. Both parameters hint at a difficult decision-making process in the largest princely state within British India.

Playing for independence from India and Pakistan, Maharaja Hari Singh deliberately postponed any decision about accession to either side. The story of the 1947-48 Kashmir war, of the interference of troops from Pakistan and India, and of an UN-negotiated peace treaty has repeatedly been retold.[11] The first Kashmir war broke out shortly after Independence in 1947 and the Indian army, as well as the Pakistan army, was commanded by British high-ranking officers. The confrontation created a paradoxical situation: in October 1947 Field Marshal Sir Claude Auchinleck was the Commander-in-Chief of both the Indian and Pakistan Armies. Some authors suggest that this fact led to the early involvement of the United Nations in peace negotiations.[12] According to Lamb (1994: 71-72) "the opinion of

Language groups 1981

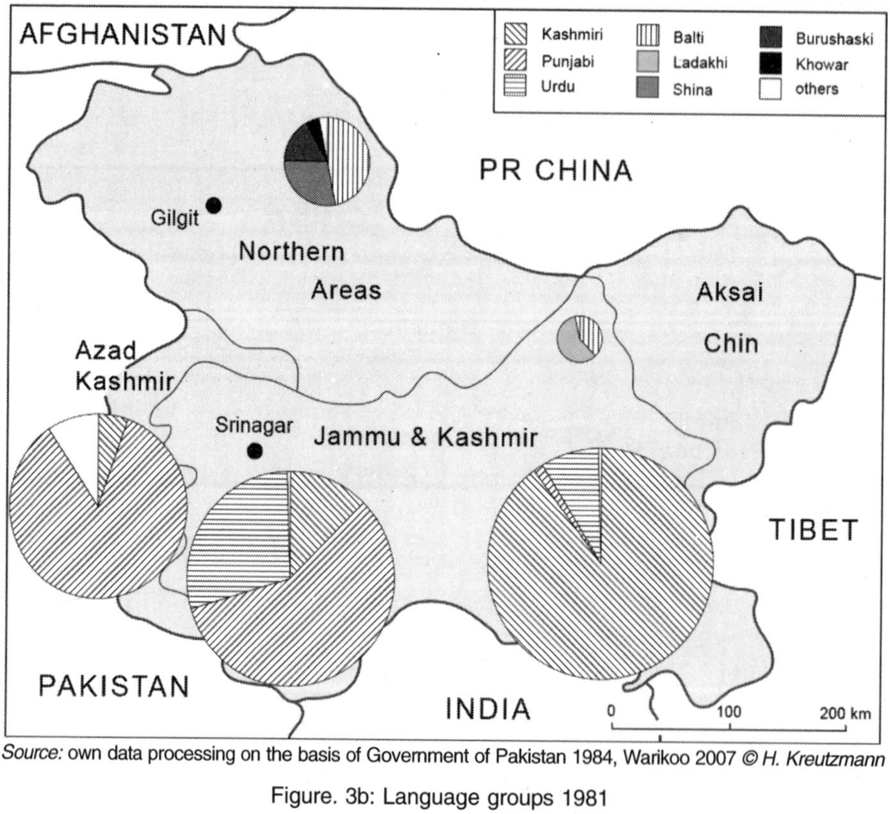

Source: own data processing on the basis of Government of Pakistan 1984, Warikoo 2007 © *H. Kreutzmann*

Figure. 3b: Language groups 1981

most [contemporary] British observers [...] was that the best solution lay in a partition of the old state of Jammu and Kashmir, essentially with Ladakh and much of Jammu going to India and the rest to Pakistan." The concept of Partition was reiterated by the British UN representative in 1950 but rejected by India and Pakistan in favour of a unitary plebiscite throughout Kashmir.

This confrontation resulted in a cease-fire line separating Indian and Pakistani spheres of influence. With minor deviations it survived the 1965 war, which saw Kashmir as the major military theatre. The Tashkent agreement of January 1966, negotiated by the Prime Minister Lal Bahadur Shastri and President Ayub Khan under the mediation of the Soviet Prime Minister Aleksei Kosygin, confirmed the status quo and the retreat of troops behind the actual line of control. During the third Indo-Pakistan war of 1971, Kashmir played a secondary role, and the 1972 Simla Conference further prolonged the status quo. Since then all demands for an impartial and internationally supervised referendum/plebiscite on the future status of Kashmir have been postponed. The population distribution pattern in the disputed region has significantly changed (Fig. 3b), which is not only a result of population growth and its administrative make-up but also reflects the impact of migration and mobility.

Since the mid-1980s, fighting for control of the Siachin glacier region between specially trained army units has taken place every summer (Fig. 4). The barren tracks of the uninhabited Siachin region form a challenging and remote battleground where both sides fight because of the shortcomings of land surveyors' and diplomats' ambiguity in a commercially unproductive territory. Unfortunately, servicemen from the mountain regions who are adapted to, and who are able to survive at altitudes above 5,000 m have become the victims of this senseless fight year after year. During the Kashmir wars, huge amounts of money were tied up in armoury in the Northern regions for a battle between two independent states where local residents are pawns in a competition neither side might ever totally win. In 1999, another war between the two competing parties was barely averted when the 'Kargil Crisis' led to military encounters, territorial gains and losses, and to numerous victims among the soldiers who mainly originated from the mountain regions.

A second point should be emphasized because of its importance in related disputes: the extent of the state ruled by the former Maharaja of Kashmir and its status under international law are inconsistent in the demands of all the parties concerned. To the present day, the whole of Gilgit-Baltistan—the former Gilgit Agency including the principalities of Hunza at the time, Nager, and the governorships of Punial, Yasin, Kuh, Ghizer and Ishkoman, the Chilas and Baltistan Districts—are marked on Indian maps as part of Indian Kashmir. According to this point of view, Kashmir borders in the west with Chitral (North-West-Frontier Province/Khyber-Pukhtunkhwa)[13] and in the north with the People's Republic of China (Fig. 4). But India also repudiates the present frontier with China. This delineation originates from the 1963 Pak-Chinese Treaty, which involved a settlement over 8,800 sq km of disputed territory of which Pakistan has since controlled 40 per cent. In addition, Chinese claims for Aksai Chin, which followed

Figure 4: Disputed territories and constitutional peculiarities of Western High Asia

the construction of the Xinjiang-Tibet road through this uninhabited territory in 1956, are unacceptable for India. Consequently, Indian maps indicate that Aksai Chin is within its national boundaries.

Pakistan's views have changed over time. In the aftermath of the local uprising causing the abolition of Dogra rule in Gilgit and Baltistan, a short-lived 'Independent Republic of Gilgit' was established on 1 November 1947, preparing the way for the unanimously accepted accession to Pakistan.[14] Consequently, the official version of the Pakistan Government distinguishes between Kashmir on the one hand and the Gilgit Agency (Northern Areas/Gilgit-Baltistan) on the other. This viewpoint is supported by a lengthy historical investigation and legal interpretation based on colonial files regarding the status of certain territories in the Gilgit Agency. In 1941 an internal, administratively binding decision summarized the results of a previous discussion about the two principalities in question: "Hunza and Nagir [Nager]: – Though these are under the suzerainty of the Kashmir State, they are not part of Kashmir but separate states".[15]

This deliberate uncertainty in the formulation of the legal status is one of the obstacles to a negotiated solution. The Government of Pakistan has treated the Northern Areas and Kashmir as separate entities, which is reflected in different constitutional configurations (Fig. 5).[16] Azad Kashmir (AK) is governed by its own President elected from an assembly composed of the AK Parliament and the AK Council. By contrast, Gilgit-Baltistan is granted neither provincial status within Pakistan nor a similar semi-autonomous parliamentary setup like that of Azad Kashmir (Fig. 2, 3b). In recent years there have been attempts by Azad Kashmir politicians to link Gilgit-Baltistan to their issue of pressing for a plebiscite to be held throughout Kashmir. Although this move seems to enhance their chances of a

Constitutional set-up of Pakistan

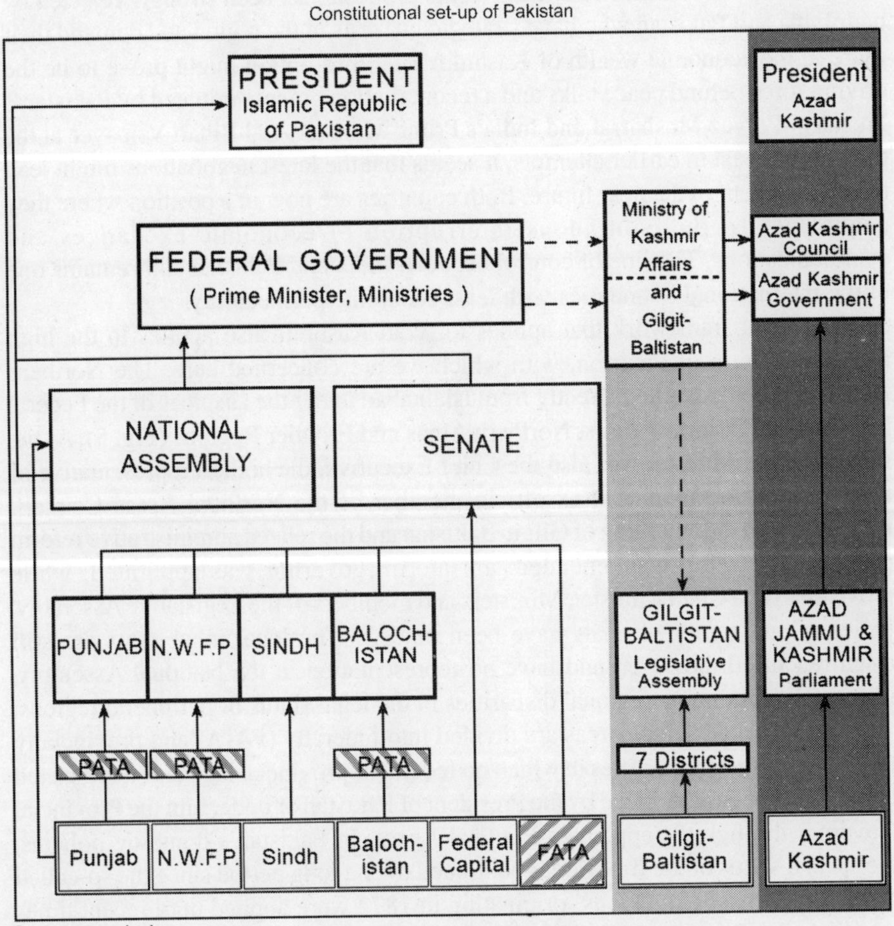

Source: own design

Figure 5: Constitutional set-up

vote in favour of Pakistan—in the event where a referendum on the aspired affiliation to either side is ever held—outspoken representatives of Gilgit-Baltistan refuse to cooperate. In their opinion, the struggle for independence has succeeded in dismissing any link with Kashmir. In recent times the federal government has made moves to unify both regions. These plans have been rejected given that there are no ethnic and regional similarities, no traffic link, and no economic exchanges. The inhabitants of Gilgit-Baltistan (approx. 0.87 million according to the latest population census of 1998) once again fear domination by Kashmiri bureaucrats as the population of Azad Kashmir falls to around three million.[17] Therefore they advocate an independent province with similar civil rights and representation to those in the other provinces, not a separate constitution as in Azad Kashmir.

Both India and Pakistan claim to be the rightful representatives of the people of Kashmir, yet in recent years Kashmiri nationalists have promoted the creation of an independent Kashmir composed of Pakistan-controlled Azad Kashmir and Indian-held Jammu and Kashmir Province.[18] This proposal has been strongly rejected by both India and Pakistan who have strategic interests in the region and demand their share in the economic wealth of Kashmir. The third option might prove to be the driving force behind peace talks and a reconciliation process initiated by Pakistan's President Pervez Musharraf and India's Prime Minister Atal Bihari Vajpayee at the time. In contrast to earlier attempts, it seems that the latest negotiations might lead to an agreement in the near future. Both countries are now in a position where they cannot afford the continuous interruption of economic exchanges and communications. The line of control between India and Pakistan still remains one of the international boundaries with least economic permeability.

The legal framework that applies to Azad Kashmir also applies to the high mountain region of Pakistan, with which we are concerned here. The Northern Areas had been governed directly from Islamabad under the auspices of the Federal Minister for Kashmir Affairs, Northern Areas and Frontier Regions (Fig. 5). At the same time the Minister was also the Chief Executive, the highest representative of the Northern Areas and an ex-officio member of the Northern Areas Council. Recently, with the emerging of Gilgit-Baltistan and the related administrative reform package, the setup was amended: an interim governor was appointed, while provisions for a Chief Minister, Ministers and members of the Legislative Assembly, judiciary and senior officials have been made.[19] The local inhabitants are still disenfranchised, however, and have no representation in the National Assembly, illustrating continuing regional disparities in the legal status of peripheral regions.

The so-called Tribal Areas are divided into federally (FATA) and provincially (PATA) administered entities in which no federal or provincial legislation is enacted unless it has been authorized by the President of Pakistan or under him the Provincial Governor through the appointed Political Agent. In Pakistan's domestic policies, the special status of the tribal areas continued over a long period since the so-called 'Frontier Crimes Regulations' originating in 1872 were applied until recent times and since the Government left internal affairs to the tribal leaders (*malik*, *sardar*). Renewed attention has been drawn to the Tribal Areas after the Pakistan Government started operations against Taliban supporters and their strongholds in clean-up operations mainly focussing on Waziristan. The fierce fighting is ample evidence for the limited control the administration and army can execute in this border region to Afghanistan.

To sum up, the mountain regions of Pakistan in the Hindukush, Karakoram and Himalayas are characterized by a state of uncertainty marked by a special legal status, direct and indirect rule, and a limited validity of certain civil rights. All these peculiarities are linked to colonial and geopolitical legacies. On the other hand, huge subsidies have been allocated to these regions, which fare much worse than the rest of the country when a comparison is made of average provincial incomes. Azad Kashmir and Gilgit-Baltistan have been allocated substantial funds for regional development. These aspects need to be highlighted when it comes to a discussion on participation, governance and civil society.

Central administration and regional conflicts in Northern Pakistan

Gilgit-Baltistan is neither represented in Pakistan's National Assembly nor in its Senate. The territories strongly claimed by all governments are administered directly from Islamabad via the Ministry of Kashmir Affairs and Gilgit-Baltistan Administration. So far no minister from Gilgit-Baltistan has ever been appointed. Hence, the constitutional situation in Gilgit-Baltistan is still clearly different from the one in the centrally governed Tribal Areas (FATA) and in Azad Kashmir (AK). The Tribal Areas send representatives to both houses of parliament and have assumed a major role in forming the government.[20] The Pakistan-administered parts of Kashmir have been granted their own constitution and autonomous government. Nominally, an independent president represents Azad Kashmir vis-à-vis other countries.

For years the political representatives of the Northern Areas attempted to put an end to the discrimination committed against them and to obtain the same civil rights on the basis of Pakistan's constitution. The administrative reforms that were heralded on the occasion of President Zulfiqar Ali Bhutto's visit in 1972 and subsequently implemented abolished the authority of the traditional élites in Northern Pakistan, leaving a power vacuum. From the start, the representatives of the national government in Islamabad were confronted with the local population's demand to be granted the status of a province (Bhutto 1972: 173). As an interim solution, a *Resident and Commissioner* took charge of the newly created districts of Gilgit, Baltistan and Diamir. In 1974 Hunza was the last princely state to be incorporated within the Northern Areas; at the same time two new districts—Ghizer and Ghanche—were created by subdividing Gilgit and Baltistan. An assembly of representatives—the Northern Areas Advisory Council—with 16 elected members became the highest regional body, a function it kept until 1999, when it was renamed Northern Areas Legislative Council (NALC) and its membership and portfolio were expanded.[21]

During the military dictatorship of Zia-ul-Haq (1977-1988) this administrative structure was modified, and the Northern Areas were reorganized into three districts.[22] The administrative centralism practised in the Northern Areas was further strengthened under martial law. In response and parallel to the democracy movement at national level, a Movement for the Determination of the Constitutional Position of Northern Areas was formed, calling for the constitutional integration of the Northern Areas into Pakistan and electoral rights for the population. Among the committee's demands was the release of imprisoned demonstrators and of judges and public officials who had gone on strike, equal pay for civil servants, and equal grants for students from the region.[23] The Kashmiri citizens preferred—and still prefer—a union with Azad Kashmir, because the Northern Areas are expected to vote for Kashmir to join Pakistan in a future referendum in accordance with the UN resolution.[24] In addition, a Kashmiri[25] majority population anticipates economic advantages and administrative jobs in an amalgamated and upvalued province. A resolution presented by a prominent representative of the Northern Areas to President Zia-ul-Haq in 1987 demanded independence and equal opportunities, as well as general infrastructural and economic improvements:

- "The determination of political status
- Appointment of a local advisor for Northern Areas
- Reservation of all Northern Areas posts for the people of Northern Areas
- Participation at local, national and international level
- Special provision for the participation of people of Northern Areas in all services
- Establishment of Secretariat like Azad Kashmir in Northern Areas
- Right to appeal in High Court and Supreme Court
- Increase in the allocation of seats in professional colleges
- Reservation of seats in public and other educational institutions for the students of Northern Areas
- All the services of Northern Areas should be brought at par with Federal Services […]
- Discouragement of the people responsible for creating religious tension"[26]

Yet so far these demands have been only partially met. During Benazir Bhutto's first term of office (1988-1990), plans to grant provincial status to the Northern Areas became more concrete. After the internal regional subdivision had been increased again to five districts in November 1989, the plan for the Northern Areas was to introduce a similar model to the one in the Federally Administered Tribal Areas (FATA).[27] The presidentially decreed change in government in Islamabad in August 1990 did not engender any progress. The plan was put on hold. The call for constitutional recognition of the Northern Areas continues, backed by protest demonstrations and strikes that lend momentum to the formation of a regionalist movement.[28] Progress was only made after the creation of a three-month interim government tolerated by the army and led by Prime Minister Moeen Qureshi following the summer riots in 1993. The official agreement was:

> "The federal government has upgraded the administrative and judicial set up in the Northern Areas *without changing the constitutional status of the area.*

> The Northern Areas will have the *chief executive with a status of federal minister*. The Judicial Commissioner will be appointed for the area, who will be a serving or a retired judge of the high court. The Northern Area's administration will also have the *executive authority like a province*.
>
> [...] a greater degree of administrative and judicial autonomy to the Northern Areas administration had been given in order to ensure full participation of the people in the affairs of the country and the area.
>
> The Northern Area's council has been expanded from its present 21 members to 26 members. The Chief Executive of the area would be the chairman of the council, who would appoint three members of the council as his advisers. The administration of the area would be delegated the powers of the provincial government.
>
> [...] the law enforcing agencies in the area would also be reorganised and the Chief Commissioner office would also be restructured.
>
> [...] the *Northern Areas administration* would be attached with the federal government and it *would be answerable to the Federal Government*."
>
> (*The Muslim* 5.10.1993, italics by HK)

Demands from earlier petitions were at least partially satisfied. On the one hand, this restructuring gave the Northern Areas similar rights to those of a province; on the other hand, the constitutional *status quo* was maintained with respect to electoral rights and central administration.[29] All amendments were laid down in the Northern Areas Legal Framework Order of 1994, which stated that the Chief Executive would be the Federal Minister for Kashmir and Northern Areas affairs in a personal union. Further modifications were applied in 1999 when the Supreme Court urged a solution for the unconstitutional status of the Northern Areas.[30] Consequently, the NALC discussed an 'Interim Constitution Act' which would enable the Government to assign the Northern Areas the status of a province including all associated institutions. Modelled after the Azad Kashmir constitution, an elected president would represent a government consisting of prime minister, minister and legislative assembly.[31] Even when an administrative reform was executed in Pakistan to install the 'nazim' system, the Northern Areas were spared. In 2004, the army still justified the presence of 'Army Monitoring Teams' in the Northern Areas, although they had been abolished in the rest of Pakistan two years earlier. The Northern Areas Deputy Chief Executive Fida M. Nashad legitimated their existence because "the Northern Areas were not a constitutional part of Pakistan".[32] Nevertheless, the Northern Areas have not yet been awarded a provincial status. The announcement of a 'constitutional package' for the Northern Areas on October 23, 2007 during the visit of President Pervez Musharraf to Gilgit came as somewhat of a surprise. The NALC was given the status of a "legislative assembly with powers to debate and pass its budget. The existing council has 36 seats, 24 elected and 12 reserved (six for women and six for technocrats)" (*Dawn* 24.10.2007). The post of deputy chief executive was renamed "chief executive with full administrative and financial authority and the existing chief executive (a federal minister) would be the chairman of the Northern Areas government".[33] The amendment to the Legal Framework Order (LFO) for the Northern Areas included the transfer of administrative and financial power from the Ministry of Kashmir and Northern Areas (KANA) to the newly named Northern Areas Legislative Assembly (NALA). Although a budget of 7.5 billion PRs was henceforth transferred from the federal government to the Northern Areas for further allocation, the main issue of constitutional status was yet to be solved. The prescribed terminology within the Government of Pakistan was given in 2004 as: "The areas constitute an integral part of Pakistan, but is not a federating unit".[34]

Consequently, despite cosmetic reforms and amendments of the Legal Framework Order, the Northern Areas' demand for a provincial status remained unfulfilled. The 1999 Supreme Court verdict ruling that the Government of Pakistan ensure that the people of the Northern Areas enjoy their fundamental rights, namely, that they are governed through their chosen representatives and have access to justice through an independent judiciary, among other things, for enforcement of their Fundamental Rights guaranteed under the Constitution, has so far not been implemented. The Northern Areas have remained in a constitutional limbo because the issue of the Northern Areas' allegiance to Pakistan has been intrinsically tied up with resolving the Kashmir dispute and therefore deferred. Consequently, enfranchisement still figures at the top of the agenda of demands and has gained symbolic value over time. Gaining equal status with other Pakistani citizens is judged in light of what is perceived as a legitimate democratic demand by the residents of the Northern Areas. The right to vote in general elections has been repeatedly denied the inhabitants of the Northern Areas. The diachronic reconstruction of the 'uplifting' of Northern Areas' citizens to the status of their fellow-Pakistanis has shown that democratization is sidelined in Gilgit and Baltistan in the debate about the constitutional status (Photograph 1).

From Northern Areas to Gilgit-Baltistan—Name Change or Move Forward?

The demand for political integration and the call for economic improvement of the Northern Areas gained new momentum in 2009 when another so-called 'reform package' was announced.[35] It was yet another cosmetic change by altering the nomenclature of the region: official notification was given that the Northern Areas would cease to exist and that the territory would be called Gilgit-Baltistan in the future. Prime Minister Yousaf Raza Gilani on his visit to Gilgit announced: "The people of Gilgit-Baltistan would have their own governor, chief minister, an independent judiciary and all institutions which came under the new system."[36] Prior to his visit a transitional plan was designed that was constituted by the President's Secretariat. Despite continuing disenfranchisement, this move is recognized as a major step to delink Gilgit-Baltistan's future from the Kashmir stalemate, for transferring substantial responsibilities to Gilgit, and for raising the status of Gilgit-Baltistan closer to that of other provinces in Pakistan. Consequently, it is not surprising that the headline in an article on 12 November 2009 read: "Polling got underway for 23 legislative seats in the Gilgit-Baltistan province."[37] Even if it

Photograph 1: Demand for equal citizenship: "Vote is our right". Photo by H. Kreutzmann.

is only a small step and sometimes blurred in semantic opaqueness, the widening gap in the nexus of the Kashmir issue on the one side and the Gilgit-Baltistan strive for provincial autonomy on the other provides some consolation for the latter.

The Northern Areas' location is often mistaken for the North-West Frontier Province, especially when Western media report about suicide attacks and military clashes in 'Northern Pakistan'. Therefore the name change was attributed quite high priority by local and regional representatives and stakeholders.

Borders on the Move, Changing Regionalization and the Role of Stakeholders

In the former princely state of Chitral, which was separated from Gilgit in 1896 and incorporated into the newly created North-West Frontier Province (NWFP) in 1901, the Pakistani administration gradually restricted the *mehtar*'s authority from 1953 onwards by appointing a Political Agent. Up until 1969, Chitral, Swat and

Dir were under direct control as Federally Administered Tribal Areas, before the traditional rulers lost their last privileges in the course of reforms of social structures. Later, these areas were given their own districts within Malakand Division. In the 1973 constitution, the basic elements of which are still valid today, Chitral is a Provincially Administered Tribal Area (PATA) within the NWFP.[38] Both administrative areas have been granted special status in the constitutional reality of Pakistan. These phenomena are primarily due to the persistence of colonial administration principles in border areas rather than to a determined Pakistani nationalities or minorities policy. The territorial borders of administrative units primarily followed pre-colonial and colonial patterns of regional power.

The administrative setup is still quite clear. The Hindukush regions of Chitral are an integral part of Pakistan as they are included in the North-West Frontier Province, now renamed Khyber-Pukhtunkhwa. Gilgit-Baltistan continues to hang in limbo since provincial status has not yet been awarded. For bureaucrats and administrators this is a straightforward concept. Nevertheless, in recent years international and national donor agencies treat the mountain areas of Pakistan differently. They regard the northern mountain fringe ranging from Chitral to Ghanche as an entity they have identified as Northern Areas and Chitral (NAC)[39] or more recently as Gilgit-Baltistan and Chitral (GBC; Fig. 6).Under pragmatic considerations they have rekindled a discussion that goes back to colonial times.

Developmental practices by Non-Governmental Organizations (NGOs) have created new perceptions and handling of space. International actors defy the

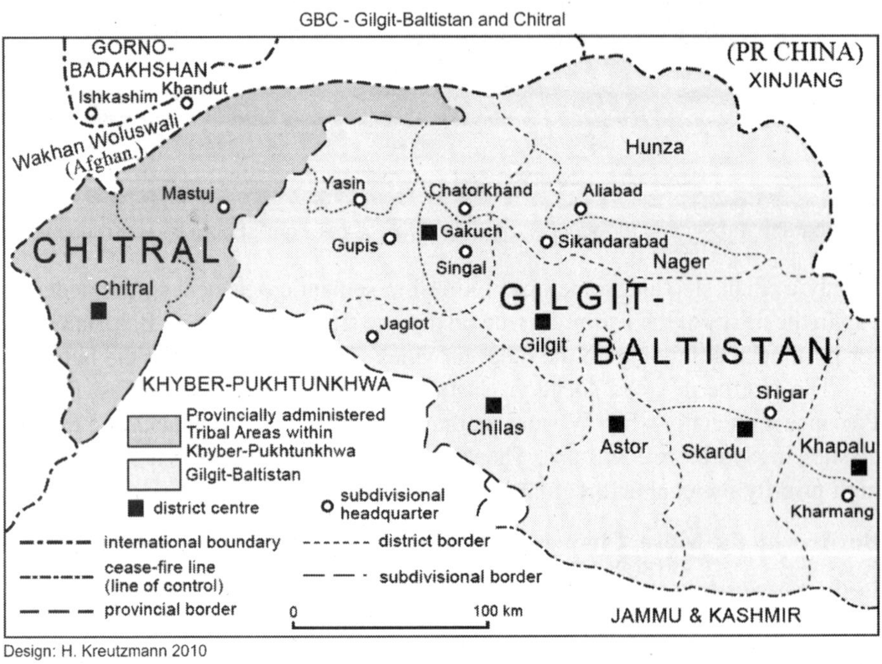

Figure 6: GBC—Gilgit-Baltistan and Chitral

Photograph 2: Graffiti of Balawaristan National Front along Karakoram Highway.
Photo by H. Kreutzmann.

administrative separation for the purpose of defining project areas for the implementation of their development packages. Superseding existing Pakistani border demarcations by development agencies is for the betterment of mountain people. In terms of living conditions and participation in economic affairs, GBC (Gilgit, Baltistan, Chitral) can be regarded as one entity. Civil society groups in Chitral, Gilgit and Skardu are discussing their demand about unifying the areas as a 'mountain province'.

From a rather different perspective, 'mountain nationalists,' striving under the banner of an entity called 'Balawaristan' (Photograph 2), combine Chitral and Northern Areas and extend their claim eastwards towards Ladakh. Here is an endeavour to justify historical roots in a common cause linked to the mountainous regions. The protagonists of the 'Balawaristan National Front' constantly repeat their mantra that Northern Pakistan remains a 'Pakistan-occupied' territory; if the Northern Areas are not covered by the Pakistan Constitution, then they should be permitted to create the independent state of 'Balawaristan'. So far the following of this Gilgit-based movement is rather limited in contrast to the claims they promote via Internet.[40] Indian activists and diplomats who interpret the term indiscriminately as an historical spatial entity are backing the Balawaristan case.[41] Therefore, by constant repetition the newly coined term Balawaristan has become familiar in public discourse.

Development and governance in Gilgit-Baltistan

We have discussed the reasons for the Pakistani government not giving full provincial status to Gilgit-Baltistan. While economic support was increased manifold during the last decade of the Cold War in the 1980s, democratic rights and equal participation in decision-making were denied. Dual governance structures have been permitted to cover-up these deficits while the citizens are deprived of basic democratic rights. Local people and regional representatives are restricted from active participation in democratic governance. NGOs have created their own entities for practical purposes. In fact, they have created a parallel structure, e.g., village organizations (VO) and Local Support Organizations (LSO) enacting infrastructure development and governance that was initially assigned by successive Pakistani governments to appointed and/or elected bodies. NGOs somehow act as congruent units at district, sub-divisional and union council levels as strong players, often more efficient and embedded in society.[42] In some respect, NGOs can be far more effective in developmental projects than a labyrinth government. Every government—civilian or military—has experimented with its own designs for incorporating the mountain regions into the fold of mainstream Pakistan. The pioneering community development project based on a Village AID Programme in 1953 was followed by Ayub Khan's so-called 'basic democracies' that created the still-valid structure of the Union Council system. Zulfiqar Ali Bhutto transformed the 'basic democracies' into his 'people's work programme,' and subsequently introduced an 'integrated rural development programme'. In the process he abolished 'princely rule' and incorporated former independent princely states into the new structure of Northern Areas. Zia ul-Haq experimented with 'community basic services' and implemented the so-called 'local bodies and rural development' that amalgamated the concepts of his predecessors. The change in terminology did not improve the basic conditions of the Northern Areas and its people. At the same time the effect of the completion of the Karakoram Highway was felt in the Northern Areas, encouraging international NGOs to launch activities in the field of rural development. Benazir Bhutto addressed the development deficits in the Northern Areas through a 'social action programme' that left its mark by opening schools in remote settlements. Pervez Musharraf provided a 'poverty alleviation fund' for decentralized development. The administrative reforms that reached the provinces and districts of Pakistan did not leave their mark on the Northern Areas. The current presidency of Asif Ali Zardari has completed the process by rejecting the regional term 'Northern Areas' and replacing it with 'Gilgit-Baltistan'. The new reforms promise to restructure the Northern Areas in a province-like unit, and offer relief measures under the Benazir Income Support Programme (BISP).[43] Small successes in reform have kept the discourse on equal rights and democratic participation in Gilgit-Baltistan alive.

Naming games in the mountain belt

Historical roots are made responsible for a number of names that have been suggested for the Northern Areas. Bolor is often put forward as a possible term for finding a consensus. Bolor has the advantage that its regional definition is rather ambiguous and vague.[44] A variety of authors locate it in a wide-ranging area from the Pamirs and the Karakoram to Kashmir and Ladakh. Somehow they neglect the synonymous use of Bolor for Kafiristan (land of the infidels) in early sources.[45] Another such term is Dardistan, the favourite denomination of Leitner who commissioned several maps on which the area is depicted.[46] Somehow Gilgit seems to be the centre of a proposed Dardistan. Again the ambiguity of the delineation provides ample space for territorial claims. Some recent players in the naming game in the Northern Areas have suggested acronyms which incorporate certain alphabetic characters or syllables from existing toponyms. Constructions such as Baqahistan—incorporating 'baqah' (Arabic for life) with the '-stan' suffix for area or the newly created toponym Kuhimir—a combination of the Persian words 'kuh' for mountain and 'mir' for ruler—were interpreted in two ways. Kuhimir could be translated as 'mountain leader' as well as be understood as a semi-acronym made up of 'ku' for Karakoram, 'hi' for Himalaya and Hindukush, with the second syllable from Pamir completing the new term.[47] These experiments and suggestions try to avoid ethnic and/or religious connotations. For the purpose of a peaceful solution to the naming conflict they refer to orographic entities which were introduced from Turkic and Indic languages. Attributions such as Karakoram go back to the Schlagintweit brothers, who were explorers in the second half of the nineteenth century.

Others would like to portray the Northern Areas as a country of flowers (Arz-i-Gulistan). Again this is to avoid offending any community that has a stake in the naming game and any political representatives on the Northern Areas Legislative Council. Regional activists and lobbyists strongly suggest names such as Boloristan, which is mainly promoted by residents from Baltistan, Burushal as promoted by speakers of the Burushaski language, Dardistan would be the equivalent for Shina speakers, Sargin being the term preferred by residents of Ghizer district.[48] Some inhabitants feel that Karakoram would be the most appropriate term for the whole region, but then others who perceive that they live in the Pamir, Hindukush or Himalayas reject this exclusionary term. The naming game acts as a proxy for more severe rifts within mountain societies. In their effort to pacify communal fighting and to make sectarian clashes less likely the senior political representatives opted for creating a commission to name the region. A name-finding commission appointed by the Northern Areas Legislative Assembly found a consensus in December 2008 in Gilgit-Baltistan. Finally the President's directive followed that suggestion in May 2009. The conflict over a name reflects the power struggles and regional tensions that are prevalent in Gilgit-Baltistan.

Naming conflicts occur at all levels. Since Pervez Musharraf announced the creation of a new district comprising Hunza and Nager the debate has flared up, not only regarding the seat of administration. At the same time the search for a proper name—Hunza-Nager, Brushal, Kanjut—has created enmity within the region. Naming games substitute socio-economic competition for influence and power. They reflect the power struggles and regional tension that are prevalent in the Northern Areas. Sometimes they function as an alternative playground to disguise the contrasting viewpoints and serious rifts between different factions and lobby groups.

Communal conflicts

Ethnic conflicts have led to recurrent destabilization and political unrest that supports separatist forces in Pakistan. The religious composition of the mountain regions is different from the rest of Pakistan where about four fifths of the population identifies itself as Sunni Muslims. In Gilgit Town which is the focal point of confrontation in the North the denominational formation is tripartite. The three major denominational Muslim communities—Ismailiya, Sunna, Twelver Shia—are roughly of equal size in Gilgit Town. In rural areas regional variation is ubiquitous. While Ghizer and Hunza are dominated by Ismailis, Astore, Baltistan and Nager are predominantly Twelver Shia, whereas both Chilas and Chitral are majority Sunni regions. In addition, the Nurbakhshia of Ghanche in Baltistan need to be accounted for as the sole Islamic sect which is only to be found in Baltistan.[49]

Ethnic conflicts in the Northern Areas are primarily due to communal and sectarian rivalries between the majority Shia and the minority Sunni groups. In the 1980s armed clashes frequently broke out, coming to a head in 1988 when many died (Table 1).[50] In this crucial year a death toll of more than 500 was registered. In the aftermath, the subsequent death toll was estimated by the local authorities to be 373 between 1988 and 2005. Since then, about 110 people have lost their lives.[51] Public perception is that this is a clash between Shia and Sunni factions, very much in tune with similar confrontations in the rest of Pakistan. This obvious rift has developed since the 1980s, during Zia-ul Haq's regime and his Islamization policies.[52] By endeavoring to contain the conflict and to mediate, the administration arranges regular meetings for conflict resolution between different communities. These representatives are the Sunni Anjuman-e-Ahle Sunnat and the Anjuman-e-Imamia, which acts for the Twelver Shia population of Gilgit.

Table 1: Conflict constellations in Northern Pakistan

Year	Region	Groups in conflict	Effects
1982	Chitral	Sunni-Ismaili	August: Ismaili property destroyed in Chitral Bazaar, burning-down of buildings, militant fights causing eight deaths
1983	Gilgit	Sunni-Twelver Shia	October: militant encounters on Muslim holidays (tenth of *muharram*); injured persons
1983/84	Nager, Hunza	Twelver Shia -Ismaili	October: militant encounters on Muslim holidays (tenth of *muharram*); injured persons
1985	Gilgit	Sunni-Ismaili	May: accusation of heresy behavior against Ismailiya and Agitation against the Aga Khan Rural Support Programme by Sunni preachers from Punjab, ban on such preachers and eviction from Northern Areas
1986	Chilas	Sunni-Ismaili	March: fire set to Ismaili *jamaat khana*
1988	Gilgit Town & environs	Sunni-Twelver Shia	January: bomb attacks in Gilgit BazarMay: massacre (Sakwar, Jalalabad) and damage to villages and infrastructure, militant fights with > 500 dead persons), closure of Karakoram Highway

1989	Gilgit	Twelver Shia-Sunni	October: during holiday *Eid-e-Millat-un Nabi* demonstrations and fighting, arrests and curfew imposition
1989	Chitral	Sunni-Ismaili	October-November: campaign against the Aga Khan Rural Support Programme, accusation of being a pioneer institution for the creation of an Ismaili State
1990	Jaglot	Sunni-Twelver Shia	February-March: Kidnapping of a minibus with Twelver Shiite and Ismaili passengers, murder of all passengers in front of a Sunni clergyman; culprits and suspects escape verdict and punishment: 9 persons dead, temporary arrest of more than 1,500 persons
1991	Gilgit, Ghizer	Sunni-Twelver Shia	September: murder in cold blood of 6 Nagerkuts in Ghizer and of six other persons in Gilgit October: murder of two prominent Sunni politicians in Gilgit, ban on down country preachers from Northern Areas
1992	Gilgit	Sunni-Twelver Shia	May-June: militant skirmishes in Gilgit Bazaar: 15 persons dead, 38 arrests and 23 days of curfewAugust-September: militant encounters in Gilgit Bazaar: 7 persons dead, 20 arrests, house searches, substantial confiscation of weaponry, announcement of night curfew
1992	Chitral	Sunni-Ismaili	August: heresy accusations by Sunni clergymen, challenge with regard to theological debates, arrest of 25 Sunni and Ismaili preachers
1993	Gilgit, Nager, Skardu	Sunni-Twelver Shia	August-September: militant encounters in Gilgit Bazaar (> 25 dead persons), murder of two Sunni truck drivers from Hazara in Nager, curfew, house searches for weapons, arrest of local clergymen and ban on preachers from down country, blockage of Karakoram Highway
1996	Gilgit	Federal Government-local people	June: Killing of a person seeking employment in the Northern Scouts triggered-off a brawl in Gilgit, causing substantial destruction of public buildings. One person dead, numerous injured followed by arrests
1999	Chitral	Sunni-Ismaili	August: anti-Ismaili demonstrations in several settlements of Chitral in the aftermaths of the murder of a Sunni JUI leader, subsequent threats against employees of AKRSP, temporary closure

			of AKRSP offices in Chitral and retreat of several Ismailis from Central to Upper Chitral
1999	Gilgit	Sunni-Shia	After the introduction of new Islamiat textbooks from the Punjab Textbook Board Aga Zhiauddin Rizvi, Shia leader in Gilgit, claims that non-equivocal and indisputable books need to be introduced. Shia and Sunni positions needed to be considered. Government authorities consent to look into the matter
2001	Gilgit	Sunni-Shia	Clash of Shia and Sunni students in a Gilgit high school about the textbook issue initiates a spread of protests and strikes all over the Northern Areas where Sunni and Shia communities live
2002		Sunni-Twelver Shia	January: following the national ban on sectarian organizations such as Tehrik-i-Jafria Pakistan (TJP), Sipah-i-Sahaba Pakistan (SSP), Lashkar-i-Taiba, Jaish-i-Mohammad and Tehrik Nifaz Shariat-i-Mohammadi a close-down of all their offices in the Northern Areas is announced, TJP is represented in political bodies with 12 district councilors in Gilgit and BaltistanMarch: reform of national curricula and their publication
2003	Gilgit	Sunni-Twelver Shia	April: new Islamiat text books are commissioned on order of the Ministry of EducationJune: violent protests against Islamiat curricula, several injured persons and arrests
2004	Gilgit and environs, Baltistan	Sunni-Twelver Shia	May: Shia scholar and clergyman announces ultimatum for the implementation of new Islamiat curricula. To mitigate the dispute the bureaucracy is challenged June: 12 days of curfew in Gilgit following public unrest, 4 persons dead, educational institutions in Gilgit Town and Skardu are temporarily closed, numerous arrests of Shia demonstrators, prosecution of alleged crimes such as arson and violent attacks against public buildings and civil servants, securing of unlicensed weaponry, more than 1,000 persons demonstrate in several settlements of Baltistan
2005	Gilgit	Sunni-Twelver Shia	January: murder of the high-ranking Shia clergyman Aga Zhiauddin Rizvi, violent

			protests and retaliation: 12 persons dead, substantial property damage, imposition of curfew, by July more than 30 persons dead March: murder of the recently deposed Superintendent of Police and four accompanying persons on their way to hand-over responsibilities October: 12 persons dead and more than 100 injured civilians following violent encounters between security forces and Shia demonstrators, imposition of prolonged curfew, blockages of KKH and protest in several villages Between 1988 and 2005 it was estimated by government authorities that 373 persons were killed in the Northern Areas during sectarian clashes
2006	Gilgit	Sunni-Twelver Shia	The toll of the riots which started the previous year rose to more than 100 dead and many injured. During the year the situation calmed down, curfews were relaxed, but control posts and patrols remained in place
2008	Gilgit	Sunni-Shia	December: Assassination and attempted murder on leading officials in government institutions were interpreted as sectarian-based attacks
2009 2010	Gilgit Gilgit	Sunni-Shia Sunni-Shia	September: Eight persons were killed in a bomb attack, followed by gun attacks between rival groups July-August: Two people in Gilgit in sectarian strife in Kashrote muhallah in Gilgit Town

Source: compiled on the basis of Abdul Malik and Izhar Ali Hunzai. 2005; *Dawn,* 15.1.2001, 1.6.2003, 21.4., 14.5., 4.6., 6.6., 7.6., 16.6.2004, 9.-14-1., 21.7., 15.10., 17.10., 30.10.2005, 24.1., 5.6., 24-25.10.2007, 16.9.2008, 30.8., 12.9., 30.9., 11.11.2009; *Dawn Magazine* 2004; Herald 1999, 2005; Kreutzmann 1996: 246; Zaigham Khan 1996: 142-143 and own interviews.

In Chitral the confrontation lines that have sporadically led to violent clashes run between the mainly Sunni in the south and the dominantly Ismaili in the north of the district.[53] As in Gilgit, Chitral's rapidly expanding district capital and the trading centres with their changing population structure are often arenas of conflict. In rural areas, important groups are no longer formed on the basis of criteria such as language, origin and social status, but according to membership of specific religious denominations. Hence, spatial factors are of decisive importance in questions of group formation, and changing coalitions have to be considered.

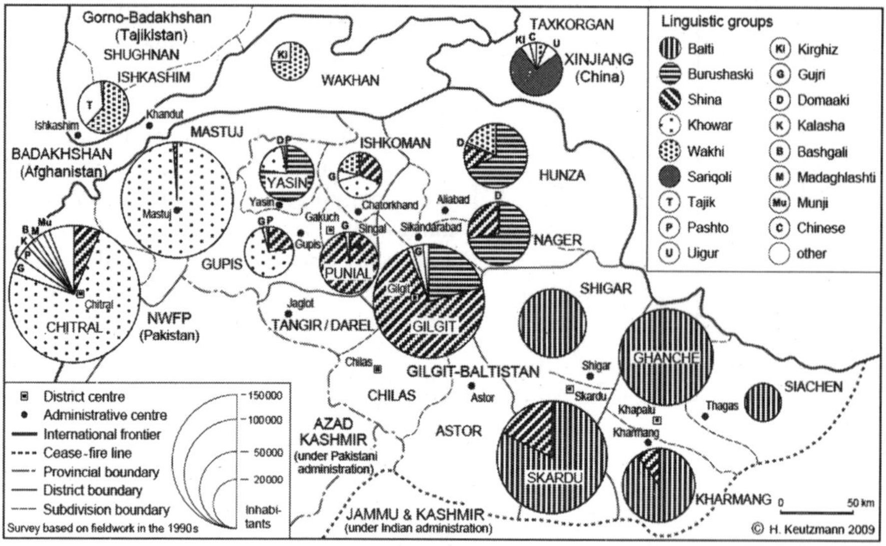

Figure 7: Linguistic variegation in Gilgit-Baltistan

To give one example: from the perspective of the Wakhis in Northern Pakistan, these questions are important because Wakhis belong to a linguistic and religious minority. In Pakistani censuses and surveys there is no breakdown of these criteria, so no data on group composition or on statistical groupings are available to representatives of the communalistic factions or to government employees.[54] With regard to religion, the only distinction made is between Muslims and other beliefs.[55] This procedure is practised throughout Pakistan and contributes to the fact that group sizes and data about members of religious groups are often manipulated for political reasons. Shia representatives overestimate their proportion of the population, just as the Sunnis give unrealistic figures for the groups' proportional representation.[56]

The language factor poses similar difficulties: in linguistic terms, the Census of Northern Areas—the last was conducted in 1998—included the great majority of the autochthonous population groups in the category 'Other Languages', which, after all, accounted for 97.11 per cent of inhabitants in 1981; in 1998, they amounted to 96.9 per cent in the Gilgit District and 98.3 per cent in Baltistan.[57] The first census, conducted in 1951 after Pakistan's independence, had distinguished between the local language groups (Government of Azad Kashmir 1952). Since then, the category 'Others' has not been further differentiated. Only in our own language survey at the beginning of the 1990s could we establish reliable data about the linguistic variegation in Northern Pakistan and adjacent regions (Fig. 7).[58]

The primary characteristic of this major group of local languages is that it is not written; hence, it differs from the category of national languages, for which extensive printed literature is available. Attempts to develop appropriate scripts for local vernaculars have been made mainly by non-local linguists and local language associations. The past few years have seen an increase in academic and private initiatives in this field.[59] Because Pakistan's aim was to standardize rather than diversify its language spectrum after independence, government campaigns have not been launched to create writing systems or to introduce support measures. Schools in mountain regions do not teach local languages; English and Urdu are the accepted medium of instruction, as elsewhere in Pakistan.

In recent years, the dispute has taken on a new dimension. Tensions between denominational groups have found a new arena. The curriculum taught in schools in the Northern Areas has evoked a textbook dispute about the display of religious history and schools of thought.[60] Representatives of the Shia community allege that the 'textbook boards' of Pakistan over-emphasize the Sunni version of Islamic history. The academic and didactic dispute about contents in religious interpretation and Islamiat textbooks escalated in 2005 and subsequently more than 100 people lost their lives in the Northern Areas.

In terms of education and literacy Gilgit-Baltistan fares much better than expected (Table 2). Efforts by government and non-government institutions have contributed to the fact that primary net enrolment rates in Gilgit-Baltistan not only outnumber Azad Kashmir, but at the same time the values also exceed all averages for the four provinces. Within Gilgit-Baltistan, Gilgit and Ghizer districts rank at the top, while only Diamir District fell below the national average in 1998. As regards education, Gilgit-Baltistan has covered ground in recent years and has finally joined mainstream Pakistan. The combined female and male adult literacy rates increased between 1991 and 2005, going from 36 to 59 per cent in Gilgit-Baltistan.[62] In economic terms, Gilgit-Baltistan continues to lag behind although this progression has gained pace. A recent report on socio-economic trends states: "The incidence of poverty has sharply declined [...] from a high baseline of 67 per cent in 1991 to 25 per cent in 2005" (Aga Khan Rural Support Programme 2007: xii). Nevertheless, the so-called 'food inflation' over subsequent years has detrimentally modified the situation that aggravated living conditions throughout Pakistan and as well in Gilgit-Baltistan. The rockfall at Atabad in the Hunza Valley in January 2010 that interrupted traffic on the Karakoram Highway and cut-off 20,000 people from supplies aggravated the situation further.[63] Whenever a crisis occurs, the government is blamed first followed by repeated demands for a provincial status that is expected to ease the handling of any crisis.

Conclusion

The process of territorial reorganization and creation of spatial entities distinguishes Gilgit-Baltistan as a region striving for political and administrative autonomy. The danger of separatism is less virulent here than in Balochistan, Seraiki, Khyber-Pukhtunkhwa or Sind. Gilgit-Baltistan is too small to survive on its own and as a land-locked area would need to be accepted for accession to China. At present such a move seems to be most unlikely. Nevertheless, the close proximity to China and

Table 2: Comparison of education – Gilgit-Baltistan in relation to Pakistan's provinces

<table>
<tr><td rowspan="2">Primary Net Enrolment Rate[a]</td><td colspan="3">1998</td><td>2005-06</td><td></td><td></td></tr>
<tr><td>Total</td><td>Male</td><td>Female</td><td>Total</td><td>Male</td><td>Female</td></tr>
<tr><td>Pakistan</td><td>42%[b]</td><td>47%[b]</td><td>37%[b]</td><td>66%[c]</td><td>73%[c]</td><td>59%[c]</td></tr>
<tr><td>Punjab</td><td>44%[b]</td><td>47%[b]</td><td>40%[b]</td><td>68%[c]</td><td>70%[c]</td><td>66%[c]</td></tr>
<tr><td>Sindh</td><td>41%[b]</td><td>47%[b]</td><td>35%[b]</td><td>67%[c]</td><td>77%[c]</td><td>57%[c]</td></tr>
<tr><td>NWFP</td><td>39%[b]</td><td>47%[b]</td><td>30%[b]</td><td>66%[c]</td><td>81%[c]</td><td>51%[c]</td></tr>
<tr><td>Balochistan</td><td>36%[b]</td><td>44%[b]</td><td>28%[b]</td><td>40%[c]</td><td>47%[c]</td><td>33%[c]</td></tr>
<tr><td>Azad Kashmir</td><td>45%[c]</td><td>46%[c]</td><td>44%[c]</td><td>57%[c]</td><td>58%[c]</td><td>56%[c]</td></tr>
<tr><td>Gilgit-Baltistan</td><td>51%[d]</td><td>58%[d]</td><td>43%[d]</td><td>83%[c]</td><td>100%[c]</td><td>66%[c]</td></tr>
<tr><td>Gilgit</td><td>68%[d]</td><td>73%[d]</td><td>62%[d]</td><td colspan="3" rowspan="5">Notes:
[a] age 5-9, class 1-5.[b] Government of Pakistan. 2006. PSLM 2004/05, p. V; data given for Financial Year 1998-99.[c] Government of Pakistan. 2008. EFA Mid Decade Assessment Country Report, p. 95, 99; own calculations; data given for FY 2001-02/ FY 2005-06 resp.[d] Government of Pakistan. 2001. District Census Reports 1998 (various issues), Table 14; own calculations.</td></tr>
<tr><td>Ghizer</td><td>64%[d]</td><td>71%[d]</td><td>57%[d]</td></tr>
<tr><td>Baltistan</td><td>45%[d]</td><td>55%[d]</td><td>33%[d]</td></tr>
<tr><td>Ghanche</td><td>58%[d]</td><td>66%[d]</td><td>49%[d]</td></tr>
<tr><td>Diamir</td><td>29%[d]</td><td>36%[d]</td><td>22%[d]</td></tr>
</table>

Source: Data compilation and calculations by Andreas Benz[61] (2009) based on Government of Pakistan 2000, 2001a, b, c, d, 2006.

its importance as a vital section of the transit corridor from Khunjerab Pass to Gwadar Port at the Arabian Sea draw special attention to the region that gives some leverage for political demands.

The general claim and desire for enhanced democratic participation is disguised in secondary issues of bureaucratic reforms and splitting up existing divisions. Subdividing administrative units creates jobs in public service. Therefore the struggle for equal participation and adequate representation of Gilgit-Baltistan and its status reflects the negotiations for sharing public resources and civil rights. The disputes about districts and divisions, about the location of district headquarters and hospitals and schools follow the same track. That disenfranchisement in National Assembly elections continues to the present day in Gilgit-Baltistan serves as a reminder that legacies of boundary-making have long-lasting effects and sometime seem to be invariants.

Boundaries, both known to us and delineated on maps, were introduced in the Hindukush-Karakoram-Himalaya ranges in the late nineteenth century. The results of the 'Great Game' have created major sources of conflict and war. Kashmir, the Siachin Glacier and Gilgit-Baltistan are prime examples of virulent disputes which are still burning issues today. The status of the Gigit-Baltistan is a legacy from the British Raj and is rooted in the overall Kashmir conflict. Present competition for more autonomy and representation in federal institutions is linked to the country's uneven participatory approach that is reflected in spatial entities confronted with extra-constitutional status, the application of colonial rules, and the implementation of indirect rule in tribal areas, as well as the direct administration of disenfranchised people by the centre of power. The process of slowly adjusting and changing the position of Gilgit-Baltistan has created conflicts and confrontation along different lines. Borders are the identifiable results of conflict constellations and administrative attempts to resolve them. Besides international conflicts over space there are internal disputes at different levels of society. Contested administrative setups, demands for equitable quota systems and legitimate representation in public services, regionalism and sectarian clashes form the spectrum of social, spatial and political boundary-making in Gilgit-Baltistan.

The history of territorial reorganization and the creation of spatial entities reveal that boundary-making has been seen as a strategy to mitigate prevalent encounters between competitors and/or to solve pending disputes among different stakeholders. In a society that has limited resources for redistribution and in which certain groups enjoy outstanding privileges, the popular movements for territorial readjustments can be interpreted as a way of searching for an equilibrium. High costs are borne by communities and the people in Northern Pakistan.

REFERENCES

Abdul Malik. 2005. 'Village-based Development in the High Mountains of Pakistan: Lessons from the Aga Khan Rural Support Programme (AKRSP)', in P. Gyamtsho *et al.* (dir.), *Sustainable rural development in mountainous regions with a focus on agriculture in the Tibet Autonomous Region*, Feldafing, InWEnt, pp. 113-125.

Abdul Malik and Izhar Ali Hunzai. 2005. 'Promise and Challenge of Pluralism for Sustainable Development in Mountain Areas: Experience and Encounters from Northern Areas and Chitral' (accessible as PDF on site: http://www.akdn.org/mountains/11 - Pluralism and Sustainable Development (Eng).pdf).

Aga Khan Rural Support Programme. 2007. *An assessment of socio-economic trends in the Northern Areas and Chitral, Pakistan (1991-2005)*, Gilgit, AKRSP.

Aitchison, C. U. 1909. *A Collection of Treaties, Engagements and Sanads Relating to India and Neighbouring Countries Vol. XI: The Treaties &c, Relating to the North-West Frontier Province, Baluchistan, Jammu and Kashmir, Eastern Turkistan and Afghanistan*. Calcutta, Superintendent Government Printing.

Alder, G. J. 1963. *British India's Northern Frontier 1865-95. A Study in Imperial Policy*, London, Longmans.

Arif Hasan. 2009. *The Unplanned Revolution. Observations on the process of socio-economic change in Pakistan*, Karachi, Oxford University Press.

Asif Ezdi. 2009. 'The Gilgit-Batistan bungle', *The News International*, 10.9.2009: 6.

Aziz Ali Dad. 2007. 'Renaming the Northern Areas', *Dawn*, 11.9.2007.

Aziz Siddiqui. 1994. 'Too Little, Too Late', *Newsline*, 6 (5): 49-50.

Beg, F. A. 2007. 'Proposed Names for the Northern Areas', *Dawn*, 21.9.2007.

Bhutto, Z. A. 1972. *Speeches and Statements, July 1, 1972–September 30, 1972*, Karachi, Department of Films and Publications.

Dani, A. H. 1989. *History of Northern Areas of Pakistan* (Historical Studies [Pakistan], Series 5), Islamabad, Sang-e-Meel.

Dawn. 3.11.1999; 30.11.1999; 15.1.2001; 24.10.2002; 1.6., 9.7.2003; 21.4., 14.5., 4.6., 6.6., 7.6., 16.6., 4.7.2004; 9.-14.1., 21.7., 1.8., 15.10., 17.10., 30.10. 2005; 20.9.2006; 24.1., 5.6., 24.-27.10.2007; 16.9.2008; 30.8., 12.9., 30.9., 11-12.11., 30.11.2009.

Dawn Magazine. 2.5.2004.

Dichter, D. 1967. *The North-West Frontier of West Pakistan. A Study in Regional Geography*, Oxford, Clarendon Press.

Durand, A. 1888. *Report on the Present Military Position in Gilgit*, Simla (IOL/P&S/7/57).

– 1899. *The Making of a Frontier.* London (reprint: Graz, Akademische Druck- und Verlagsanstalt 1974; Karachi, Indus Publications 1977).

Elias, N. 1886. *Report of a mission to Chinese Turkistan and Badakhshan in 1885-1886.* Calcutta (IOL/P&S/20/A 27).

Frontier Post, The. 4.-6.11, 3.12, 5.12.1989.

Government of Pakistan. 1984a. *Pakistan Statistical Yearbook 1984.* Karachi.

– 1984b. *1981 District Census Report of Northern Areas* (published by Population Census Organization, Statistics Division). Islamabad.

– 1990. *The Constitution of the Islamic Republic of Pakistan* (published by the Ministry of Law, Justice and Parliamentary Affairs, Law and Justice Division). Islamabad.

– (probably 2004). *Northern Areas n.d: Brief on Northern Areas.* Gilgit.

– 2000. *1998* District *Census Report of Gilgit* (published by Population Census Organization, Statistics Division). Islamabad.

– 2001a. *1998 District Census Report of Baltistan* (published by Population Census Organization, Statistics Division). Islamabad.

– 2001b. *1998 District Census Report of Diamir* (published by Population Census Organization, Statistics Division). Islamabad.

– 2001c. 1998 *District Census Report of Ghanche* (published by Population Census Organization, Statistics Division). Islamabad.

– 2001d. *1998 District Census Report of Ghizer* (published by Population Census Organization, Statistics Division). Islamabad.

– 2006. *Pakistan Social and Living Standards Measurement Survey 2004-05. Provincial/District* (published by Statistics Division, Federal Bureau of Statistics). Islamabad.

– Ministry of Education. 2008. *Education for All. Mid Decade Assessment. Country Report Pakistan.* Islamabad.

– President's Secretariat. October 28, 2009. *Letter from the President's Secretariat containing the Executive Summary on Transition to Self-Governance and Development of Gilgit-Baltistan.* Islamabad.

Grare, F. 2007. 'The evolution of sectarian conflicts in Pakistan and the ever-changing face of Islamic violence', *South Asia. Journal of South Asian Studies*, 30 (1): 127-143.

Herald. October 1999, April 2005.

Holdich, T. H. 1909. *Indian Borderland 1880-1900*, London, Methuen & Co.

Holzwarth, W. 1998. 'Change in Pre-colonial Times: An Evaluation of Sources on the Karakorum and Eastern Hindukush Regions (from 1500 to 1800)', in I. Stellrecht (ed.), *Karakorum-Hindukush-Himalaya: Dynamics of Change* (Culture Area Karakorum, Scientific Studies, 4/II), Cologne, Rüdiger Köppe, pp. 297-335.

India Office Library & Records. *Crown Representative's Records—Indian States Residencies—Gilgit, Chilas, Hunza and Nagir Files (Confidential)*, IOR/2/1086/303.

– *Files Relating to Indian States Extracted from the Political and Secret Letters from*

India 1881-1911, IOL/P&S/7/132/455.
– *Files Relating to Indian States Extracted from the Political and Secret Letters from India 1881-1911*, IOL/P&S/7/143/468.
– *Files Relating to Indian States Extracted from the Political and Secret Letters from India 1881-1911*, IOL/P&S/7/189/1100.
– *Departmental Papers: Political and Secret Separate (or Subject) Files 1902-1931*, IOL/P&S/10/279, 286, 588, 826, 924, 973.

Iqbal Khattak. 2009. 'New frontiers', in *Himal Southasian*, October-November: 18-19.

Jettmar, K. 1975. *Die Religionen des Hindukusch.* (= Die Religionen der Welt 4/1), Stuttgart, Kohlhammer.

Khan, A. W. 2005. 'Gilgit bleeds yet again', in *Dawn* 15.1.2005.

Khan, Mahmood Hasan. 2009. *Participatory rural development in Pakistan. Experience of Rural Support Programmes*, Karachi, Oxford University Press.

Kreutzmann, H. 1989. *Hunza - Ländliche Entwicklung im Karakorum* (Abhandlungen - Anthropogeographie. Institut für Geographische Wissenschaften der Freien Universität Berlin Band 44), Berlin, Dietrich Reimer.
– 1995. 'Globalization, Spatial Integration and Their Impact on Sustainable Development in Northern Pakistan', *Mountain Research and Development*, 15(3): 213-227.
– 1996. *Ethnizität im Entwicklungsprozeß. Die Wakhi in Hochasien*, Berlin, Dietrich Reimer.
– 2002. 'Streit um Kaschmir', *Geographische Rundschau*, 54(3): 56-61.
– 2005a. 'The Karakoram Landscape and the Recent History of the Northern Areas', in S. Bianca (ed.), *Karakoram. Hidden Treasures in the Northern Areas of Pakistan*, Turin, Alberto Allemandi, pp. 41-46.
– 2005b. 'Linguistic diversity in space and time: a survey in the Eastern Hindukush and Karakoram', *Himalayan Linguistics,* 4 (http://www.uwm.edu/Dept/CIE/HimalayanLinguistics/Journal_2005/Kreutzmann_HLJ4.pdf).
– 2008. 'Kashmir and the Northern Areas of Pakistan: Boundary-making along contested frontiers', *Erdkunde*, 62 (3): 201-219.
– 2010a. 'An inevitable disaster', in *Himal Southasian* 23 (6): 24-25.
– 2010b. 'Pakistan's Hunza Valley. Another paradise lost?', in *Newsline* 22 (12) June 2010: 38-43 (http://www.newslinemagazine.com/2010/06/pakistans-hunza-valley-another-paradise-lost/).

Kreutzmann, H. and S. Schütte. 2009. *Three years after. Evalution of the GRC/ICRC livestock programme in the earthquake-affected areas of Pakistan-administered Kashmir* (= Berlin Geographical Papers 36), Berlin, ZELF.

Kumar, R. 2005. *Making Peace with Partition*, New Delhi, Penguin.

Lamb, A. 1991. *Kashmir. A Disputed Legacy 1846-1990*, Hertingfordbury, Roxford.
– 1994. 'The Indian Claim to Jammu & Kashmir: Conditional Accession, Plebiscites and the Reference to the United Nations', *Contemporary South Asia*, 3 (1): 67-72.

Leitner, G. W. 1891. 'Hunza, Nagyr, and the Pamir regions', *The Imperial and Asiatic Quarterly Review*, N. S. III: 53-76.
– 1893a. 'Dardistan', *The Imperial and Asiatic Quarterly Review*, N. S. VI: 202-207.
– 1893b. 'Dardistan in 1893 and the Treaty with Kashmir', *The Imperial and Asiatic Quarterly Review*, N.S.VI: 422-425.
– 1894. *Dardistan in 1866, 1886 and 1893*. Woking, Oriental University Institute (reprint: New Delhi 1978, Karachi 1985).

Lockhart, W. S. A. and R. G. Woodthorpe. 1889. *The Gilgit Mission 1885-86*, London, Eyre & Spottiswoode.

Masoodul Mulk. 1991. *A Microcosm of Farmers Strategies in Chitral* (AKRSP Occasional Papers 1), Gilgit, AKRSP.

Mehmood Khan. 1996. 'Prime Minister's Reform Package for Northern Areas', *The Muslim*, 18.5.1996: 14-15.

Muslim, The. 31.5, 20.6.1985; 27.11.1989; 27.8.1991; 5.10.1993.

Prescott, J. R. V. 1965. *The Geography of Frontiers and Boundaries*. London, Aldine Pub. Co.

– 1975. *Map of Mainland Asia by Treaty*, Carlton, Melbourne University Press.

– 1987. *Political Frontiers and Boundaries*, London, Allen & Unwin.

– H. J. Collier and D. F. Prescott. 1977. *Frontiers of Asia and Southeast Asia*, Carlton, Melbourne University Press.

Raman, B. 2005. *Unrest in Gilgit-Baltistan* (South Asia Analysis Group Paper No. 1241; http://www.saag.org/papers13/paper1241.html, retrieved on 26.10.2007).

Rieck, A. 1995a. 'Sectarianism as a Political Problem in Pakistan: The Case of Northern Areas', *Orient*, 36 (3): 429-448.

– 1995b. The Nurbakhshis of Baltistan: Crisis and revival of a five centuries old community', *Die Welt des Islams*, N.S. 35 (2): 159-188.

Salaria, A. A. 1989. 'Decision on Status of Northern Areas Shortly', in *Dawn*, 12.3.1989.

Samaddar, R. (ed.). 2002. *Space, Territory and the State. New Readings in International Politics*, Hyderabad, Orient Longman.

Schwartzberg, J. E. 1997. 'Who are the Kashmiri people? Self-identification as a vehicle for self-determination', *Environment and Planning*, A 29 (12): 2237-2256.

Sherullah, Beg. 1976. *An Introduction to Northern Areas of Pakistan*, Gilgit (unpublished brochure).

Singh, C. 1998. *Natural premises. Ecology and peasant life in the Western Himalaya 1800-1950*, Delhi, Oxford University Press.

Sökefeld, M. 1997. 'Jang Âzâdî: Perspectives on a Major Theme in Northern Areas' History', in I. Stellrecht (ed.), *The Past in the Present. Horizons of Remembering in the Pakistan Himalaya* (CAK Scientific Studies 2), Cologne, Rüdiger Köppe, pp. 61-81.

– 1999. 'Bâlâwaristân and Other Imaginations: A Nationalist Discourse in the Northern Areas of Pakistan', in M. van Beek, K. B. Bertelsen and P. Pedersen (eds.), *Ladakh: Culture, History, and Development Between Himalaya and Karakoram* (Recent Research on Ladakh 8), Oxford, Aarhus University Press, pp. 350-368.

– 2005. 'From Colonialism to Postcolonial Colonialism: Changing Modes of Domination in the Northern Areas of Pakistan', *Journal of Asian Studies*, 64 (4): 939-973.

Stöber, G 2007. 'Religious Identities Provoked: The Gilgit 'Textbook Controversy' and its Conflictual Context', *Internationale Schulbuchforschung* 29: 389-411.

Tanveer Kayani and M. A. Khan. 1998. 'Reconstitution of Provincial Boundaries in Pakistan: An Appraisal and Suggestions', in Israr-ud-din (ed.), *Studies in Pakistan Geography*, Peshawar, Department of Geography, pp. 263-282.

Tariq Ali, 1983. *Can Pakistan Survive? The Death of a State*, Harmondsworth, Penguin.

Usman Malik and A. Schimmel. 1976. *Pakistan. Das Land und seine Menschen. Geschichte, Kultur, Staat und Wirtschaft*, Tübingen and Basel, Erdmann.

Warikoo, K. 2007. *Language and Politics in Jammu and Kashmir. Issues and Perspectives* (http://www.koshur.org/Linguistic/11.html, retrieved on 25.10.2007).

Wood, G., Abdul Malik and Sumaira Sagheer. 2006. *Valleys in transition. Twenty years of AKRSP's experience in Northern Pakistan*. Karachi and Oxford, Oxford University Press.

World Bank. 2002. *The Next Ascent. An Evaluation of the Aga Khan Rural Support Program, Pakistan*, Washington, World Bank.

Zaigham Khan. 1996. 'Saviours of the lost park', *The Herald*, September: 142-143.

Notes:

1. The manuscript is based on the talk given during the CNRS Conference on 'Territorial Changes and Territorial Restructurings in the Himalayas' in Paris in December 2008. The material presented draws on archival research and fieldwork in Pakistan generously supported by the Deutsche Forschungsgemeinschaft (DFG) between 1984 and 2008. An extended version was published in the journal 'Erdkunde' in Autumn 2008. Since administrative changes in 2009 and recent developments have changed the situation significantly, an updated version is presented here.
2. The term 'state' is charged with meaning and is used here to meaning 'sovereign bounded communities' (see Samaddar 2002: xii).
3. Numerous publications in academic journals, scientific books and fiction refer to Partition. New literature which addresses a future that is "making peace with partition" (Kumar 2005) has emerged only recently.
4. Tanveer Kayani and Khan 1998.
5. The term Gilgit-Baltistan was introduced as the official name of the former Northern Areas of Pakistan in September 2009 (Government of Pakistan, Ministry of Kashmir Affairs and Northern Areas, *An order to provide greater political empowerment and better governance to the people of Gilgit-Baltistan*, Islamabad, September 9, 2009, to be published in the next issue of the extra-ordinary gazette of Pakistan Part-II, 61 p.). In the course of this chapter Gilgit-Baltistan and Northern Areas describe the same spatial entities in different time frames and can be used interchangeably in certain contexts. More specifically, the term Northern Areas is only applicable for administrative purposes between 1974 and 2009.
6. Durand 1899.
7. See Alder 1963; Prescott 1965, 1975, 1987; Prescott, Collier and Prescott 1977. Prescott (1965: 9-31) grounded concepts of boundary perception by referring to Ancel, Holdich, Mackinder and Ratzel. The lively debate in France, Great Britain and Germany has strongly been influenced by these protagonists.
8. Holdich 1909. In the context of historical geography the question of congruence between political and environmental entities arises. Chetan Singh discussed the aspect of 'natural frontier' in his thought-provoking book on 'natural premises' (Singh 1998:10).
9. Elias 1886: 71-72.
10. See Durand 1888; Elias 1886; Lockhart and Woodthorpe 1889.
11. See Kreutzmann 1995, 2002; Lamb 1991 for further references.
12. Tariq Ali 1983: 65; Lamb 1994: 69.
13. The North-West Frontier Province was renamed Khyber-Pukhtunkhwa in 2009.
14. See Kreutzmann 1989, Sökefeld 1997.
15. India Office Library and Records: Crown Representative 's Reco rds –Indian States Residencies –Gilgit, Chilas, Hunza and Nagir Files (Confidential): IOR/2/1086/303. Nagir is another name for Nager.
16. The latest configuration of the separate constitutional entities within Pakistan includes the name changes of North-West Frontier Province into Khyber-Pukhtunkhwa and Northern Areas into Gilgit-Baltistan.

17. See Kreutzmann and Schütte 2009.
18. See Schwartzberg 1997; Sökefeld 2005.
19. See Executive Summary on Transition to Self-Governance and Development of Gilgit-Baltistan. 2009. *Letter from the President's Secretariat*. October 28.
20. Similarities between the Tribal Areas and the region which became known as the Northern Areas in the 1970s only emerged in the field of political participation. The 'Frontier Crimes Regulations ' prohibited activities of political parties in Gilgit and Baltistan until 1972, see Rieck 1995a: 439.
21. Government of Pakistan (n.d, probably 2004): 6; Sherullah Beg 1976.
22. During the period of martial law (1977-1985), the Northern Areas were governed as a separate *Martial Law Zone* (Dani 1989: 422). Representatives of the Northern Areas merely had observer status even in the *Majlis-i-Shora* created by Zia-ul-Haq as the predecessor of the current National Assembly.
23. Salaries of non-local 'conscripted 'employees were 25 per cent higher than those of local clerks, and study grants for Northern Areas students were less than half the amounts received by students from the FATA; see *The Muslim* 31.5., 20.6.1985.
24. The third option under discussion was to make the Northern Areas constitutionally equal to Kashmir, i.e. to allow their own constitution, a parliament, and representation by a President of the Northern Areas. However, this option was soon rejected; see Lamb 1991: 158-181 (on the role of the UN); Salaria 1989.
25. The term Kashmiri is used here to identify citizens of Kashmir, not speakers of the Kashmiri language who are a minority in Azad Kashmir.
26. The resolution was presented by Dr Sher Zaman in December 1987; quoted according to Dani (1989: 426-427). It is characteristic of the dictatorship period that the call to give the Northern Areas the right to vote was not included in the list. Nevertheless, a lot of graffiti was written on the walls of public buildings, especially during this period (Photograph 1).
27. See *The Frontier* Post 5.11., 3.-5.12. 1989, *The Muslim* 27.11.1989. Besides, Benazir Bhutto had appointed advisors from the districts of the Northern Areas, hence fulfilling one of the above-listed demands.
28. In 1985, many people had already been arrested and imprisoned during a general strike in Gilgit. Subsequently, this means of protest was frequently used, with similar consequences. The movement was advocated mainly by judges and lawyers. See *The Muslim*, 27.8.1991.
29. Further concessions —including the transfer of budgetary and administrative powers to the *Northern Areas Council* and lower-level executive organs, as well as the creation of a *Chief Court* in Gilgit —were announced in April 1994 (Aziz Siddiqui 1994).
30. *Dawn,* 9.7.2003; Khan 2005.
31. In addition, a full-flung judicial setup including a supreme court and high court, as well as public services and election commissions, were included. "The proposed Interim Constitution Act, 2003, for Northern Areas, comprises 85 articles and 5 schedules" (*Dawn,* 9.7.2003).
32. *Dawn,* 4.7. 2004.
33. *Dawn,* 24.10.2007, 25.10.2007. In addition, the creation of a new district consisting of Hunza and Nager was decreed on the same occasion. In 2004 the new district of Astor was constituted (*Dawn* 1.8.2005). Up until today the dispute about the location of the district headquarters — either in Eidgah or Gurikot — has not been solved and was referred to the courts (Dawn 20.9.2006; 24.1.2007). The constitutional practice

was immediately challenged; whether the Northern Areas are a part of Pakistan and whether the Supreme Court is cognisant.

34. Government of Pakistan (n.d., probably 2004): 1. This point of view is reflected in the building of Basha-Diamir dam on the river Indus. The dam construction site was identified at Basha which is located just a few kilometres south of the Northern Areas territory, whereas most of the water will be stored in Diamir district in the Northern Areas. It seems that the Government of Pakistan wanted to play safe and selected a dam site within the area covered by its constitution and accepted by international law. This stance did not convince the World Bank that refrained from any involvement in funding the project.
35. The fourth 'package ' announced in 2009 has created a "mixed reaction of the local people" (Asif Ezdi 2009) or is perceived as an "eyewash reform" (Iqbal Khattak 2009: 19).
36. Dawn, 30.11.2009.
37. Dawn, 12.11.2009. In the same article it is mentioned that "the total number of registered voters in the province is over 700,000 …"
38. See Dichter (1967: Fig. 1); Government of Pakistan (1990: 176); Masoodul Mulk (1991: 17).
39. See the evaluation reports by the World Bank and regular reporting from development agencies such as the AKDN (see Abdul Malik 2005: 114; Abdul Malik and Izhar Ali Hunzai 2005; World Bank 2002).
40. Reference is given to http://www.balawaristan.net where the aims and objectives of the movement are given; see Sökefeld 1999, 2005 for an interpretation of aspirations, motives and strategies.
41. Raman 2005.
42. See Arif Hasan 2009: 1-26; Khan 2009: 94-147 ; Wood, Abdul Malik and Sumaira Sagheer 2006.
43. See Executive Summary on Transition to Self-Governance and Development of Gilgit-Baltistan. *Letter from the President 's Secretariat.* 2009.
44. One cartographic example for identifying Bolor with the 'Massif des Ts'oung-ling (Bolor)' (Onion Mountains) is the map entitled 'Carte chinoise du Si-Yu ou Asie Centrale pour accompagner le mémoire intitulé récit officiel de la conquête du Turkestan par les Chinois (1758-1760)' where the 'Massif des Ts'oung-ling (Bolor)' is located in an area south of Kashgar and east of Wakhan. Holzwarth (1998) has discussed and interpreted local and regional sources of information. In reference to his in-depth study, Bolor and Boloristan comprise an area which some authors project on to today's Chitral, Gilgit and Baltistan, other s trace a much narrower space.
45. Holzwarth (1998: 300-301).
46. The maps by E. G. Ravenstein 'The skeleton map of the countries between Kashmir & Panjkorah including Chilas, Kandia & other districts of Dardistan' to a scale of 1: 500,000, published in 1875, and the 'Map of the Pamirs' to a scale of 1: 4 million, published 1892 in the Asiatic Quarterly Review; see Leitner 1891, 1893a, b, 1894.
47. See Beg 2007 who promoted such place names which he identified as un-biased due to their constructive character. Bahqahistan can be broken down into "Ba" for "bam-e dunya" (roof of the world), "qa" for Karakoram and "hi" for Hindukush and Himalayas. These attempts follow similar interpretation models which have been applied to the country's name, Pakistan.
48. Aziz Ali Dad 2007. For all these terms, some historical references are given going

back as far as the work of Greek historiographers, chroniclers from different periods and colonial writers; see Dani 1989, Holzwarth 1998, Leitner 1893a, b, 1894.

49. See Kreutzmann 2005a, b; Rieck 1995a, b. The vast majority of residents, more than 99 per cent, profess the Muslim faith in the Northern Areas. For the Gilgit District, the latest census recorded 99.74 per cent Muslim, for Baltistan 99.77 per cent (Government of Pakistan. 2000, 2001a). Only very few Ahmadi, Bahai, Buddhists, Christians and Sikhs live there and are registered.
50. See Kreutzmann 2008.
51. See Kreutzmann 2008.
52. Sunni sectarian organizations were founded in the 1980s (see Grare 2007). On their agenda were excommunication strategies directed mainly towards Twelvershia groups.
53. See similar circumstances and motives of clashes between the two conflicting parties during the colonial period: In 1901-1902 large groups of Ismailis converted to Sunna (especially in Mulkho and Turkho) after agitation by clerics; similar events were reported in 1906 (India Office Library & Records: IOL/P&S/7/132/455: Gilgit Diary 23.2.1901; IOL/P&S/7/143/468: Chitral Diary 18.2.1902; IOL/P&S/7/189/1100: Chitral Diary 23.5.1906). In 1917, the Ismaili pir Bulbul had to leave Chitral and go into exile after he had stirred up an oppositional group from Mastuj against the Mehtar (India Office Library & Records: IOL/P&S/10/973: 75: Gilgit Diary December 1927). In the winter of 1924-1925, Ismailis started to flee from Chitral to seek asylum in Gilgit Agency, causing the colonial administration to intervene.
54. Since the 1931 Census of India, no specific data have been published on religious groups, castes and other communities/sects either in India or in Pakistan.
55. The percentage of non-Muslim minorities in the Northern Areas amounts to about 0.2 per cent of the total population (Government of Pakistan 1984a: 18; 2000: 26; 2001: 26). In Chitral (NWFP) the 4,000 Kalasha amount to 1.5 per cent. The Kalasha inhabit the Hindukush valleys of Bumburet, Birir and Rambur and are a prominent group of the *kafir* (infidels), who do not follow any of the religions of the Book. Meanwhile, almost 50 per cent of them have converted to Islam. Studied by ethnographers at an early stage (see Jettmar 1975), the Kalasha are now being marketed as an exotic ethnic group and tourist attraction.
56. See the data in Usman Malik and Schimmel 1976: 205.
57. Government of Pakistan 1984a, 2000: 43; 2001a: 26.
58. See Kreutzmann 2005b.
59. See Kreutzmann 1995: 213-227; 2005a: 41-46; 2005b.
60. See Stöber 2007.
61. I am grateful and indebted to Andres Benz for providing material from his Ph.D. dissertation on the impact of education on development in Gilgit-Baltistan.
62. Aga Khan Rural Support Programme. 2007. *An assessment of socio-economic trends in the Northern Areas and Chitral, Pakistan (1991-2005).* Gilgit: xi.
63. See Kreutzmann 2010a, b.

1.1.2.

Politics of Territoriality: Indigeneity, Itinerancy and Rights in North-East India

Sanjib Baruah

On the evening of 7 February 1944, with the Second World War still raging, Robert Reid, a recently retired officer of the Indian Civil Service, addressed a meeting of the Royal Geographical Society in London. The subject of his lecture was the Excluded Areas of Assam. This colonial administrative term—the Excluded and Partially Excluded Areas of Assam to be exact[1]—referred to territories located in part on the borderlands between North-East India and Burma and to which the war had drawn significant public attention. While Reid emphasized the area's heterogeneity, the people in those borderlands, he said, have one thing in common: "neither racially, historically, culturally, nor linguistically they have any affinity with the people of the plains, or with the people of India proper." If they were "tacked on as an Indian province" it was only a matter of historical accident and "a natural administrative convenience" (Reid 1944: 19).

'Hill people' was one of the few general terms Reid used to describe the people living in those borderlands. This is not surprising. In this part of the British Empire, the hills-plains binary more or less coincided with a legal distinction made between 'tribal' people that were governed by customary law and other colonial subjects that were governed by general law. However, the presence of 'tribal' people in the plains or 'non tribal' and non-native tribal people in the hills caused considerable difficulty for colonial ethnic classification.

Reid's lecture was an introduction to the region and its peoples: he guided the audience through a map, and showed photographs of people living in those lands. He spoke with great authority. Reid, after all, was the Governor of Assam from 1937 to 1942. He was forthright about placing those people on a 'civilizational' scale, and he confidently expressed his likes and dislikes. Thus the "Dufflas, Akas and Miris," he said, are "very primitive peoples, who respond hardly at all to the influences of civilization." While the Nagas of the Tirap Frontier Tract were "rather a degraded, backward type," in their "abode proper" they were "frank and independent by nature, often a cheerful and hospitable disposition." Indeed the qualities of "those picturesque people" appealed to "the men who work there"—i.e. colonial administrators—so much that they "become devotedly attached to them" (Reid 1944: 19-21).

Phrases such as "abode proper" and the "backward and degraded type" point to the peculiar logic of colonial racial and ethnic classification: the fixing of 'tribes' to their supposed natural habitats and, to borrow Paul Gilroy's phrase, a "bio-cultural" notion of ethnic traits as "fixed, solid almost biological" and inheritable (Gilroy 1987: 39). This notion of ethnicity and the mixing of ethnicity with territory—what I would call a frame of ethno-territoriality—made it necessary to distinguish between so-called pure and impure types to account for those that stray from their assigned physical spaces, or do not conform to ethnic stereotypes.

The colonial discourse of race, caste and tribe in India, as Crispin Bates reminds us, was no "mere faltering steps on the road towards the formulation of a purer science of Indian sociology"; it was "in many ways the Peacock Throne of British India, carried off by the new Constitution of 1950, but still greatly missed by many" (Bates 1995). Indeed Reid made little effort to hide his political preferences. Whatever the precise form of India's future political dispensation, he was sure that for the peoples of the excluded areas, it would mean less "protection," and perhaps even "the abolition of protection." He therefore reminded his compatriots that: "We are responsible for the future welfare of a set of very loyal, primitive peoples who are habituated to look to us for protection and who will get it from no other source." Reid believed that this factor, along with the demonstrated military importance of these borderlands that saw action during the war, pointed toward only one policy option: Britain retaining those areas as "a civil administrative unit comprising the Hill Areas along the north and east frontiers of Assam and taking in as well the similar areas in Burma itself" (Reid 1944: 27-28). Historians know this idea as the crown colony scheme that came up during the last days of British rule in the region and was rejected by both Indian and Burmese nationalists.

Reading this text more than sixty years later, it is obvious how wrong Reid was about the region's political future. In the new post-World War II political balance, there was little chance of a 'crown colony' on the Indo-Burmese border coming into being. Reid thought that the people of the Excluded Areas "are not by a hundred years, ready to take their place in a democratic constitution, or to compete with the sophisticated Indian politician for place and power" (Reid 1944: 27). Yet not long after India's independence in 1947, the people in these areas began participating in India's democratic institutions. Today a number of districts of the old excluded areas are full-fledged constituent units of federal India. Despite Reid's fears that protection would be abolished, the postcolonial Indian Constitution retained most of those safeguards placing them under the supervision of elected bodies.

Is there nothing more to Reid's text than a tired attempt by an unapologetic imperialist at justifying colonial rule in the name of civilization and the 'primitive's' need for tutelage? Such a nationalist-triumphalist reading of the text, I would suggest, underestimates the damage that the colonial discourse has caused, and obscures the continuities between colonial and postcolonial practice.

The excluded areas along with the colonial province of Assam to which they were "tacked on", as Reid put it, now constitute a major part of North-East India and parts of Bangladesh. The focus of this paper is on North-East India. Colonial Assam is now divided into five states: Assam and four other states that were once part of the excluded areas: Arunachal Pradesh, Meghalaya, Mizoram and Nagaland. Apart from these five states North-East India also includes Manipur and Tripura that were 'native states' in colonial times. Since 2003 Indian official usage has included an eighth state: Sikkim, once an independent Himalayan kingdom, and part of India since 1973. For the purpose of this paper however, Sikkim is not included in North-East India.

The Argument

Territoriality—"the spatial strategy to affect, influence, or control resources and people, by controlling area" (Sack 1986)—has been a recurrent theme in the postcolonial politics of North-East India. Territorial changes and territorial restructurings take the form of both political demands emerging from society and a policy tool used by the state. A number of factors explain the saliency of the politics of territoriality: (1) retention of the colonial frame of ethno-territoriality primarily through the distinction still maintained between subjects governed by customary law and subjects governed by general law, and between 'hill tribes' and 'plains tribes'; (2) the morphing of safeguards for the customary practices of hill tribes into protective discrimination for ethnically defined groups turning the excluded areas into *de facto* ethnic homelands; (3) a constitutional-legal context in which the list of groups entitled to protective discrimination, and the boundaries of territorial units remain permanently open to modification and change; and (4) the frontier character of the region where massive immigration makes territoriality an idiom of resistance by 'indigenous' ethnic groups against what is seen as a process of minoritization.

Indigeneity is a persistent theme in North-East India's politics of territoriality. However, the dominant discourse of the local and the outsider is far too dissonant with the actually existing political economy of the region. The framing of some people as outsiders produces recurrent episodes of ethnic violence and displacement, and a permanent crisis of citizenship. Nor are the politics of territoriality limited to conflicts between groups that call themselves indigenous and those labelled outsiders. 'Indigenous' groups with competing agendas also find themselves in conflict with one another. For instance, the demand for a new *de facto* ethnic homeland, or efforts to bring together segments of an ethnic group scattered across many units into a single homeland implies the break-up of an existing unit. In such a situation those favouring boundary change and those defending the territorial integrity of an existing unit are inevitably pitted against each other.

The rest of the paper will elaborate this argument by emphasizing a few of the key elements.

1. Institutional Legacies of Colonial Knowledge

The 'hill tribes' in North-East India, as indicated above, were so called because of the colonial ethnographer's inclination to fix them to their supposed natural habitats. As anthropologist Matthew Rich has shown, a particular problem that the British colonial scholar-administrator was confronted with in North-East India was how to make sense of the 'egalitarian' mores and habits of many of the peoples—for instance, the absence of caste in the hills. Given their notion of India as an essentially 'hierarchical' civilization, they had to figure out whether these people were outside or inside the racial unity of India. There was no easy answer since the ethnic kin of the same egalitarian people also performed Hindu-like rituals a short distance away. Such facts had to be either assimilated into the master principle of caste, or categorized as "external to the caste system yet internal to a racially defined unity of India." The hills and plains therefore became the master oppositional binary in the colonial solution to this 'problem' (Rich 2006). The peoples of North-East India were therefore classified as belonging either to the hills or to the plains. It is this history of colonial ethnography that explains the contemporary discourse of territorialism in North-East India. Thus communities like the Koch Rajbongshis that are in the plains and the non-tribal side of the divide now seek Scheduled Tribe (ST) status, and communities like the Bodos that are recognized as ST, but are on the plains side of the divide now seek Sixth Schedule status once only available to 'hill tribes.'

The colonial distinction between hill tribes and plains tribes is still retained in the Indian census. Thus of the 23 STs in Assam, 14 are hill tribes and 9 are plains tribes. Since the census still counts tribes only in their supposed natural habitats, it produces the absurdity of the number of people being classified as plains tribal being zero in the hills, and those classified as hill tribal being zero in the plains. Thus if one goes by the Indian census, the number of 'hill tribals' living even in metropolitan Guwahati, is zero (Prabhakara, 2003).

The Inner Line, originally implemented in 1873, is another colonial institution that has survived. It originally separated territories where colonial administration was firmly established, and areas where British authority remained nominal. Today Indian citizens, as well as foreigners, require permits to enter the states of Arunachal Pradesh, Mizoram and Nagaland. Since only certain ethnic groups legally belong to these territories thanks to the legacy of the colonial ethno-territorial framework, it is not difficult to see why they would develop a stake in the Inner Line. After all, it becomes a legal way of excluding outsiders from the *de facto* homeland set aside for particular groups. However, the logic of the commodity economy and everything that is being done in these areas for the sake of development, mean that outsiders do come and work across the Inner Line. But a permanent distinction is maintained between those that can legally belong and those that cannot. It is hardly surprising that there are growing demands by ethnic activists for extending the Inner Line to other parts of North-East India. Occasionally there are calls for the Inner Line even in the plains districts of Assam.

2. Constitutional-Legal Context and the Politics of Territoriality

Certain characteristics of India's postcolonial constitutional order facilitate these politics of territoriality. India's Constitution leaves open the question of which groups are entitled to what preferences constitutionally and politically (Weiner 1983: 46). Its demos-enabling federalism (Stepan 2001: 338-39) puts few constraints on central government's power to make and break states. To create a new state by changing the political boundaries of an existing state, it has to barely consult the elected legislature of the state concerned.

Preferential policies, as Myron Weiner observed, tend to create a particular political process affecting the ways in which groups are organized, the demands that are made, the issues that constitute policy debates, and the way coalitions are formed. By facilitating group mobilization in support of new preferences or the extension of existing preferences, preferential policies create political struggles over how the state should allocate benefits to ethnic groups, generating a backlash on the part of those ethnic groups excluded from benefits, intensifying the militancy of the beneficiaries, and reinforcing the importance of ascription as the principle of choice in allocating social benefits and facilitating mobility (Weiner 1983: 49).

If this is true about India as a whole, in the North-East the institutional legacy of the colonial ethno-territorial framework gives these struggles a territorial character. The Indian Constitution empowers the president of India to specify by public notification the "tribes or tribal communities or parts of or groups within tribes or tribal communities which shall for the purposes of the Constitution be deemed to be Scheduled Tribes." According to Marc Galanter, a major scholar of Indian law, the ethnic communities that are listed on the schedule were "defined partly by habitat and geographic isolation, but even more on the basis of social, religious, linguistic and cultural distinctiveness—their 'tribal characteristics.' Just where the line between 'tribals' and 'non-tribals' should be drawn has not always been free from doubt" (Galanter 1984: 150).

In this constitutional-legal context it is hardly surprising that in North-East India new groups constantly demand ST status, those who have ST status seek the protection of the Sixth Schedule or the Inner Line, and those having Sixth Schedule status ask for full-fledged states. Thus the Bodos, a 'plains tribe,' now have sixth schedule status—so far only available to 'hill tribes.' As I shall develop in the next section, in the former excluded areas there are now elected state governments under the control of ST politicians, and a visible and well-to-do ST elite. This has captured the imagination of ethnic activists in the entire region and beyond. There is a perception that the STs in the states with the most comprehensive protective discrimination regimes and rules of exclusion have done well economically and have been relatively successful in protecting themselves from being swamped by immigrants. While a homeland has become an aspiration for ethnic groups that do not have one, ethnic activists of the existing homelands have become zealous defenders of what they see as their entitlements.

The ease with which states are made and unmade has reinforced the idea that any demand for an exclusive ethnic homeland might be successful if backed by sufficient evidence of political support, including the capacity for violence. This is a major factor in the persistent politics of territoriality in North-East India. In Manipur, the fear that significant parts of its territory can be bargained away in closed-door negotiations between the central government and leaders of the Naga independentist movement is a major source of anxiety.

3. From Space of Customary Law to Ethnic Homeland

The Indian Constitution originally used the phrase "Tribal Areas of Assam" to deal with the old excluded areas. Thus while the rest of the country also has ST populations and the Constitution makes provisions of protective discrimination for them, the 'Tribal Areas of Assam' were treated differently. While the Fifth Schedule covers tribal areas in other parts of the country, the Sixth Schedule applies to these areas.

The Sixth Schedule provided for autonomous districts and autonomous regions. These districts were to have elected councils which enjoy powers to levy some taxes, to constitute courts for the administration of customary law—justice involving tribals—and law-making powers on subjects including land allotment, occupation or use of land, regulation of shifting cultivation, formation and administration of village and town committees, appointment of chiefs, inheritance of property, marriage and social customs.

The process of forming Autonomous District Councils, however, did not quite proceed the way Constitution-makers had anticipated. The outbreak of the independentist Naga rebellion meant that the political conditions for holding elections to the Naga Hills District Council were lacking. Instead, in 1963 the state of Nagaland was created. According to article 371A of the Constitution that applies only to Nagaland, without explicit resolutions of the Nagaland Assembly, laws passed by the Indian parliament that impinge on Naga customary practices—including matters of ownership and transfer of land—do not apply to Nagaland. The North East Frontier Tracts where the Sixth Schedule was eventually supposed to be implemented also underwent a different process of institutional transformation following the Indo-China war of 1962.

Beginning with the creation of Nagaland in 1963, a number of new political units were created in the region culminating in the creation of the North Eastern Council in 1971. By now they are all full-fledged states. Four of these states including Nagaland were part of the excluded areas of colonial Assam. These states acquire a de facto ethnic homeland character because the lion's share of public employment, business and trade licenses, the rights to land ownership and exchange, and even the right to seek elected office in these states are reserved for members of the those STs with which the territory is iconically associated. Table 1 gives the number of reserved seats in the state legislatures of northeastern states and also gives the percentage of the ST population. In the Legislative Assemblies of Arunachal Pradesh, Mizoram and Nagaland all but one seat are reserved for STs. In Meghalaya 55 of the 60 seats are reserved.

Table 1: North-East Indian states: Reserved seats for Scheduled Tribes in state legislative assemblies

States	ST as % of population*	Leg. Assembly total members	Leg. Assembly seats for STs	Leg. Assembly unreserved seats
Arunachal	63.7	60	59	1
Assam	12.8	126	16	102**
Manipur	34.4	60	20	40
Meghalaya	86.6	60	55	5
Mizoram	94.8	40	39	1
Nagaland	87.7	60	59	1
Tripura	31.0	60	20	33***

* based on 1991 census data
** 8 reserved for scheduled castes
*** 7 reserved for scheduled castes

Through a constitutional amendment the balance between reserved and unreserved seats in the assemblies of Arunachal Pradesh, Meghalaya, Mizoram and Nagaland has been frozen in order to ensure that delimitation of constituencies in light of demographic changes never changes the current ethnic balance. Thus these territories remain ethnic homelands in the sense that there is a guarantee that their political class would always have a clearly defined ethnic character following the colonial ethno-territorial frame. This of course says little about who actually resides in these territories even historically, but especially since the commodity economy and the development process have brought many outsiders to these sparsely populated areas. The fact that outsiders can never become full citizens is built into the very structure of these homelands.

4. A Land of Immigrants

For more than a century North-East India has been a frontier region attracting massive immigration from the rest of the subcontinent. From 1880 onwards, according to the historian David Ludden, "when statistics appear for the first time" permanent cultivation expanded over a hundred years at extremely high rates in North-East India—"faster than almost anywhere else in South Asia." Much of this expansion was the result of lowland agriculturalists "investing in land at higher altitudes." Indeed until about 1960, "additional increments of agricultural production in South Asia" were achieved entirely by means of the physical expansion of cultivated farmland (Ludden 2003: 17). It is quite clear from the census data that this expansion of permanent cultivation has meant massive immigration that has significantly increased the population density of North-East India over time. The process continues till this day. Indeed it has barely begun in some of the sparsely populated old excluded areas.

That colonial Assam was seen as a frontier—an area with vast tracts of 'wasteland'—and that the story of tea in Assam begins with this 'wasteland' being made available to European entrepreneurs, is relatively well known. But land allocated to tea is not the only area that the colonial state viewed as 'wasteland'. As the colonial rule was consolidated, apart from tea plantations, other vast tracts of land were also settled upon by people from the rest of the subcontinent. Most dramatic was the flow that began in the 1920s, of land-hungry peasants from densely populated East Bengal.

This migration continued after the Partition of 1947 giving it an international and even illegal dimension. However, it is hardly surprising that Partition could not suddenly change the logic of a frontier, and divert the flow of people from one of the subcontinent's most densely populated areas to a relatively sparsely populated region. From the point of view of North-East India, the effect of Partition was mostly to intensify migration pressure from East Bengal with significant numbers of Hindu refugees now being added to the flow. The 'Bangladeshi' question, however, is only part of the broader question regarding the demographic transformation of the region. Migration from the rest of India and from Nepal and Burma (in the case of Mizoram) is also a source of tension.

However, frontiers are not natural, they are man-made. It is unequal political power, and often conquests, that turn territories into frontiers for other people. It is not surprising therefore that political resistance on a frontier typically makes an appeal to the principle of self-determination. Indeed resistance is often a sub-text to North-East India's politics of territoriality. On the other hand, the indigenous elite also benefits from the long-term changes in land relations that are part of the extension of permanent cultivation and the increase in population density. In the space of customary law these changes take place mostly beyond the eyes of the law. Immigrants also provide various kinds of cheap labour that benefit the better-off classes. So there is also an element of denial involved in the anti-immigrant rhetoric that is so much part of the region's politics of territoriality.

5. Territoriality Versus the Logic of Political Economy

This politics of territoriality has produced a major structural dilemma for the postcolonial practice of citizenship in North-East India. This particular form of territoriality, to borrow Mahmood Mamdani's words, penalizes those that the commodity economy dynamizes. Those who are mobile are either penalized by being defined as outsiders (Mamdani 2005) or mobility is discouraged because preferences that go with indigenous status are made specific to habitats to which particular groups are linked.

The plight of the descendants of tea workers brought as indentured labourers to Assam in the nineteenth and early twentieth centuries has become a major issue in Assam recently. Tea workers posed a classificatory problem for the census of Assam as early as 1891. The "aboriginal tribes of central India" were explicitly excluded from the "forest and hill tribes" in that census, and instead were classified simply as labourers (Rich 2006). Today many among them who trace their roots to Munda, Oraon, Santhal and other people of the Jharkhand, like to call themselves adivasis. According to some estimates there are as many as 4 million adivasis in

Assam—more than half of Assam's tea labour community. Adivasi activists argue that since their ethnic kin in their places of origin are recognized as STs, they should have the same status in Assam. Rather strikingly, adivasi activists today use the bow and arrow as an ethnic symbol.

That a group that provided the muscle for the nineteenth century capitalist transformation of Assam today demands recognition as STs status as well as the framework within which the debate is being conducted, draws attention to the dangerous residues of colonial knowledge that remain in India's political culture.

The then tribal affairs minister P.R. Kyndiah had said without any sense of irony that ST status for adivasis would involve examining the case using the criteria of "tribal characteristics, including a primitive background and distinctive cultures and traditions." Ethnic activists opposed to the adivasi claim cite Home Minister Shivraj Patil's statement with approval that the adivasis have "lost their tribal characteristics". They also argue that the adivasis are not "aborigines of Assam".

Seen through the prism of the global political economy, the tea labour community of Assam is part of the same nineteenth century migration that took Indian labourers to plantations in various parts of the British Empire, such as Fiji, Guyana, Mauritius or South Africa. India now celebrates the Indian Diaspora. The *Pravasi Bharatiya Divas* honours descendants of those migrants to far-away shores, some of whom have risen to become presidents and prime ministers of their countries. Viewed from this perspective, the efforts of the descendants of those who remained within India's borders to claim ST status—using the idiom of adivasi and the symbolism of the bow and arrow—is quite extraordinary. Surely, given the contribution of the tea labour community in blood and in sweat to creating modern Assam, no other group has a better claim to full citizenship rights and compensatory justice than them. Yet in a political culture that is infused with a colonial framework of ethno-territoriality, there is little space for articulating their demands except through a borrowed language of remembered tribal-hood.

The tension between the politics of territoriality and the logic of political economy plays out in other ways as well. The 2005 Arunachal Pradesh *Human Development Report* features an unlikely class of agricultural revolutionaries—migrant sharecroppers who bring the technology of wet rice cultivation and introduce settled cultivation to many parts of the state. The bullock-driven plough used by them, according to the report, is the main instrument for extending settled cultivation and is therefore the symbol of Arunachal's "agricultural modernization." They are now "an important segment of the peasantry" extending settled cultivation to uncultivated land. Even though heralded as agricultural modernizers, the legal status of this group is quite vulnerable. The contract between sharecroppers and landlords says the report, "is only short-term and eviction may take place any time." (Government of Arunachal Pradesh 2006). Since access to land in Arunachal is the domain of customary law, the leases that allow them to live and cultivate are merely oral agreements. Indeed, even the residential rights of most 'outsiders' in Arunachal are severely restricted under the Inner Line Permit regime. The tensions between the idea of democratic citizenship and the logic of development on the one hand, and North-East India's politics of territoriality on the other could hardly be more obvious.

6. The Assam Movement: The Dangerous Discourse of Illegality

Except for two hill districts, most of the present-day state of Assam was the plains and non-excluded part of the colonial province of Assam—the space of general law as opposed to customary law. Apart from the mostly unsuccessful attempt to implement the Line System, separating 'tribal belts' and to prevent the transfer of land from 'plains tribals' to immigrants, there are no restrictions on the settlement of outsiders in the plains of Assam. Indeed settling immigrants in the 'wastelands' of Assam was the explicit goal of colonial policy. Assam therefore has experienced the most intense pressure from immigration, leaving aside Tripura which was a native state in colonial times and thus not subject to the colonial bifurcation between the spaces of general and customary law. Resistance to immigration shook up the politics of colonial Assam in the 1930s and the 1940s, and profoundly shaped Assamese attitudes towards the Partition of 1947.

From 1979 to 1985 Assam saw extraordinary mobilization against illegal immigration and the enfranchisement of non-citizens. The political upheaval became a model for other 'anti-foreigner' movements in the entire region. It ended with an agreement signed between the Indian government and the leaders of the movement. Based on various "cut-off dates" of entry into India, some foreigners were to be given Indian citizenship, some were to be disenfranchised temporarily, and more recent illegal immigrants were to be deported. In addition, there was a government commitment to making significant investments in Assam's development, and to finding ways to protect the embattled public cultural face of Assam.

But given the difficulties of identifying foreigners, the Assam Accord was impossible to implement. In that sense the Assam Movement was a failure. When in July 2005, two decades after the end of the Assam movement, the Indian Supreme Court spelt out the dangers that Assam faced from continued illegal immigration from Bangladesh, its language was remarkably like that used by the supporters of the Assam movement.

In any political system, a powerful social movement that brings millions of people to the streets for an extended period and then appears victorious at the negotiating table, ultimately failing to achieve its goals would have notable consequences. But the Assam Movement failed in other ways as well. One of its unintended consequences was that it brought to the surface serious fault-lines in Assam's social and cultural fabric. Despite the movement's leadership constitutionally and legally framing the issue as one of illegality—both of immigration and enfranchisement—labels such as Bangladeshis and foreigners made many long-term residents of Assam feel uncomfortable and insecure. Things came to a head in Nellie and a number of other places in February 1983 when amidst tensions produced by an election that campaigners portrayed as Assam's last struggle for survival, and boycotted on grounds that it was based on electoral rolls filled with names of non-citizens, hundreds were killed in group violence. In Nellie, in a climate of deep anxiety amidst rumors of 'foreigners' finding shelter in Bengali Muslim villages, neighbours were suddenly transformed into dreaded 'foreigners' (Kimura 2009) and the Assam's so-called 'last struggle for survival' came dangerously close to a civil war.

It was not only relations between Hindus and Muslims that were destroyed by the Assam movement. Soon, a number of tribal groups, notably Bodos, launched movements seeking autonomy or even separation from Assam, partly because of another unintended consequence of the Assam movement. Following an extended period of political mobilization that privileged the distinction between the legal and the illegal, people suddenly expected the actual situation on land officially labelled as a reserve forest or set aside for other forms of public use, to conform to the formal legal labels. Thus the law suddenly began to define anyone occupying such land as encroachers, and even tried evicting them. That included even those with claims to being the most 'indigenous' of Northeasterners. This factor contributed significantly to the tribal revolt of the years following the Assam Movement. However, the most enduring legacy of the failure of the Assam movement is surely the United Liberation Front of Assam or Ulfa. It came into existence in 1979, the same year as when the Assam movement began. Ulfa's position regarding the issue of foreigners is accommodative, but it views the failure of the Assam Accord as one more piece of evidence of Assamese interests being disregarded in the pan-Indian political dispensation.

7. Indigenous, Non-indigeneous Conflicts in the Space of Customary Law

The plot of Siddharth Deb's novel *The Point of Return* is based in a nameless North-East Indian hill state. It can be read as a portrait of the consequences of the politics of territoriality in the space of customary law. In this fictional hill state, the refugees of the Partition of 1947 are seen as interlopers. Thus after leaving "their homes forever to try and find themselves within the nation" they discover that their journey is not over. "The hills that appeared beyond the horizon were only another mirage, their destination just another place that would reject them." The narrator on a visit to his "hometown" remembers the "life time of fear" that the protagonist felt, and from which the son, the narrator, ran away to escape. A hill-town that "drummed in the message of death" to the protagonist, he imagines, must have seemed "like a lost spot on the map of the nation, its remote beauty and even more remote violence surfacing in the national newspapers only as little single-column reports of 'disturbances'" (Deb 2004: 292, 295).

Moving from the fictional to the real, according to a scholar who examines human security issues in the state of Meghalaya, the outsider discourse is so dominant in the state that it tends to exclude any human security issues even from the agenda of civil society organizations. "Unless a cause of insecurity fits the 'outsiders' discourse, it is marginalized. And sometimes human security issues are co-opted into the 'outsiders' discourse" (McDuie-Ra 2006).

8. Conflicts Between the Space of Custom and Space of Law

Creating four new states out of the colonial state of Assam has produced a whole set of new territorial conflicts. The disputes are mostly over forestland located on the border. In colonial times, in the interest of better management of reserved forests many of these tracts were transferred from across the Inner Line to the plains districts. Hence there was "an administrative boundary, a forest boundary, and a political boundary, making for a somewhat confusing situation" (Verghese 2005). Once these lines became inter-state boundaries, the fact that many of these *de jure* reserve forests were by then settled by a variety of 'encroachers,' and because land claims can be based on ethnicity in the space of customary law, but not in the space of general law, various forms of boundary disputes have flared up from time to time. This however, is not the only dimension of these border disputes. For instance, the demand for Nagalim—'greater Nagaland'—also marks the rejection by Nagas of the unilateral determination of the boundaries of Nagaland without taking into account the wishes of the people.

The political consequences of the two different legal regimes in the two kinds of space became very apparent in the summer of 2007 when suspected Bangladeshi nationals were expelled from Arunachal Pradesh by vigilante groups. They were emboldened by the Inner Line Permit regime and the rules of customary law that restricts *de jure* access to land—no matter what the *de facto* situation. Not surprisingly, it resulted in an exodus of these so-called Bangladeshis to Assam. The debate in Assam that followed underscores the difference between the space of customary law and the space of general law. Many ethnic activists in Assam urged the state government not to allow those displaced from Arunachal Pradesh—Bangladeshis in their words—to settle in Assam. On the other hand, the Congress-led state government described them as residents of Assam. The leader of a political party that speaks for minority interests called them Bengali speaking Indian Muslims and said that only a judicial authority can determine the citizenship status of each individual.

What is to Be Done?

Mahmood Mandani has observed in the context of Africa that defining "political identity, political rights, and political justice first and foremost in relation to indigeneity had turned the colonial world upside down, but it did not change it" (Mamdani 2005). That the genocide in Rwanda was ultimately the product of the Hutu and Tutsi being constructed as native and settler (Mamdani 2001) should serve as a warning about the dangers of politics of territoriality shaped by the institutional legacy of colonial knowledge.

Elsewhere I have put forward a proposal for multi-level citizenship—citizenship both of India and of a state—as an alternative to the ethnic homeland model. Multi-level citizenship is not unknown in federal systems. Instead of effective local citizenship being determined by state monitored customary rules that define ethnic belonging (that prevails in four of the seven Northeastern states), state-level citizenship could introduce a civic principle, and give the right to define the rules of inclusion and exclusion to territorially defined political communities. Such a provision could be extended to all Northeastern states.

My premise is that citizenship discourse is based on a different grammar than the discourse of territoriality grounded in colonial knowledge. In principle, most countries recognize three ways of becoming a citizen: birth within the territory of a country (*jus soli*), descent from a citizen (*jus sanguinis*) and naturalization. If *jus sanguinis* incorporates the principle of citizenship gained through blood ties, the other two principles can incorporate the ethnically or culturally different outsider. On the other hand, current rules in North-East India in effect guarantee the ethnic composition of the political class governing ethnic homelands, with few ways of incorporating the outsider except when the customary practice of a ST fixed in that space allows it; only through the mediation of the *jus sanguinis* principle. Countries of course vary in terms of how much of the *jus soli* principle is applied to the claims to citizenship of children of immigrants born in a country, and on the degree of difficulties that are involved in obtaining citizenship through naturalization. Yet inherent in the grammar of the citizenship discourse is that new members can enter the political community as full members, unlike the rules of inclusion and exclusion inherent in the idea of ethnic homelands shaped by the legacy of the colonial ethno-territorial framework. It is difficult within the discourse of citizenship not to recognize the right to citizenship of second- or third- generation immigrants. Outsiders and their descendants cannot remain foreigners for ever.

The obvious advantages of introducing state-level citizenship in this frontier region would be that it could define political communities in civic terms and introduce a dynamic element of incorporating new members. It could make a decisive break from the notion of ethnic homelands. State-level citizenship could allow elected state governments and legislatures to make rules by which an internal immigrant or his or her descendant can become a citizen of the state, and a full member of the local political community.

Conclusion

Ethnic activists in North-East India often allude to ancient kingdoms to justify contemporary territorial claims, but the exclusionary territorial imaginaires that shape their political projects are firmly grounded in the colonial ethno-territorial framework. It is easy to miss in this particular discourse of territoriality the break in spatial dynamics that colonial rule represents. Colonial occupation, as African intellectual Achille Mbembe reminds us, was a matter of seizing, delimiting, and asserting control over physical geographical area—of writing on the ground a new set of social and spatial relations. The writing of new spatial relations (territorialization) was, ultimately, tantamount to the production of boundaries and hierarchies, zones and enclaves; the subversion of existing property arrangements; the classification of people according to different categories; resource extraction; and finally, the manufacturing of a large reservoire of cultural imaginaires (Mbembe 2003: 25-26).

In North-East India today it is becoming vitally important to take stock of the local practices of space, and to recognize the violent break in spatial dynamics made by colonial occupation. It is not hard to find evidence of a few old, local spatial practices that still persist today, despite colonial disruption. But in order to do so, one has to step away from colonial archives. For instance, only about 40 miles from the city of Guwahati, the Jonbil mela still takes place every year, where a descendant of the Gobha king presides over a fair in which Tiwa, Khasi and Karbi people that straddle the colonial hill-plains divide, barter edible roots in exchange for fish (Ramirez 2007: 22).

On the other hand, before North-East India became a remote and militarized border region in the mid-twentieth century, it was incorporated into the global capitalist economy in the nineteenth century along with other colonial era plantation economies such as Fiji, Mauritius and Guyana. The difficulties that the first generation of tea workers presented to a colonial ethno-territorial classification, and the contemporary efforts of their descendants to claim compensatory justice with a borrowed idiom of remembered tribal-hood underscore the need to fundamentally rethink North-East India's politics of territoriality.

By the end of the nineteenth century as tea plantations, oil wells, coalmines, and cash crops, such as jute, were changing the landscape of the plains of Assam by subverting local economic and social networks and property regimes, the Inner Line was there, more than anything else, as the security parameter of this colonial capitalist frontier. Designed partly to keep 'primitives' bound to their 'natural' space in the hills, it defined the boundaries of the 'civilizational' space that the colonial authorities were trying to create. It was also intended to stop European adventurers from straying across and from grabbing more land. The colonial state was unwilling to spend resources to 'tame' those 'wild' areas, to establish a government presence and to ensure the security of property. Yet it did not want actions by private individuals to incur the wrath of the 'primitives', and risk the safety of the fledgling enclave economy taking shape in the plains.

The memories of a real or imagined shared past, or the political *imaginaries* shaped by the colonial ethno-territorial framework, cannot be the basis for rights and entitlements in such a region. The current politics of territoriality must make room for politics that are based on an understanding of the local practices of space and a vision of a common future for those who live in the region today.

REFERENCES

Bates, C. 1995. *Race, Caste and Tribe in Central India: The Early Origins of Indian Anthropometry*, Edinburgh Papers in South Asian Studies, No. 3, University of Edinburgh.

Deb, S. 2004. *The Point of Return*, New York, HarperCollins.

Galanter, M. 1984. *Competing Equalities*, Berkeley, CA, University of California Press.

Gilroy, P. 1987. *There Ain't no Black in the Union Jack: the Cultural Politics of Race and Nation*, London, Hutchinson.

Government of Arunachal Pradesh. 2006. *Human Development Report 2005*, Itanagar, Department of Planning, Government of Arunachal Pradesh.

Kimura, M. 2009. 'Agency of Rioters: A Study of Decision-Making in the Nellie Massacre, Assam, 1983', in *Beyond Counterinsurgency: Breaking the Impasse in Northeast India*, New Delhi, Oxford University Press.

Ludden, D. 2003. *Where is Assam? Using Geographical History to Locate Current Social Realities* (CENISEAS Papers 1), Guwahati, Assam, Centre for Northeast India South and Southeast Asia Studies, Omeo Kumar Das Institute of Social Change and Development.

Mamdani, M. 2005. 'Political Identity, Citizenship and Ethnicity in Post-Colonial Africa', Keynote address, World Bank Conference on 'New Frontiers of Social Policy: Development in a Globalizing World,' Arusha, December 12-15.

– 2001. *When Victims Become Killers: Colonialism, Nativism, and the Genocide in Rwanda*, Princeton, NJ, Princeton University Press.

Mbembe, A. 2003. 'Necropolitics', *Public Culture* 15(1):11-40.

McDuie-Ra, D. 2006. 'Civil society Organisations and Human Security: Transcending Constricted Space in Meghalaya', *Contemporary South Asia* 15 (1): 35-53.

Prabhakara, M. S. 2003. 'Promises and Problems', *Frontline* 20 (5) March.

Ramirez, P. 2007. 'Choosing Between Hills and Plains: The Complexity of Identities in the Assamese Foothills'. Paper presented at an International seminar on 'The Politics of Belonging in the Himalayas: Democracy. Citizenship, and Belonging in the Himalayas', India International Centre, New Delhi, March 19-21.

Reid, R. 1944. 'The Excluded Areas of Assam', *The Geographical Journal*, 103, (1/2): 18-29.

Rich, M. 2006. 'Hill and Plains in the Colonial Imaginary of India's Northeast: Khasi Citizens and Tribal Subjects'. Unpublished Masters Thesis, Department of Anthropology, University of Chicago.

Sack, R. D. 1986. *Human Territoriality: Its Theory and History*, New York, Cambridge University Press.

Stepan, A. 2001. *Arguing Comparative Politics*, New York, Oxford University Press.

Verghese, B. G. 2005. *Water Resources in the Northeast: Development Options in a Cooperative Framework*, New Delhi, Government of India, Ministry of Development of Northeastern Region & The World Bank.

Weiner, M. 1983. 'The Political Consequences of Preferential Policies: A Comparative Perspective', *Comparative Politics*, 16 (1): 35-52.

Notes

1. In this paper the term Excluded Areas is used to refer to both types of area.

1.1.3.

Margins and Borders: Polities and Ethnicities in North-East India[1]

Philippe Ramirez

Both the affirmative action policies of the Indian state and the demands of ethno-nationalist movements contribute to the ethnicization of territories, a process which began in colonial times. The division on an ethnic basis of the former province of Assam into states and Autonomous Districts[2] has multiplied the internal borders and radically redefined the political balance between local communities. Indeed, cultural norms have been and are being imposed on these new territories for the sake of the inseparability of identity, culture and ancestral realms. The presence of certain ethnic or cultural minorities has become problematic. States have come into conflict over ethnic-minority enclaves on their respective territories. This underlines a major issue that we hope to document in the following pages through actual case studies: does the ethnic balkanization of the North-East correspond to the realization of old sovereignties? Did pre-colonial North-East India look like an assemblage of 'tribal countries', each with a clear-cut territory, a homogenous culture and a unique ethnicity?

Figure 1: State and district borders (2009), study area.

These issues do indeed emerge if one considers current data. In several communities in the North-East the strict correlation between culture and identity is far from being firmly established and the very question of ethnic affiliation may not arise at all. It may be argued that migratory blending over recent years has blurred traditional settings. We will make the opposite assumption, suggesting that such social forms pertain to ancient social identities which attribute less importance to ethnicity. These phenomena will be illustrated using the case of a hilly region divided between Assam and Meghalaya (Fig. 1).

Four 'tribes' historically coexist in this area: the Khasis, the Pnars (Jaintias, Synt-engs), the Karbis (Mikirs) and the Tiwas (Lalungs). All are listed as Scheduled Tribes in at least one of the seven states of North-East India. However, for each tribe, the extent to which it obtains advantages associated with this status varies from state to state, and in Assam it furthermore depends on whether this involves plain or hill dwellers. Giving accurate population figures about Scheduled Tribes might be a rather delicate issue, one of the reasons being that they are often referred to using outdated terms which are rejected by the groups in question. The problem is more complex for non-scheduled tribes and sections. Nevertheless, the following estimates may be put forward for 2001 (official Scheduled Tribes figures are given in bold):

	Meghalaya	Assam Hills	Assam Plains
'Khasi-Jaintia' (Khasi+Jaintia+Pnar)	**1,100,000**	**13,000**	
'Mikir' (Karbi)	**11,000**	**353,000**	200,000
'Lalung' (Tiwa)	1,000	10,000	**171,000**

In order to roughly set the scene, it might be said that the Khasis, Pnars and Karbis constitute fairly distinct entities in particular areas—though not everywhere—, where they have obtained 'Autonomous districts' (Khasi Hills, Jaintia Hills, Karbi-Anglong); their population however extends well beyond these districts. As for the Tiwas, they largely coexist with the Khasis, the Karbis and the Assamese castes, the main group in the Plains. Now, cultural belonging as well as identities are far less clear-cut within a large region at the interface between these ethno-linguistic areas. Here, often within the same village, one may come across speakers of different languages, different descent systems and barely expressed ethnic identities. We shall try to show how the atypical character of these communities pertains to particular relationships between culture, territory and political affiliation in which ethnicity plays a minor role.

It should be quite clear in the following pages that we consider the ethnic group as a real entity, exclusively founded however on identity assertions. Its limits are simply defined by the sum of individuals who claim to belong to it.[3] Indeed, many north-eastern groups satisfy this definition: it does not imply that belonging to an ethnic group is compulsory; and it does not imply that the representations associated with the ethnic groups are real.

Every winter, Jonbil Fair (*Jonbil melâ*) is held 50 km east of Guwahati. The Fair is famous among the Assamese for being one of the last places where barter is found. The event takes place three days before Magh Bihu, a key event in the

Assamese ritual year.[4] *Jonbil melâ* is extremely rich in anthropological meaning. Our focus here will only be its geographical and political context. At certain times of the day the *melâ* looks rather like the usual, modern fair where neighbouring villagers come to purchase household goods and to have a ride on the merry-go-rounds. However, a number of other activities take place, some of them typical of Magh bihu celebrations throughout Assam (collective fishing, cockfights), while others are more out of the ordinary, like the famous barter. Hill-dwellers come to exchange edible roots (taro, turmeric, ginger …) for dried fish and sweet pancakes (ass. *Pithâ*),[5] the typical delicacy eaten at Bihu (photo 1). The reasons for this bartering are obviously ritualistic rather than economic, as most visitors would find the same products at a similar price on markets closer to home. What takes place at *Jonbil melâ* might be the staging of a time when such goods were at the very heart of exchanges between the hill people and the plain people.

The fair is sponsored by a local ruler, the Gobha Râjâ, nowadays considered to be the 'King of the Tiwas'. However, it is neither strictly a state ritual nor a territorial ritual. It is neither essentially related to a single tribe, although the geographical setting of the *melâ* assigns a prominent role to the hosts, the Tiwas and the Gobha Râjâ. Visitors present the *rajâ* with various free contributions which are considered as 'taxes' (ass. *kar*). Not all of them identify themselves as Tiwa and, as we shall see, their status of 'subject of the *rajâ*' is a matter of interpre-tation. They originate from an area hardly definable in either ecological or administrative terms, straddling the borders of Assam and Meghalaya, as well as the plains and hills. Neither do visitors to the *melâ* seem to correspond to any clear cultural community: if languages alone are to be taken into account, they are speakers of Khasi, Tiwa, Karbi and Assamese. As for ethnicity, their heterogeneity is all the more puzzling, with some villagers even asserting no particular ethnic identity: when questioned about their 'tribe', they give a clan name.

The ritual territory defined by the origin of the barterers appears to be complex. How do we go about uncovering part of its foundations? We will first turn to the mythology and historical data related to *Jonbil melâ*. We will then examine the villages of origin of visitors in order to check to what extent the heterogeneity of the public attending the Fair does not simply reflect the anthropological heterogeneity of the region.

Photo 1: Hills and plains: roots (turmeric) exchanged for (dried) fish. *Jonbil mela* 2004. (Photo by Samiran Boruah)

The political significance of the Fair has evolved considerably over the last twenty years with the rise of the Tiwa movement whose political and community bodies have become the true patrons of the event.[6] The symbolic characters staged nowadays are Gobha Rajâ, a number of his own vassals, and the heir to the Ahom sovereigns (*svargadeu*), who ruled over Assam. Everyone recalls that only a few years ago, the *rajâ*s of Khyrim and Jaintia were also present at Gobha Rajâ's side. The reason for the visit by these two hill rulers is less obvious. Was it only a question of diplomacy or simple courtesy? Local representations depict Gobha, Jaintia and Khyrim as much more than mere neighbours. A number of narratives draw a triangle of special relationships between them.

A common story portrays the *rajâ*s of 'Gobha, Jayatâ and Khoirâm' (Gobha, Jaintia and Khyrim) as three brothers born on a mountain called 'Thin Makhlang' or 'Timophlang', the first from the ground, the second from a stone and the third from a hollow tree.[7] Very few people in the plains know the exact location of this place, but according to the inhabitants of the Tiwa hills, it would be situated in Meghalaya, on the ancient border between the three states.[8] Similar narratives show the three characters jumping out of a pond. Gobha is sometimes the eldest, sometimes the youngest brother. In some instances, Jayata and Khoiram come to *Jonbil melâ* to find the Bihu pancakes, in order to celebrate Rangsi, the corresponding festival in the hills.

Among Hill Tiwas, there are still traces of meetings between Khyrim and Gobha kings. At Kutusi Mokoidharam, not far from the three borders, a set of twelve monoliths (photo 2) stands right on the trail down from Nartiang (Jaintia hills).[9] Some Hill Tiwas explain that when Rajâ Khrem (Khyrim) came to pay a visit to Gobha Rajâ, he used to stop at this place where a market was held. Such a narrative is in fact of valuable historical importance in that it underlines the fundamental role of trade in the former chieftainships. One may even argue that the *raison d'être* of these states was the protection of trade and markets. It is well known that the Khasi states drew most of their revenue from market taxes (kh. *ka khrong ka dan*)[10] and most disputes between states, as well as with the British, concerned the control of or access to markets.[11]

Photo 2: Mokoidharam, site of an ancient market.
(Photo by Ph. Ramirez)

The image of chiefs coming and going over the hills to ensure the smooth running of markets and of them taking their share might give an idea of the nature of the Hill states. It is not at all clear whether very defined and close borders separated the 'territories' of these chiefs.

The oral tradition itself sometimes confuses various rulers: it is quite common to hear Tiwas speaking of 'Jotya-Khairam', i.e. 'Jaintia-Khyrim', a generic term referring to the higher hills of the plateau. And this pattern seems to have been used more widely than merely around Gobha. According to one of the myths of origin for neighbouring Dimoria, the latter's rajas are supposed to have descended from one of four brothers: 'Gubha, Dimorua, Khoiram and Milim', whose cradle was 'Thimuflong'. Another narrative published by Gohain conjures up for us the components of the former political system.

> When the Khorang clan (Khoiram) of the Jaintias wanted a bullock for their religious festival, they would go to the Magro clan of the Lalungs living in the Jaintia habitat with a betel-nut and liquor [...] Next morning, the chief [of the Magro] would hand over a rope to them [...] They would go to the Gobha raja with liquor and a betel-nut and the latter would say, 'Go to the field and select a young bull'. [They used to pay] one rupee and four annas to the cowherd, but no price was to be paid for the bull.[12]

Proper nouns must not be taken for granted for they appear to confuse clans and states. Instead, it is the structural features which are noteworthy here. One rapidly perceives the social landscape of this paradigmatic narrative: a network of economic and ritual relationships among clans, and between clans and states, which transcended the limits between hills and plains or even made possible exchanges between hills and plains. This is in tune with the patterns above which focused on the association of several chiefs all placed on the same footing.

Historical documents about the region portray a less egalitarian and peaceful image of the relationships between these three rulers. However, they confirm the existence of a regional politico-economic system based on trade between the hills and plains. For three centuries, two major powers competed against each other in the area: the Ahom rulers, dominant in the plains, and the Jaintias controlling the eastern Meghalaya plateau. The region is mentioned in the Assamese chronicles (ass. *buranjî*) dating from the seventeenth century onwards. At that time, and until the advent of British Rule in 1826, the Ahom sovereigns almost entirely controlled the Brahmaputra Valley, yet hardly any of the highlands. They sent some military expeditions into the hills (in 1707 especially, in the Jaintia hills) but never occupied them. They nevertheless maintained multiple trade relationships with the hills and mountains, whether with Bhutan or Tibet, the Naga, Khasi or Garo Hills.[13] As a matter of fact, they entered a system and networks which had pre-existed them long before. In this system, the 'gates' leading to the hills, the *duar*s, were much sought after strategic points.[14] The *duar*s themselves provided access to major trade routes, but the nearby hills also concealed some resources valued by merchants from the plains, such as salt, lime, lac, wax ...[15] Ahoms attempted to control the *duar*s by gaining the allegiance of local leaders, to whom they granted the status of *râjâ*. In this venture, they competed with other states, namely Jaintia and Khyrim, which also claimed suzerainty over the same chiefs.[16] In the early eighteenth century, the affiliation of the rulers of Gobha, Neli and Khala shifted from one to the other. They seem to have seldom been fully independent, but to have played on rivalries to protect themselves, possibly lending simultaneous allegiances to two or more suzerains.[17]

When the British arrived on the scene, Khyrim and Jaintia had been in mutual conflict for several years, a situation which affected exchanges through Gobha.[18] The image of a peaceful coexistence between the three kings was nevertheless maintained, unless it actually only emerged afterwards. The intimacy among the three kings, which is tangible in the folk narratives, would reflect a very fluctuating situation in which the area fell either alternatively or simultaneously under the authority of several states. In 1829, when accounting for the difficulties in preparing a map of the Khasi hills, David Scott provided a very significant clarification: "throughout these mountains peculiar spots are to be found belonging to one chief, although surrounded with the territory of another, and that two or more of them are occasionally found exercising authority in the same village."[19] One point in particular must be stressed; the possibility that an area may have been simultaneously subjected to several authorities. It could have led to an intermingling of different cultural models imposed or inspired by the respective elites of the dominant states.

The region where the three rajas mingled roughly consists of the north-eastern corner of the Meghalaya plateau: low hills (200-800 m) covered mainly by slash and burn regrowth vegetation and bamboo groves. To the west and south lies Meghalaya. Formerly part of Assam, this Indian state was created in 1972 for the benefit of Garo and Khasi-Jaintia Scheduled Tribes (ST). Except for its capital Shillong, the whole state is classified as a 'Tribal area' under the 6th Schedule of the Indian Constitution. The task of settling customary affairs, as well as a section of local government, has been handed over to three Autonomous District Councils: Garo Hills, Khasi Hills and Jaintia Hills. To the East lies the state of Assam, and more precisely the Karbi-Anglong Autonomous District (1970), realm of the Karbi ST, under the same 6th Schedule. To the North, the plains of Assam, which on this side do not include any official "Tribal area", but where the Tiwa (plain) ST is demanding its own Autonomous District.[20]

This territorial arrangement stems from a series of processes initiated on the arrival of the British and which today continue along the same lines.[21] The conviction among numerous observers is that these political territories no doubt correspond to ancient territories inhabited 'from the beginning' by stable, well defined and unique populations. It is unclear whether such a representation existed in pre-colonial times, but it seems just as firmly rooted among the ethno-nationalist elites as it was among British rulers. Thus, the oriental part of Meghalaya, known since colonial times as the 'Khasi and Jaintia hills', is believed to be the ancestral territory of the Khasis and Jaintias. Associated with the image of a specific ethnic territory is that of a homogenous population. Today this perception is dominant among people claiming a Khasi and/or Pnar identity. Certainly, the Khasis do acknowledge some cultural differences among themselves. The most common classification recognizes four localized groups: in the highlands, the Khynriam, to the south the War, to the east the Pnar and on the northern and eastern fringes, the Bhoi. However, for a majority of the first two groups, there is no doubt about their original unity.[22] They are all thought to descend from the 'Seven huts' (Hynniew trep). Among certain ethno-nationalists, the mere evocation of differences among the Khasis is perceived as an unacceptable attack on the unity of the group.[23] Among the Pnars and the Bhois, the matter is more a subject of debate, although the Khasi identity is widely accepted.

As early as the beginning of the twentieth century, Gurdon—author of the first monography on the Khasis—portrayed a very ambiguous image of the northern-belt people, whom the 'other Khasis' called 'Bhoi' and considered with some disdain. Gurdon claimed they were mostly Mikir—these days the common designation for Karbi—and that the term Bhoi was "a territorial name rather than tribal".[24] Several decades later, despite the area opening on to the outside, Bhoi's specificity does not seem to have diminished. It was formally recognized in 1992 with the creation of Ri-Bhoi district (lit. 'Bhoi country'). The recent attempt to rechristen it 'Northern Khasi District' and the strong reaction which ensued prove that a century later the classification of Bhois as Khasis remains problematic, both for external observers and the Bhois themselves.[25]

Let us consider the north-eastern corner of the Meghalaya plateau. The anthropological complexity of this area makes any description arduous. Cultural, linguistic and ethnic diversity is not only widespread, but neither does it assume the same aspect nor does it appear to the same degree everywhere. A comparison of two zones might be made. In the first zone, villages are generally monolingual and mono-ethnic. In the other, several languages are spoken in the same village and ethnic identities are either complex or barely determined. On the one hand a compartmentalized mosaïc, on the other a continuum of blends. (Fig. 2)

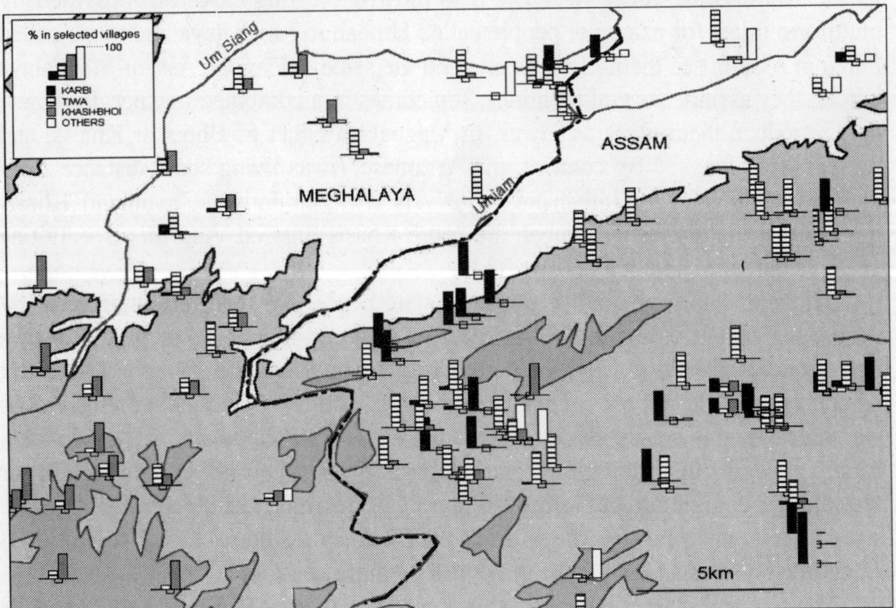

Figure 2: Ethnicity of names in villages of the Assam-Meghalaya border, as perceived by a Tiwa speaker. Names were taken from the 2006 Assam and Meghalaya Electoral Rolls. (© Ph. Ramirez, CNRS).

We do not intend to go into the reasons for these differences which may have been shaped by the interplay of Christianisation and Scheduled Tribe policies.[26] Suffice to say that, in some parts, the limit between the two zones follows the border between Ri-Bhoi and Karbi-Anglong, i.e. between Meghalaya and Assam. Thus, in this area, the Assam-Meghalaya border neither corresponds to a limit between two ethnic territories, nor two mono-cultural areas, but between different types of heterogeneity. We will presently consider the first zone, the eastern part of Ri-Bhoi district. Here, the linguistic landscape is more multicoloured than in other parts of Meghalaya. Although the local dialect of Khasi, Bhoi Khasi,[27] prevails both as the mother tongue and *lingua franca*, Karbi and Tiwa are spoken in some fifteen villages, to which one must add an atypical Assamese parlance used ironically not on the Assamese border but well inside the hills, in the Marngar locality. One must bear in mind that these languages belong to three entirely distinct families: Mon-Khmer (Khasi), Tibeto-Burmese (Tiwa, Karbi) and Indo-Aryan (Assamese).

Similarly, clanic affiliations are also more complex than elsewhere in Meghalaya, a fact that has a very significant impact on ethnic identity. As in many parts of India, ascribed identity is most generally a result of the perception of patronyms, or 'titles'. In tribal systems, these correspond to clans or to local clan segments. The general paradigm in North-East India is that a particular title indicates one and only one ethnic affiliation. This is not the case, however, in the area we are dealing with, where the ethnic connotation of names varies according to several factors, depending on the informant and the local or social context in which information is given. Thus, the same group might be referred to as either Bhoi, Khasi, Khasi-Bhoi, Karbi or Tiwa. The most interesting cases are provided by multiform titles: for example, people called Umbah in Meghalaya and those called Puma in Assam, see themselves as one and the same. Along the Assam-Meghalaya border, they assume several identities, sometimes in a combined manner. In Assam they introduce themselves as Tiwas, in Meghalaya either as Bhois or Khasis, and sometimes Lalungs.[28] By contrast, the Assamese Tiwas living some distance from the border claim that the Umbah are exclusively Tiwa; similarly the 'mainland' Khasis might insist that the same Umbah are 'pure Khasis' instead. And this is only one example among many others (Fig. 3).

Might one further qualify the inter-state border, in this particular area, by suggesting that it corresponds far less to a real cultural break or gap than to a transition between two ways of perceiving the intermediate groups? Dominant perceptions would impose a clear break where cultural variations are elusive. As we shall see, this would be accurate only if two further details were provided. Firstly, it seems that the cultural variations in question do not evolve in a linear way along a continuum, but through a mosaic of micro breaks. Secondly it must be stressed that dwellers in the border zones do not adhere easily to the dual categorizations emanating from the politico-ethnic centres.

The fundamental social unit in this part of Ri-Bhoi, as in the entire Khasi Hills Autonomous District, is a community of villages called (kh.) *raid*. This administrative unit is part of the 'traditional' political system in force within the

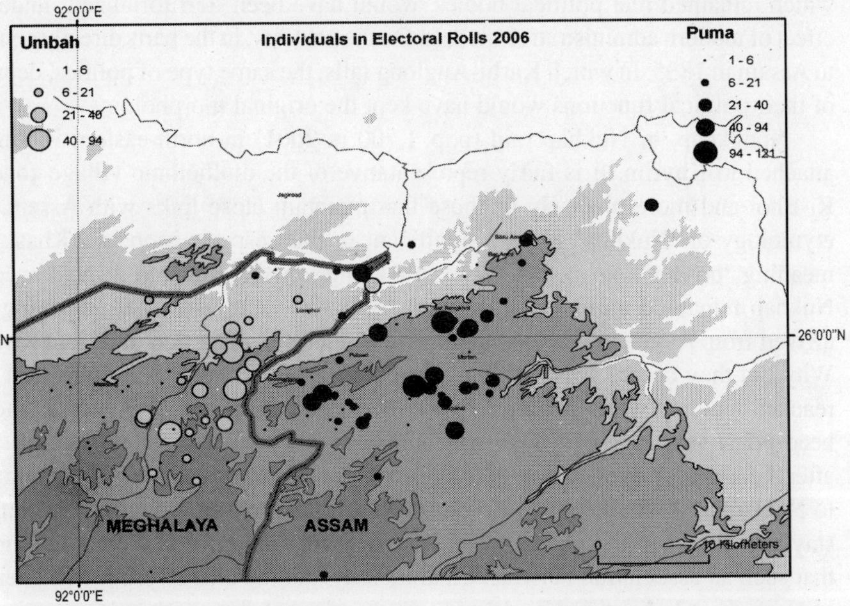

Figure 3: Borders, spellings and identity: the example of Puma (black) versus Umbah (grey), Meghalaya and Assam Electoral Rolls, 2006. (© Ph. Ramirez, CNRS).

Autonomous District: the states, (kh.) *hima*, under the authority of a kingly person, (kh.) *syiem*, are divided into a number of *raids*. *Raids* are ruled by representatives of clans, (kh.) *basan*, and in Ri-Bhoi they are presided over either by a vassal chief called *syiem raid* or a (kh.) *lyngdoh* priest. However old and whatever the origin of its inhabitants, the *raid* is one of the main references in social identity. Every *raid* formerly came under the authority of the Tiger-god Ryngku (Khla Ryngkhu) who punished crimes and to whom an annual sacrifice was offered. According to the *syiem* of Khyrim, all the *raids* in north-eastern Ri-Bhoi placed themselves willingly under his predecessors' authority in order to seek protection from the Jaintia tyranny.[29] As a matter of fact, it seems that Jaintias occupied the region several times during the seventeenth and nineteenth centuries. When the British arrived, it was under Shyllong and when the latter split a few years later (1853), the local *raids* were divided up between Mylliem and Khyrim along a line running north-south.

The present *raids* display several attributes of modern administrative units, including a delineated territory. However, one might question how long this form has actually been in existence. The *raid* should be compared with other comparable institutions in the vicinity. In Karbi-Anglong, whether in the Karbi-speaking or Tiwa-speaking areas, some politico-ritual networks are found to be spatially discontinuous (see infra). In other words, the sum of links between a centre and its dependencies does not form a continuous spatial entity but rather a network interlaced with other similar networks. We suggest that this open and discontinuous pattern is older than the form the *raids* display nowadays. The Khasi hill *raids*, which remained real political bodies, would have been 'territorialized' under the effect of modern administrative models. On the contrary, in the parts directly annexed to Assam in 1835, in which Karbi-Anglong falls, the same type of polities, deprived of their political functions would have kept the original morphology.

Nongkhap, or Nukhap raid (pop. 1,700 in 2001) in north-eastern Ri-Bhoi is attached to Khyrim. It is fairly representative of the multiethnic village zones of Ri-Bhoi and more precisely of those that maintain close links with Assam. The etymology of 'Nukhap' is very significant in this respect: *khap* is a Khasi term meaning 'border', *nu* means 'new' in Assamese. According to a local account, Nukhap remained uninhabited up until "300-400 years ago", before immigrants arrived from Assam and placed themselves under the protection of Syiem Khyrim. Why did they come? It is said that the first inhabitants were Karbi and that they reached Nukhap while fleeing 'the Bodo king of the plains'.[30] According to the head-priest (kh. *lyngdoh*), his own family arrived from the Jaintia hills in 1618 after the Jaintias had overcome a 'local tribe'. Another informant described migration to Nukhap as a return to their origins: "We are people from the hills. We couldn't stay any longer in the plains". Without going into detail, it is worthwhile noting that such an association between tribal identity and the hills is gaining momentum in North-East India, though a large majority of tribals live in the plains.

In keeping with a widespread model in Meghalaya, the existing villages of Nukhap claim to have come from a common founding village, Nongbah (kh. 'great village').[31] From Nongbah Nukhap came the founders of the *raid*'s four villages. They were met by immigrants from Tiwa-speaking villages situated on the Assamese side. Nongbah Nukhap was finally abandoned after 'the conversion'. The *raid* experienced large-scale Christianisation in the first half of the twentieth century.[32] All villages are now exclusively Presbyterian or Catholic. Following an initial period, when Christians were expelled from villages and regrouped, the remaining 'pagans' (kh. *jyntir*) had to choose in turn between conversion and exile. Beside the distribution of settlements, conversion had an obvious impact on culture and ethnicity. The influence of catechists preaching in the Khasi language and using Khasi books might have been decisive. Culturally, Christianisation might have taken the same path as a large movement of 'Khasi-isation' generated by other factors such as the creation of Meghalaya state and mass schooling. It might not be a generalised phenomenon in Ri-Bhoi, but in Nukhap the linguistic situation has evolved significantly over the last two generations. Bhoi Khasi has now become the only mother tongue, even though twenty years ago Karbi and Tiwa were still spoken in some houses. Today, nowhere in this *raid* are these languages understood. By contrast, just a couple of kilometres to the east, still within Meghalaya, in a fringe devoid of roads and teachers, Tiwa is the only language known to everyone.

In Nukhap, Khasi-Assamese bilingualism is associated with exchanges both upstream and downstream. Though a dependency of Khyrim, the largest Khasi state, Nukhap seems to be much more oriented towards Assam than Meghalaya. People only go to the district headquarters, Nongpoh, to settle administrative matters. Before the Topatuli-Umsning road had been built (1975), leading to Shillong and the Upland, villagers used to exclusively venture to markets in the Brahmaputra Valley: to Topatuli and Jagiroad for rice and areca nuts, and once a year to *Jonbil mela* to barter roots for fish. Today, after each monsoon, Nukhap people still go down to Hariaburi, famous for its dried fish that have a long conservation period.

In border communities naturally turned towards the plains, the creation of Meghalaya has paradoxically weakened the links with Khasi states. The *syiem*s had an obvious economic interest in the hill-plain borders: they used to exploit forest resources and to collect taxes on markets in the foothills, both through the local agents they appointed. The complete closure of forests and the drawing of a new boundary, together with the building of inner roads have noticeably reduced the Upland states' business opportunities in this area, with there being no land tax in the Khasi hills. Thus, the little we know of Nukhap's history tells us mostly about outside contributions and exchanges.

The main feature which distinguishes the Bhoi *raid*s from both Assam and the Upper country is the feeble assertion of any ethnic belonging. Nukhap inhabitants are hardly expansive on this topic. The label 'Bhoi' is accepted by everyone. This acceptance may be a recent phenomenon, because in the early twentieth century the term was deemed pejorative in the eyes of the Bhois themselves and later remained a synonym of 'ignorant' in the Khasi language.[33] More than elsewhere, the social context of speech is the decisive factor in assessing identity: when speaking with a Khasi speaker from the Upper country one calls oneself 'Khasi Bhoi'. On meeting an Assamese person, Karbi or Tiwa origins will be stressed instead. And 'Bhoi' seems to be the name favoured when dealing with foreign anthropologists. There is no hidden strategy behind such opportunism. It truly reflects the social representations, which in this area are characterised by the conjunction of multiple ethnic affiliations. In Nukhap, social identity could be summarised in the following manner: one belongs first to a *raid*, within which one's original ethnic belonging is still remembered, yet not deliberately displayed. In the *raid*, language and customs are fairly uniform and are linked at the upper level to a large entity, 'Bhoi', which is primarily defined by stressing the differences with the 'Khasis', i.e. the 'other Khasis': Khynriams, Pnars and Wars. On the other hand, cultural dissimilarities with groups residing in Assam, which are objectively more obvious, are not often pointed out. The reason might be that the arena in which ethnic identity is of relevance for the people of Ri-Bhoi is first of all Meghalaya, to which political belonging is widely assumed; in this respect cultural differences with Assam are not problematic. Finally, belonging to a traditional state (kh. *hima*), be it Khyrim or Mylliem, does not seem to correspond to any sense of identity, and this seems the case in the larger part of the Khasi hills.

The anthropological features of Nukhap are typical of many but not all *raid*s. To the north, the border *raid*s are, like Assam itself, compartmentalised into villages or hamlets asserting a single identity: 'Khasi', 'Karbi' or 'Tiwa'—but not 'Bhoi', as far as we know. Often, but not always, the original languages have survived with the Bhoi language. Here we reach the historical fringes between the highland and plain states, where political allegiances have been complex. In Mylliem state for example, the geographical specificity of this belt has been acknowledged, and is still recognized by the Khasi Autonomous Council. According to law,[34] Mylliem is divided into two parts: the 'Highlands' (kh. *ri-lum*: 'land of the hills') and the 'Lowlands' (*ri-them*), which come under different administrative structures. In Ri-Bhoi, the Lowlands are distributed among sixteen chiefs: three *lyngdoh raids*, eight *syiem raids* and five *bangthe raids*, each recruited from a particular clan. Two of the *syiems* are entitled '*syiem kuba*', which might well refer to the family of the Gobha raja, considered nowadays in Assam to be the "Tiwa king". One of the *raids*, Marngar, forms a linguistic isolate surrounded by entire Khasi and Karbi-speaking zones, where a very peculiar form of Assamese is used.[35] Local people also assert a peculiar identity, both Khasi and Marngar, this latter ethnonym being found exclusively in these nine villages. Finally, the Bangthe Raids, as their name suggests, are run by Karbi headmen (kb. *bangthe*). These *bangthes* preside over villages where the inhabitants—though not all—bear 'Karbi' titles, speak either Karbi, Khasi or Assamese as their mother tongue and whose Christian (educated) sections are keener to assert their Karbi identity than the so-called 'Hindu' (i.e. non-Christian) sections.

Historically, Bangthe Raids overlapped Dimoria principality, a strategic marketplace in the plains long coveted by the Jaintias, the Khasi states and the Ahoms.[36] The villages of Dimoria are emblematic of the anthropological complexity of North-East India and have been—and still are—subjected to multidirectional migrations. Lokkikok, nowadays a Tiwa-speaking village just inside the foothills, provides a good example of the context of these migrations:

> We came down from Murji Kunji, near Marjong in the hills, to search for land. We settled first in Jagiroad Deosal, then in Lokkikok in 1942. The place had been inhabited by Karbis. They had stolen and shared among themselves the buffaloes belonging to the Dimoria King. A woman, angry about not getting a share, went to the king, who cursed and expelled them. They left after cursing the place and settled in Belkhuri, 2 km downstream. When we arrived in the area looking for a possible site, one of the Karbis showed us Lokkikok. At first we suffered from numerous diseases. We finally found out that a local spirit was responsible, the Tortoise Eater Lord (Kaso Khua Gohain). A cult was established and the troubles stopped.

These few words sum up the common historical aspects of the peopling of this region: shifting cultivation, political conflicts, morbidity, setting up of new cults and moves from the hills to the plains. Only banishment for incest is missing here. The route taken starts from Marjong, in the Tiwa hills, then makes a first stop at the important Deosal shrine, in the plains near Gobha, and next, in a very typical move, runs along the foothills for 20 km to Lokkikok. Narratives emanating from upland villages, like those from Bangthe Raids, describe routes leading from the plains to the hills, a migratory movement which is nowadays rare but which seems to have been common till half a century ago.

It is attested that in the nineteenth century, slash and burn was the sole method of farming in the northern part of the Meghalaya plateau.[37] Yet amazingly, the need for land is rarely mentioned as a factor of migration in local histories. Either land was plentiful enough to avoid moving villages, or the practice was regular enough not to be specially noticed. Another factor which immediately comes to mind because regularly mentioned in the Assamese chronicles is epidemics. The scale of some pandemics was such that it certainly had a decisive impact on the demographic map of Middle Assam and the Khasi-Janitia hills. Between 1891 and 1901 the localities in the foothills described here lost a third of their population, following several epidemics such as the 1897 *kala azar* (leishmaniasis) epidemic and the 1900 cholera.[38] However, given the present state of knowledge, the effects are still difficult to assess, and we have no trace of significant waves of migration towards the hills in this particular period.[39]

The relationship between ritual practices and migrations is rather unexpected. According to some indigenous accounts, the abolishment of human sacrifices has made survival impossible in some localities. Soon after Independence, the Assamese police were ordered to do away with human sacrifices in the former 'excluded' zones in which the British did not apply the general law. This was the case of Sonaikuchi. Some villagers offered goats instead, others ceased to worship. Mysterious deaths started to occur and were attributed to the wrath of the local gods. Some families, terrified, fled to the Bhoi hills.

Finally, one of the major causes of migrations has obviously been the development and decline of economic opportunities. Two particular cases may be mentioned: in the early twentieth century, hundreds of families came down from the Jaintia hills, attracted by recent developments involving lac cultivation and trade. Most of them left when the prices collapsed after the Second World War. Similarly, in the early twentieth century, thousands of Nepalis settled in Meghalaya, including Ri-Bhoi, where they specialised in milk production. They have almost disappeared from the Khasi hills since the xenophobic riots of 1987.[40]

The spatial mobility of Bhoi groups must be put into some perspective along with some features of their social structures. Firstly, the local polities (villages or *raids*) have the capacity to integrate newcomers, whatever their culture. As an example, the basic political unit of the Khasi states, the *raid,* is first and foremost an association of lineages, which may be from all sorts of cultural backgrounds, as we have seen in Nukhap.[41] This does not exclude Khasi cultural dominance or in some cases internal disputes along ethnic lines, but both the rules for admission and the functioning of the polity do not take into account ethnic affiliation.[42]

Exogamous marriages are almost absent from anthropological literature on the region. However, marriages between distant localities as well as between different ethnic groups are now widespread, at least in Ri-Bhoi, Morigaon and West Karbi-Anglong. They may be associated with the fast growing matrimonial networks now developing throughout North-East India, yet it seems that they were already common in many localities in the past. As a matter of fact, in certain areas, such as Raid Nukhap, there have been so many marriages that the question is what should be retained as an 'interethnic marriage'. In Nukhap, as in many areas of Ri-Bhoi and Karbi Anglong, a factor which facilitates local and ethnic exogamy is ambilineal descent, i.e. the combination of matrilineality and patrilineality. Descent depends on the choice of residence, i.e. the place where the young couple will settle. Under such a regime, the system of descent adhered to by the family of the 'other' spouse is not an obstacle to marriage. Ambilineality might be considered by some as a consequence of a recent expansion of exogamy or as a result of the decline in matrilineality. Although we cannot go into detail here, we suggest that this type of descent, which is prevalent in the Tiwa hills today and which is attested to in the late nineteenth century,[43] might have been practised for a long time over a larger region and that it has not only resulted from but also facilitated cross-cultural marriages.

Another challenge to the ethnic compartmentalization in Ri-Bhoi are the trans-ethnic phratries. Although phratries have been described among the Khasis and Tiwas,[44] it has not yet been noted that they possibly comprise clans from different ethnicities, which are considered 'equivalent' and form exogamic units. The present phratries could be interpreted as mutations of Khasi and Tiwa original phratries, which would have lost their ethnic belonging as a consequence of 'local globalization'. However, I think it reasonable to allege that they are rather the remnants of an old social system dissociating ethnicity and descent.

The multi-ethnicity of phratries and matrimonial networks might have extended well beyond the actual areas where it has survived. It might have been associated with a human geography characterized by high mobility and the interlacing of village networks. The village network is a general feature of settlement organisation in this region, whether in a multi-ethnic or mono-ethnic environment. The *raid* is a true network, since, as we have seen, its inhabitants claim they all come from the same 'great village'. Similarly, certain *raid*s are considered as 'elders', as original sites from where other *raid*s emerged, such as Iapngar for the *raid*s of north-east Ri-Bhoi. The situation is similar in the Jaintia hills, where villages fall into two different categories: (kh.) *shnong barim*, ancient villages, and *shnongbri*, farm-villages which stemmed from the former. How is an ancient village in fact different? It has a *lyngdoh* priest and it houses the matrilineage deities, which means that people from farm-villages have to go to their native village to worship them.[45] A very similar arrangement is found in the Tiwa hills. Each Tiwa perceives himself and is perceived by others as belonging to one of the seven major villages, (tiw.) *krai baro*. These are the only 'true' villages, those that have a (tiw.) *loro* priest. They are supposed to be the oldest settlements, although this is not always the case due to various divisions and mergers. Villages without *loro* are called *kunji phara* or *pham*, 'extensions', and each of them is linked to a founding village.[46] The largest network centred around Bormarjong village links sixteen settlements straddling Meghalaya and Assam.

Finally, the traditional Karbi political system relies on a sophisticated fabric of village networks. Ronghang, Chingthong and Amri, which are considered to be the three original geographical cradles of the Karbis, each come under the jurisdiction of four, three and two councils of dignitaries respectively, who act as judges and supervise rituals for a set of villages.[47] These functionaries reside in a sacred village, which definitely evokes the Khasi *nongbah* although it is not considered to be a founding village. At the lower levels, villages are organized into complex clusters (kb. *longri*) governed by a hierarchy of local chiefs (kb. *habe*) emanating from specific clans. The spatial distribution of village clusters does not follow a simple juxtaposition, but often a series of interweavings.[48] Since the eighteenth century at least, the Jaintia political domination has certainly contributed to the modelling of Karbi as well as Tiwa supra-village structures. A sign of this influence may be found in similarities between the titles of functionaries in the Jaintia, Tiwa, Karbi and Bhoi areas.[49] Whether they have been imposed by the Jaintia state or inspired by a common model, these institutions are obviously related to each other by a common 'politico-ritual culture'—Leach would have spoken of a 'ritual language'.[50] Ethnic revivalism has not uprooted them and, on the contrary, they have found a new legitimacy in Karbi-Anglong Autonomous District. As a matter of fact, these ancient political structures are not based on 'Karbi', 'Tiwa' or 'Khasi' principles, but on the tribal principle of local power-sharing by descent groups.

Conclusion

The new territories of North-East India are being created on an ethnic basis. The ideal model pursued is clearly an ethnically and culturally homogenous nation-state. Ethnic polities which have been established, or are sought after, are supposed to stem from timeless spatial sovereignties. Without judging the legitimacy of these claims, it is worthwhile noting that many local data point to models which are the opposite of such representations. In several areas of the North-East, local communities are not the subset of a particular ethnicity, but an association of lineages asserting various ethnicities or none at all. As a matter of fact, these local communities look upon themselves as small multi-ethnic nations. If such societies represent relics, this would mean that in part of North-East India at least, the states had no strict ethnic basis; that a change of ethnicity was common and that ethnicity itself was not compulsory. If, on the contrary, we are dealing with the product of a recent intermingling of populations, then it must at least be admitted that the boundaries between ethnic groups are far less firmly rooted that we generally suspect.

The perception of ethnicity depends both on the spatial point of view and on the spatial focus. Seen from the centre, like Shillong for the Khasi hills, one may feel that people sharing the same ethnic identities correspond to obvious cultural entities, with geographical cohesion and a historical depth. Yet as one gets closer to the margins, heterogeneity grows in linguistic as well as in cultural and identity fields, though in various forms and on different scales. This does not mean that the anthropological landscape is amorphous. Distinctions emerge among different configurations of multi-ethnicity, for example in the contrast between mixed villages and mosaics of mono-ethnic villages. Nonetheless, anthropological territories, whatever their definition, rarely coincide with current political territories. As ethnic groups, the Karbis, Khasis and Tiwas are present in significant numbers on both sides of the border between an official Khasi territory and an official Karbi territory.

Can the idiosyncrasies of these areas be explained by their location on the margins of major clearly identifiable political and cultural poles? In this case, they would be mere transitional areas, zones of compromise, of mingling, where specificities become blurred. This would in itself be a finding that would run counter to the very common image of a compartmentalized cultural landscape. However, it would be necessary to explain why this transition zone does not take the shape of a regular continuum and how discontinuities have appeared: migrations, ethnic and religious conversions, political conquest, economic change ... The local representations of history lead us on to another track, that of a true 'marginal' space, where cultures, identities and state authority constantly undergo a redefinition, a space which the 'Three Kings' or 'Three Borders' metaphor does indeed effectively illustrate. Much more than a space of contested ethnicities or of deculturation, the 'Three Borders' would be one of the last portions of a multi-ethnic and multicultural space caught nowadays between the conquering territories of states and ethnicities.[51]

The anthropological situation prevailing at the 'Three Borders' almost represents a counter-model opposed to the ethnic exclusivism which has affected North-East India over the last four decades. This movement has materialized in the form of new political territories, the states and Autonomous Districts. Yet if my presumptions are right, ethnicization has also transformed the perception of the territories inherited from the past. Ancient polities are increasingly represented as historical manifestations of the spatial rights of such or such an ethnicity. As formulated by Hobsbawm, Gellner or Smith,[52] the ruling elite imposes or inspires a cultural model among the dwellers of a territory so that finally they perceive themselves as part of a single body.

If we take the 'Three Borders' as a testimony of the past, the recent spatial compartmentalization has brought about another rupture: the territories that are being imposed consist of continuous spaces. Here also, social representations have been largely conditioned by the nation-state model. According to this model, polity might be conceptualized only as a space, not as a network. And it would only be a continuous space, i.e. not intersecting with other spaces on the same scale. We have nevertheless seen that several present-day societies in the area are still organized—or were organized not so long ago—on the basis of interwoven village networks. This form of organisation is incompatible with the administrative grid of modern states, which require linear and tight limits. When such grids are imposed, networks are being severed by new boundaries—as happened on the Meghalaya-Assam border—creating enclaves which the states are striving to reduce.[53] Paradoxically, if Ri-Bhoi margins carry the remains of vanished multi-ethnic societies, they also inspire a vast meta-tribal process which may be detected nowadays in matrimonial networks. Inter-ethnic marriages illustrate developments which go counter to ethnicization and which may provoke its complete redefinition. In the near future, they may design new social entities, both egalitarian and endogamous, transcending the ethnic and territorial compartmentalisation imposed by ethno-nationalisms. Finally, the discovery of these political forms, both archaic and modern, prompt us to reconsider the way North-East Indian societies have been approached so far, starting with the ethnic group.[54]

REFERENCES

Allen, B. C. 1902a. *Census of India 1901, vol. IV: Assam, part 1. Report*, Shillong, Assam Secretariat.

– 1902b. *Census of India 1901, vol. IV: Assam, part 2. Tables*, Shillong, Assam Secretariat.

– 1905. *Assam District Gazetteers, vol. VI: Nowgong*, Calcutta, The City Press.

– 1906. *Assam District Gazetteers, vol. X: Khasi and Jaintia hills, the Garo hills and the Lushai hills*, Allahabad, The Pioneer Press.

Bareh, H. 1967. *The history and culture of the Khasi people*, Calcutta.

Barth, F. 1970. 'Introduction', in Barth, F. (ed.), *Ethnic Groups and Boundaries: The social organization of culture difference*, Bergen, Universitetsforlaget, pp. 9-37.

Baruah, S. 1999. *India Against Itself: Assam and the Politics of Nationality*, University of Pennsylvania Press.

Baruah, S. L. 2002. *A Comprehensive History of Assam*, New Delhi, Manoharlal.

Blackburn, S. 2003. 'Memories of Migration: Notes on legends and beads in Arunachal Pradesh, India', *European Bulletin of Himalayan Research*, 25-26: 15-60.

Burlings, R. 2007. 'Languages, Ethnicity and Migration in Northeastern India', *South Asia*, vol. 30, no. 3, Dec. 2007: 391-404. (A rather similar text can be found on www-personal.umich.edu/~rburling/Language%20and%20Ethnicity.pdf)

Chattopadhyay, S. K. 1988. *The Jaintias; socio-political institutions of Jaintia hills*, New Delhi, Cosmo.

Devi, L. 1960. 'Ahom-Tribal Relations (a Political Study)', Ph.D. thesis, Gauhaṭi University.

Gassah, L. S. 1998. *Traditional Institutions of Meghalaya: A Case Study of Doloi and His Administration*, Delhi, Regency Pub.

Gellner, E. 1983. *Nations and Nationalism*, Ithaca, Cornell University Press.

Gogoi, L. 1986. *Tivâ Sanskritir Ruprekhâ (pratham khanda)* ('Outlines of Tiwa Culture, part 1'), Guwahati.

Gohain, B. K. 1993. *The Hill Lalungs*, Guwahati, A. Borooah Inst. of Language, Art & Culture, Assam.

Gurdon, P. R. T. 1914 (1906). *The Khasis*, London, Macmillan.

Ḥamilton, F. 1807. *An account of Assam: with some notices concerning the neighbouring territories* (First compiled between 1807-1814), edited by S. K. Bhuyan, Guwahati, DHAS, 1940.

Hemkos. 2006. *Natun Samskaran Hemkos* (First edition 1900), edited by H. Barua, Guwahati, Sri Debananda Barua.

Hirsch, F. 2005. *Empire of Nations. Ethnographic Knowledge and the Making of the Soviet Union*, Ithaca and London, Cornell University Press.

Hobsbawm, E. 1983. 'Introduction: Inventing Traditions', in Hobsbawm, E. *et al.*, *The Invention of Tradition*, Cambridge, Cambridge University Press, pp. 1-14.

Hunter, W. W. 1879. *A statistical account of Assam*, 2 volumes, London, Trübner & Co.

Izikowitz, K. G. 1962. 'Notes about the Tai', *Bulletin of Far Eastern Antiquities*, 34: 73-91.

Leach, E. 1954. *Political Systems of Highland Burma: a Study of Kachin Social Structure*, Cambridge, Harvard University Press.

Mackenzie, A. 1884. *History of the relations of the Government with the hill tribes of the North-East frontier of Bengal*, Calcutta, Home Dpt. Press.

Martin, T. 2001. *The Affirmative Action Empire. Nations and Nationalism in the Soviet Union, 1923-1939*, Ithaca, NY, Cornell University Press.

Mullan, C. C. 1932. *Census of India, 1931, vol. III: Assam, pt. II: tables*, Shillong, Assam Govt.

Nagaraja, K. S. 1993. 'Khasi dialects: a typological consideration', *The Mon-Khmer Studies Journal*, vol. 23: 1-10.

Nissor Singh. 1906. *Khasi-English Dictionary*, Shillong, Easter Bengal and Assam.

Nongkynrih, A. K. 2002. *Khasi Society of Meghalaya: a Sociological Understanding*, New Delhi, Indus.

Pemberton, R. B. 1835. *Report on the Eastern Frontier of British India*, Calcutta, Baptist Mission Press.

Phillimore, R. H. 1954. *Historical Records of the Survey of India, vol. III: 1815-1830*, Dehra Dun, Survey of India.

Ramirez, Ph. 2007. 'Politico-ritual variations on the Assamese fringes: Do social systems exist?', in Robinne, F. and M. Sadan (eds.), *Social dynamics in the Highlands of Southeast Asia: Reconsidering political systems of Highland Burma*, Leiden, Brill, pp. 91-107.

– 2011. 'Belonging to the Borders: Uncertain Identities in Northeast India', in Pfaff, J. and G. Toffin, *The Politics of Belonging in the Himalayas. Local Attachments and Boundary Dynamics*, Delhi, Sage Publications, pp. 77-97.

Robinne, F. 2007. 'Transethnic Social space of Clans and Lineages. A discussion of Leach's Concept of Common Ritual Language', in Robinne, F. and M. Sadan (eds.), *Social dynamics in the Highlands of Southeast Asia: Reconsidering political systems of Highland Burma*, Leiden, Brill.

Roy Burman, B. K. 1994. *Tribes in Perspective*, New Delhi, Mittal Pub.

Smith, A. D. 1986. *The Ethnic Origins of Nations*, Oxford, Blackwell.

Syiemlieh, D. 1989. *British Administration in Meghalaya; Policy and Pattern*, New Delhi, Heritage.

Notes

1. Most of the field data presented in this article have been collected within the ANR project 'Language, Culture and Territory in North-East India'. Thanks go to Morningkeey Phangcho, Raktim Pator and Belinstone Khwait.
2. States of Nagaland, Meghalaya, Manipur, Tripura, Mizoram. Autonomous Districts of North-Cachar Hills and Karbi-Anglong, Bodo Autonomous Council. On the history of these successive divisions, see for example Baruah, 1999.
3. This conception closely matches the one used by F. Barth (1970: 13-14): " To the extent that actors use ethnic identities to categorize themselves and others for purposes of interaction, they form ethnic groups in this orga nizational sense".
4. Magh Bihu corresponds in other parts of India to Makara Sankranti or Thai Pongal.
5. When a term is specific to one of the regional languages, the language is given before the term, except when obvious or within series: ass. for Assamese, kh. for Khasi, tiw. for Tiwa and kb. for Karbi.
6. Jonbil mela is run by the Tiwa Literary Society (Tiwa Mathonlai Tokhra) and the Tiwa Autonomous Council. Demands for autonomy date back to 1967 under the auspices of the Lalung Darbar.
7. Gogoi, 1986: 150. Several variants may be found with slightly different characters. One of those we have recorded associates Gobha with Jayta-Khairam and Mylliem, a state founded in the mid-eighteenth century.
8. Tini Mawphlang combines an Assamese term: 'the 3' with a Khasi compound: 'grass stone'.
9. There are actually more than twelve monoliths. This arbitrary number may give an

indication of the political importance of the place. In the indigenous histories of this part of Asia, polities are often organized on a duodeci mal basis. See Izikowitz (1962) on the Tais.

10. Gurdon, 1914: 67. According to Bareh (1967: 42), the collection of these "trifling dues" was one of the characteristics that distinguished the Khasi chief, the syiem, from the common people. In Sohra, markets were part of princesses' dowries (Bareh, 1967: 86).
11. For example see Syiemlieh, 1989, p. 67.
12. *Ibid*. "Khorang", or Khoiram as Gohain notes, would not designate a "clan" but Khyrim state; similarly, Magro is a village on the border between the hills and plains.
13. Mackenzie (1884: 9-10), Blackburn (2003: 33-35), Baruah (2002: 442).
14. On the southern *duars* during British Rule, see Syiemlieh, 1989: 9, 67.
15. See, for example, Pemberton, 1835: 215.
16. Hamilton 1807 (31-33) provided a list of foothill states around Guwahati before British Rule.
17. Devi 1960: 125, 130.
18. Pemberton 1835: 221.
19. Phillimore 1954: 52.
20. In Assam there are 23 STs, two Autonomous Districts, 6 Tribal Autonomous Councils and the new Bodoland Territorial Area Districts. Meghalaya has 17 STs and three Autonomous Districts. The administrative settings are not, however, entirely similar in both states. Contrary to the Autonomous District, the Tribal Autonomous Council does not rule any particular territory and does not fall under the Constitution of India.
21. Current tribal areas roughly correspond to the 'Partially excluded areas' established by the Government of India Act, 1935, chap. V.
22. A famous saying, the origin and age of which is not known, is frequently quoted to support this vision: u khynriam, u pnar, u bhoi, u war ki dei u paid khasi ba iar : "Khyriam, Pnar, Bhoi, War, all belong to the large Khasi group".
23. Such a stand is not limited to the Khasi elites. For a discussion on similar representations among the Garos, see Burlings 2007.
24. Gurdon 1914: 2.
25. In 2004, the very powerful Khasi Students' Union demanded that the district be renamed 'North Khasi hills', claiming that the term Bhoi was derogatory. The Confederation of Ri-Bhoi People (CORP) successfully opposed this attempt, arguing that the locals had always been referred to in this way and in provocation they retorted that the name of the state itself should be changed, ince Meghalaya is a 'foreign' name ('Abode of the clouds' in Sanskrit). 'North Khasi Hills' is still used by the KSU.
26. See Ramirez, 2011, for examples of cultural diversity in the area.
27. Bhoi Khasi differs from standard Khasi (i.e. Cherra Khasi) mainly in the position of the verb which is situated before the subject. For more on this dialect, see Nagaraja, 1993.
28. 'Lalung' is now perceived as derogatory by the Tiwas in Assam, although it is generally not the case in Meghalaya.
29. Interview with Syiem Khyrim, Feb. 2006. The exact date cannot be recalled, but it must be remembered that it was before the partition of the Shyllong state, i.e. before 1853.
30. According to present-day knowledge, the Bodo, although the largest tribe in the plains, have never formed any state, at least under this designation. 'Bodo king' might refer to either a Koch or Dimasa/Kachari ruler.

31. Nongbah's are root villages but not always the very first settlements. They are often ritual sites.
32. Most conversions took place between 1910 and 1960.
33. See the dictionary edited by Nissor Singh 1906: 11.
34. See Law passed by the Khasi Autonomous District Council in 2007: 'Appointment and Succession of Syiem Mylliem'.
35. See Ramirez 2011.
36. Syiemlieh 1989: 67.
37. Chattopadhyay 1988: 59; Gurdon 1914: 40.
38. Census of India (Allen 1902b). *Kala azar* (leishmaniasis) broke out in 1890 and 1897; cholera in 1900; smallpox in 1898.
39. The great pandemics of the late nineteenth century did not entirely spare the hills. Moreover in 1898-99, they particularly suffered from the consequences of the great 1897 earthquake. (Allen 1902a: 24, 1906: 56-57). Hunter (1879 II: 189) note that in 1834-35, most Karbi villages moved their location following cholera epidemics.
40. The Census of India 1931 (Mullan 1932: 265) counts 6,932 Nepalese in the Khasi states with an increase of 2,939 since the 1921 Census.
41. For descriptions of the Khasi political setup see Gurdon 1907: 68-75, Bareh 1967: 39ff.; Nongkynrih (2002: 66-88). For the Jaintia political apparatus, see Gassah 1998.
42. Moreover, in Ri-Bhoi at least, there is no distinction between first and new settlers. On the contrary, Nongkynrih (2002: 69-71) points out that in several villages of the Sohra region, access to village positions was restricted to the 'founders' (kh. nongseng shnong), as opposed to the newcomers (shongthap: lit. temporary settlers).
43. Allen 1905: 83.
44. Gurdon 1907: 78, Nongkynrih 2002: 41, Gohain, 1993: 102-103.
45. Chattopadhyay 1988: 69.
46. The term pham certainly has a link with the Assamese pam: "land newly occupied at a distance from home" (Hemkos 2006: 676).
47. Interestingly enough, the head of each council is called (kb.) lindok and has many functions in common with the 'Khasi' lingdoh.
48. For a short discussion on the Karbi political system, see Ramirez 2007: 99-101.
49. Beyond lingdoh/lindok: doloi, pator, basan, sangot. Some of these terms look Indo-Aryan, but are not easily traceable in Assam or Bengal.
50. Leach, 1954: 101-102.
51. This model would be the multi-ethnic version of the 'bridge & buffer' model by which Roy Burman (1994: 81-91) explains the formation of tribal groups.
52. Hobsbawm 1983: 1-14, Gellner 1983, Smith 1986: 77.
53. The nationalities policy in the USSR provides an interesting point of comparison. With the aim of taking into acc ount the spatial imbrications of ethnicities in the Caucasus, territorial enclaves were multiplied. See Hirsch 2005: 145, Martin 2001.
54. Striking parallels may be drawn with the tribal zones of Pakistan and Myanmar: "Using self-identification as the critical criterion of ethnic identity, it should thus be perfectly possible for a small group of Pathans to assume the political obligations of membership in a Baluch tribe, or the agricultural and husbandry practices of Kohistanis, and yet continue to call themselves Pathans." (Barth 1970: 24); "Generally speaking, the perennial nature of the clans and lineages from the colonial period up to now contrasts with the reification of ethnicity involved both in the process of Christianisation and in the contemporary pan-Kachin movement …" (Robinne 2007: 284).

1.2. FEDERALISM AND ETHNIC TERRITORIES IN NEPAL

1.2.1.

Territorial Changes in Nepal: Proposals for a Federal State

Narendra Raj Khanal

Introduction

Territory is an area that a person, group, institution, considers their own and defends against others who try to enter it. It also includes an area that is appropriated by a person, group, and institution for a particular type of activity. It may be a state, a province, a district, or a politico-administrative unit or part of it used for different socio-economic activities such as tourism, transhumance, agriculture, employment, trade and transport. Territories change over time. In Sikles, a village located in the western mountain area of Nepal, changes in territories for transhumance, crops and employment have already been discussed (Khanal 2002). Here, the space for transhumance and crops has decreased drastically whereas in recent years there has been a surge in the number of destinations for foreign employment. Similarly, there has been a change in the origin and destination of the trading activities of Manangis inhabiting the area over different periods (Chapagain 2008).

There have also been changes in the country's politico-administrative boundaries over several periods of time. The present-day territory of Nepal was divided into several kingdoms or principalities before the unification of the country in the late eighteenth century. Besides the three kingdoms of Bhadgaon, Kathmandu and Patan, there existed 46 principalities (*baisi* and *chaubisi*) in the west, and three states—Bijayapur, Chaudandi and Makawanpur in the east (Shaha 2001). There were three Newar Kingdoms and ten Hangs (Limbuwan) in the east, twelve Magarats in the centre, four Sen states and many Khasa principalities throughout the country (Gurung 2005). During the unification period in the early nineteenth century, the country was extended to Kangara in the west and Tista in the east. Beyond the Mahakali River—the present western boundary—were the kingdoms of Kumaun and Garhwal, and further to the west were twelve and eighteen principalities. Similarly, by 1791 the Gorkhas had extended their kingdoms as far north as Digarchi (Shigatse), and Kerung and Kuti came under Nepalese rule between 1791 and 1792, the year which saw the end of the Nepal-Tibet/China War. After the 1814-1816 war with British India, Nepal was reduced to its present frontiers between the Mechi River on the east and the Mahakali River on the west (Shaha 2001). Prior to 1963, the country was divided into thirty-four districts—twenty-three districts in the hills and mountains and eleven districts in the Terai and the Inner Terai with objectives to ensure security and revenue collection. In 1963, the country was restructured into seventy-five districts. Currently, the country is divided into five development regions and seventy-five administrative and development districts.

During and after the people's movement, *Jana-Andolan 2,* people demanded that the country be restructured into federal states. The Interim Constitution of Nepal, 2007 makes provisions for a secular, republican, federal state. In part 17, it mentions the form of the state and local self governance. Article 138 makes provisions for the gradual restructuring of the state. Article 138.1 mentions "bring[ing] an end to discrimination based on class, caste, language, gender, culture, religion and region by eliminating the centralized and unitary form of the state, the state shall be made inclusive and restructured into a progressive, democratic federal system." No details are actually given about restructuring the state. However, article 138.2 specifies that "A High Level Commission shall be constituted to make recommendations for the restructuring of the state in accordance with clause 1. The composition, function, duty, power and terms of service of such a Commission shall be as determined by the Government of Nepal." In article 138.3, it further mentions that "the final decision relating to the structure of the state and federal system shall be made by the Constituent Assembly."

As proposed by the Interim Constitution, the High Level Commission has not yet been formed. However, some task forces have been set up to identify the basis on which to determine the number of states, their boundary demarcation as well as power sharing and governance by individuals, political parties and the federations of different ethnic groups. On 21 January 2010, the Committee for Restructuring of the State and Distribution of State Power of the Constituent Assembly of Nepal submitted a draft report including a concept paper. The Committee proposes dividing the country into fourteen federal states (Fig. 1). However, this is under discussion in the Constituent Assembly and has not yet been finalized. This paper aims to review the basis and the models for restructuring Nepal into federal states and to make an assessment in terms of the advantages and disadvantages.

Country's Background Information

With an area of 147,181 sq km, Nepal presents diverse biophysical and socio-cultural conditions. The altitudinal variation is very high. Altitude ranges from 60 m in the south to 8,848 m, the summit of Mount Everest, in the north within a distance of only 160 km. Physiographically, the country is divided into five regions: the Terai, the Chure (Siwaliks), the Middle Mountains, the High Mountains and the High Himal. The Terai plain in the south represents about 14 per cent of the country's total area, the Chure hills (Siwaliks) 12 per cent and the middle mountains including the Mahabharat lekh, the high mountains and the high Himal occupy 30, 20 and

24 per cent of the total country respectively. The advantages of the hilly and mountainous terrain can be found in hydroelectricity, tourism, medicinal herbs and horticulture, whereas the Terai offers the possibility of producing food grain.

Nearly 20 per cent of the area is under cultivation, followed by forests (37.8%), shrubs (4.6%), non-cultivated enclosures within cultivated areas (6.5%), pastures (11.8%), snow and ice (3.4%), lakes and ponds (0.1%), urban areas (0.1%) and rocky surfaces, sand and stone (15.7%) (LRMP, 1986). Common forms of cultivated land are the Terai/valley cultivation (58%), hill-slope cultivation with level terraces (25%) and sloping terraces (17%).

According to the Population Census of 2001, Nepal has a population of 23.1 million with an annual growth rate of 2.25 per cent. The estimated total population for 2007 is 26.4 million with an average life expectancy of 63.7 years. The literacy rate is still low (54.1% in 2001). The preliminary estimated GDP per capita at current prices stands at USD 383 for the year 2006/07. The annual growth rate of GDP at producer prices is 2.5 per cent for the year 2006/07. According to the Nepal Living Standard Survey 2003/04, nearly one third of the population (31.8%) lives below the poverty line. Similarly, about 40 per cent of the population falls below the minimum level of dietary energy requirements (CBS 2007). Nepal is an agricultural country. About 80 per cent of people rely on agriculture for their income and employment. The agricultural sector contributes 36 per cent to the GDP. More than 85 per cent of the total energy consumption is derived from traditional sources, mainly firewood. As far as natural sources of energy are concerned, the share of fuel wood consumption is 88.68 per cent, followed by agricultural residues (4.85%) and animal residues (6.47%) (MoF 2007).

There are more than 100 ethnic/caste groups with their own language and culture distributed over various parts of the country. According to the 2001 Population Census (CBS 2007), each of the 31 of these ethnic/caste groups has a population of over 100,000, or 0.5 per cent of the national population. All together, they represent 89.3 per cent of the total population. The reported number of ethnic/caste groups ranges from five in mountain areas, to thirty-six in the hills and fifty-nine in the Terai. There are forty-four identified ethnic groups: five from mountain areas, twenty-seven from the hills and twelve from the Terai, which together make up 36.4 per cent of the total population.[1] Similarly, together the fifty-three caste groups (Dalit and non-Dalit) account for 58.2 per cent of the country's total population (Sharma 2008). Major ethnic/caste groups in Nepal have a contiguous, relatively dense and concentrated geographical area. However, as a result of centuries of migration in the hills and of the more recent migration to the Terai, there is considerable ethnic/caste diversity even in areas that have a dominant ethnic/caste population. In other words, there are many small minority groups within the geographical areas of major ethnic/caste groups (*ibid*).

Proposed models for restructuring

A total of 50 concepts/models proposed by different authors, political parties, federations or members of ethnic groups have been reviewed (Table 1) (for details see Sharma *et al.* 2009). Among them, 20 models include maps on which proposed federal units are delineated. 13 models are not accompanied by any maps, and boundaries are only given in the text. 17 outline only the concept and criteria. Among them, 29 are from individuals, fifteen from political party manifestos and six from federations/members of ethnic groups.

Table 1: Number of concepts/models reviewed by type

Types/proposed by	Individuals	Political parties	Federations/members of ethnic groups	Total
Marked on maps	17	1	2	20
Boundary in text only	2	7	4	13
Concept and criteria	10	7	-	17
Total	29	15	6	50

In their political manifestos, political parties—such as the CPN-Maoist, the Nepali Congress, the CPN (UML), the Madhesi Janaadhikar Forum, the Terai Madhesh Loktantrik Party, Janamorcha Nepal, the Nepal Majdoor Kisan Party, the Sadbhabana party Mahato, the Nepal Sadbhabana Party-Mahato, the Nepal Sadbhabana Party-Anandadevi, the Rastriya Prajatantra Party Nepal, the Rastriya Janashakti Party, the Rastriya Janamorcha-Chitra Bahadur KC, the Nepal Communist Party-Ekikrit and the Chure Bhabar Rastrya Ekta Party Nepal—have all aired their opinion regarding the concept/criteria for restructuring Nepal into federal units. Among the political parties, the CPN (Maoist) is the only party that has presented a comprehensive geographical model for the federal republic of Nepal. Other major political parties have given some indication of the criteria to be used in delineating federal units. Rastriya Janamorcha-Chitra Bahadur KC opposes the federal system and is demanding a unitary system. The Majadoor Kisang Party recommends recognizing the fourteen existing zones as federal units. The Nepal Sadbhabana Party-Anandadevi, the Sadbhabana Party-Mahato, the Terai Madhesh Loktantrik Party and the Madheshi Janaadhikar Forum demand that Madesh (Terai) be recognized as an autonomous federal unit. Similarly, the Chure Bhabar Rastriya Ekta Party demands that, from east to west, the Chure and Bhabar be declared a federal unit.

The federations/members of ethnic groups, such as the Tharus, Tamus (Gurungs), Magars, Kirats, Limbus and Sherpas, have also demanded the recognition of their own homelands and have demanded that provisions for autonomous federal units be made in the new Constitution.

The criteria used and the number of federal units proposed by different authors and political parties are summarized in Table 2 (also see Sharma *et al.* 2009 for details). Many of the models/frameworks are based on the analysis of district and national level data on ethnicity, language, natural resources and potential per capita GDP, poverty level, revenue and expenditure, people's participation in mainstream economic, social, political, administrative activities and decision-making processes. However, only a few use VDC level data on ethnicity and language integrating spatial data with the help of Geographical Information System (GIS) tools (Sharma 2007, M.S. Manandhar *et al.* 2008, Mabuhang 2007).

Table 2: Number of proposed federal units and the criteria used

SNModels proposed by	Number of provinces /regions	Criteria	Basis for names
1. Gobinda Neupane (2000, 2006)	11 or 8	E, L	E, L
2. Harka Gurung (2000, 2006)	5/25	FN, AD	M, R, P
3. Pitamber Sharma (2007)	6/19	E, L, CA	M, R, P
4. Alok K.Bohara (2008)	4/12	E, L, CA	R
5. Bhawani Baral (2006)	10+1	E	E
6. Surendra KC (2006)	8 or 5	E, L, CA	E, R
7. Prem B. Singh (2006)	14	E, L, C	E, P
8. Shankar Pokharel (2006)	15	E, L, C, ACC	P
9. Ramchandra Acharya (2007)	4+1/13	E, L, C, CA, ACC	R, P
10. Narahari Acharya (2005, 2006)	9	ECO	R
11. Brishesh C. Lal (n.d)	4/11/5	E, L, C, ECO	P
12. Rajendra Shrestha (2006)	14	E, L, C	E, L, P
13. K. B. Gurung (2006)	11/6	E, L, C	E, L
14. Pari Thapa (2006)	9	E, L, C	E, L
15. Kumar Y. Tamang (2006, 2007)	11	E, L, C	E, L
16. Amaresh N. Jha (2006)	10	E, L, C	E, L
17. Baburam Acharya (2006)	4/15	H	H
18. Chandrakanta Gyawali (2007)	8	P, R	P
19. Mangal S. Manandhar *et al.* (2008)	13	E, L, C	E, L
20. Krishna Khanal (2007)	13 or 14	E, L, C, CA, ACC	NA
21. Lok Raj Baral (2007, 2008)	5	CA, NU	NA
22. Balkrishna Mabuhang (2007)	11	E	NA
23. CPN Maoist	13	E, L, C, TER	E, L
24. Nepal Majdoor Kisan Party	14	ZN	M, R, P
25. Nepal Sadbhabana Party-Anandadevi	3	ECO	NA

Note: E = Ethnicity; L = Language; C = Culture; H = History; M = Mountain range and peaks; R = River and watershed; P = Place name; FN = Financial resources; AD = Administration; CA = Comparative advantages and complementarities; ACC = Accessibility; NU = National Unity; TER = Territory; ZN = Zones; ECO = Ecology; NA = No details available.

Source: Sharma *et al.* 2009

The number of proposed federal units ranges from only three—the Terai, Hills and Himal proposed by the Nepal Sadbhavana Party—to fifteen by Shankar Pokharel. Of a total of twenty-five models reviewed, four models propose fewer than five federal units, ten models propose five to ten federal units and eleven models propose eleven to fifteen federal states. The number of districts/ethnic, linguistic enclaves/region within federal states would range between twelve and twenty-five. Federal states would be named on the basis of various criteria ranging from ethnic/language identity to mountain ranges and peaks, rivers, historical places, etc.

The models/frameworks/concepts so far proposed by different individuals, political parties and federations of major ethnic groups can be broadly divided into five groups based on the objectives, criteria used and structural forms of federation:

- Those with no change to the existing structure, with more decentralization and devolution of power (Rastriya Janamorcha-Chitra Bahadur K. C. and the Nepal Majadoor Kisan Party);
- Those taking into account ethnic/language/culture, territorial identity, and based on majority/plurality (Neupane 2000, Baral 2004, Gurung 2006, Tamang 2006, Tamang 2006, Manandhar *et al.* 2008, Thapa, 2006, Lal n.d., Jha 2006, Singh 2006, Pokharel 2006, Shrestha 2006, Nepal Communist Party, Maoist and federation of major janajati: Limbu, Kirat-Rai, Tharu, Magar, Tamang);
- Those based on physiography: Nepal Loktantrik Sherpa Sangh (High Himal from Mechi-Mahakali); the Chure Bhabar Rastriya Ekta Party (Chure-Bhabar); the Sadbhawan Party-Mahato; the Nepal Sadbhawan Party-Anandadevi (Himal, Pahad, Madesh); Madheshi Janaadhikar Forum (Madesh); the Terai Madesh Loktantrik Party (Terai-Madesh);
- Those based on geographical specificities (resource endowment, ethnicity/ language/culture, territory) and complementarities (Gurung 2006, Sharma 2007, Bohara 2008, R.C. Acharya 2007, N. Acharya 2005, B. Acharya 2006, Gyawali 2007);
- Proposals that are only concepts and criteria, providing no information on structural forms of federal state (R. Adhikari 2006, P. R. Ghimire [n.d.], B. P. Bhandari 2007, B. K. Mabuhang 2007, D. R. Dahal 2007, S. P. Sharma 2007, K. Khanal 2007, L. R. Baral 2008, P. Tamang 2006, M. Lawati 2006 and many political parties including Nepali Congress and UML).

Discussion and Conclusion

Many models/frameworks/concepts propose ethnicity, language and cultural identity as the major criteria in delineating the federal units. The basic difference is whether the nested approach is adopted: i) to delineate federal units, considering the comparative advantages and complementarities of different ecological zones by incorporating the Terai, the hills and mountains in each federal unit, ii) to link two large neighbouring countries—India in the south, east and west and China in the north to each federal unit, or iii) to opt for delineation based only on ethnicity, language and cultural identity. Four models, two based on the nested approach proposed by Harka Gurung (Fig. 2) and the author of this paper (Fig. 3), and two based on ethnic identity proposed by the Maoists (Fig. 4), and by M.S. Manandhar, S. Shrestha and P. Sharma (Fig. 5), are given as an example. An assessment of some of the models based solely on ethnic identity, as proposed by the Maoists, shows that even in the federal unit delineated as the ancestral homeland of the ethnic group concerned, they do not represent a majority. The percentage of a particular ethnic group in the total population within the proposed federal unit only ranges from 17 per cent in Tamuwan, to 26 per cent in Kirat, 27 per cent in both Limbuwan and Tamabasaling, 28 per cent in Magarat and 35 per cent in both Newa and Tharuwan. Moreover, more than 50 per cent of the population of the ethnic group concerned resides outside the federal units named after that particular ethnic group. For example, nearly 53 per cent of the total Rai population in the country resides outside the proposed Kirat federal unit. Similarly, 53 per cent of the total Newar population in the country resides outside the proposed Newa federal unit, 65 per cent of the total Magar population in the country resides outside the proposed Magarat and 50 per cent of the total Tharu population in the country resides outside the proposed Tharuwan region. It should be noted that mixing populations of different ethnic/language/cultural groups in different parts is largely due to former waves of migration. Due to such mixing, the maximization of both the percentage of ethnic groups concerned in the total population of a given region and the percentage in total national population in that region is not possible. These two have an inverse relationship. An ethnic/language group can only become a majority in the proposed federal unit by reducing the core area. However, the reduction in the size of the proposed federal unit for a given ethnic group excludes many people of the same ethnic group living in the peripheral area of that federal unit.

Another issue in delineating federal units merely based on ethnic/language identity without considering the spatial distribution of natural resources and their development potentials is that it increases regional imbalance in terms of food security and livelihood (Sharma *et al.* 2009). This kind of political and administrative boundary merely creates a barrier that limits the flow of resources and technology and that does not take advantage of complementary aspects of milieus which are an asset for future development.

Conflicts have already been fuelled by delineating federal units based on demands by *adivasi janajati* (indigenous ethnic groups). For example, it has been suggested by the Limbus that Sunsari, Morang and Jhapa districts located in the Terai be incorporated into the Limbuwan because their ancient kingdom, Bijayapur, is located in Sunsari district, and because their territory extended to the southern boundary of the country. The Magars have demanded that several isolated pockets—Surkhet to Tanahun, Ramechhap to Dhankuta—be incorporated in Magarat federal unit. Tharus have proposed incorporating in Tharuwan such isolated pockets from several Terai and Inner Terai regions, whereas the Madhesi parties have asked for a single Madesh.

Differences are found not only in the approaches and criteria proposed for delineating federal units but also in naming those units. Those who have adopted the nested approach have proposed naming the units after the name of a famous mountain range, mountain peaks, rivers and historical places instead of using the name of a particular ethnic group. Those who underlined ethnicity/language as the sole criteria have proposed naming the units after a particular ethnic group such as Limbuwan, Newa, Tharu, Magarat, Tamuwan (Gurung), Tambasaling (Tamang), etc.

The proposal regarding fourteen federal states as put forward by the Committee for Restructuring the State and Distribution of State Power of the Constituent Assembly of Nepal is based mainly on caste/ethnic identity. It does not consider issues such as the size of the area and the population of the states, natural resource endowment and financial capacity, geographical complementarities and even the inclusiveness of caste/ethnic groups identified so far for each proposed state. The size of the area ranges from only 900 sq km in Kathmandu to more than 17,900 sq km in Karnali. Similarly, the size of the population ranges from less than 50,000 in Jadan to 6,700,000 in Mithila-Bhojpura-Koch and Madhes. There is no particular caste/ethnic majority (50% or more) in any of the proposed states except for the Chhetris in Khaptad. Except for the Limbus in Limbuwan, the Tharus in Lumbini-Awad-Tharuwan and the Yadavs in Mithila-Bhojpura-Koch-Madesh, the other groups represent less than 50 per cent in the state designated as their area of identity. For example, more than 50 per cent of the total Newar population reside outside the proposed Newa state; 50 per cent of all Magars, Gurungs and Tamangs live outside the Magarat, Tamuwan and Tamsaling respectively. It is still uncertain whether those living outside the state designated as their area of identity will accept this proposal.

Though the final decision about restructuring the state and federal system shall be made by the Constituent Assembly, the author of this paper proposed to adopt the nested approach in line with Harka Gurung's proposal (Fig. 2) in which ethnicity/ language and cultural criteria would be used to delineate regions, with more

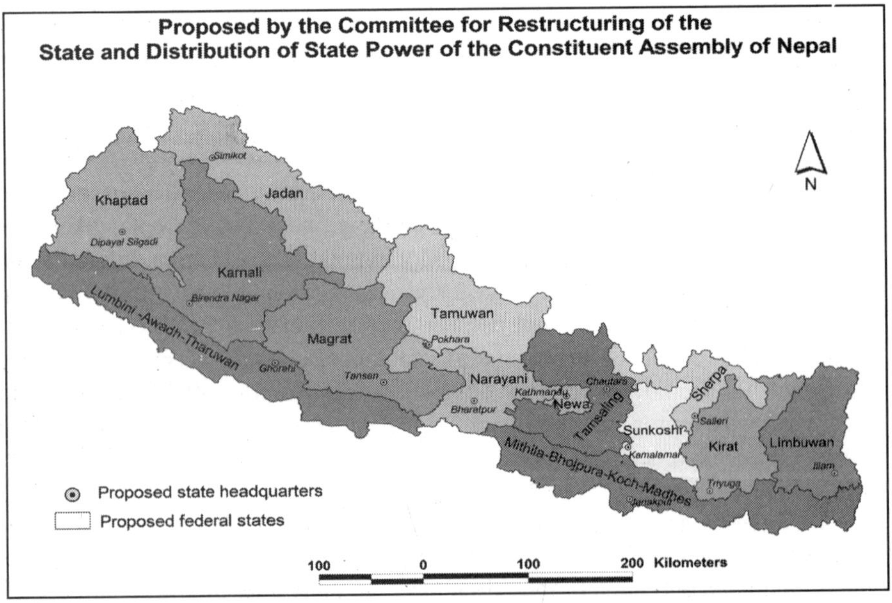

Figure 1: Map proposed in 2010

decentralization and devolution of power within each federal unit, while the geographical complementarities and cultural coexistence and cohesion in diversity would be considered to delineate federal units.

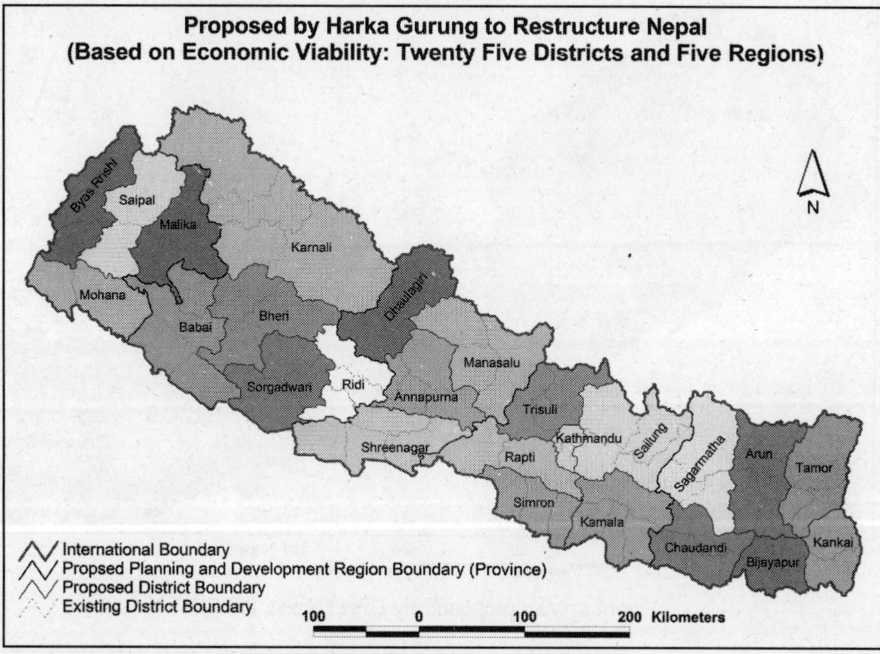

Figure 2: Map proposed in 2006

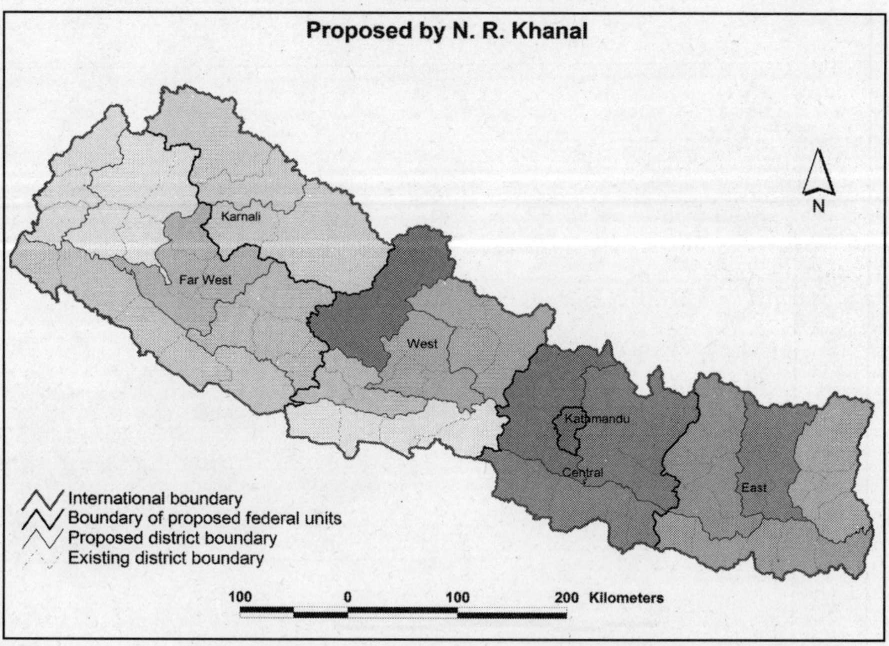

Figure 3: Map proposed by Narendra Khanal in 2008

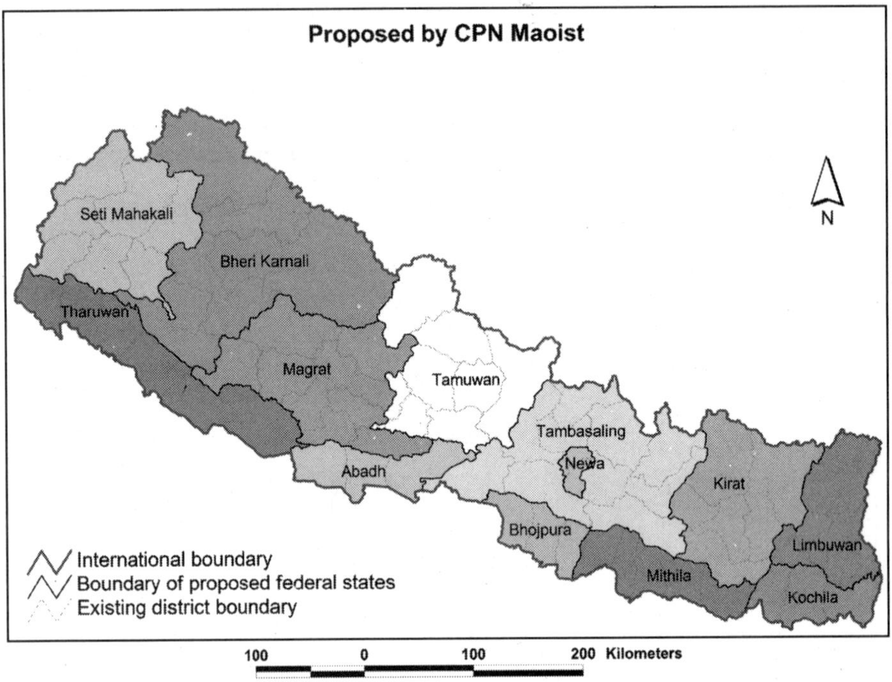

Figure 4: Map proposed by CPN Maoist, 2006

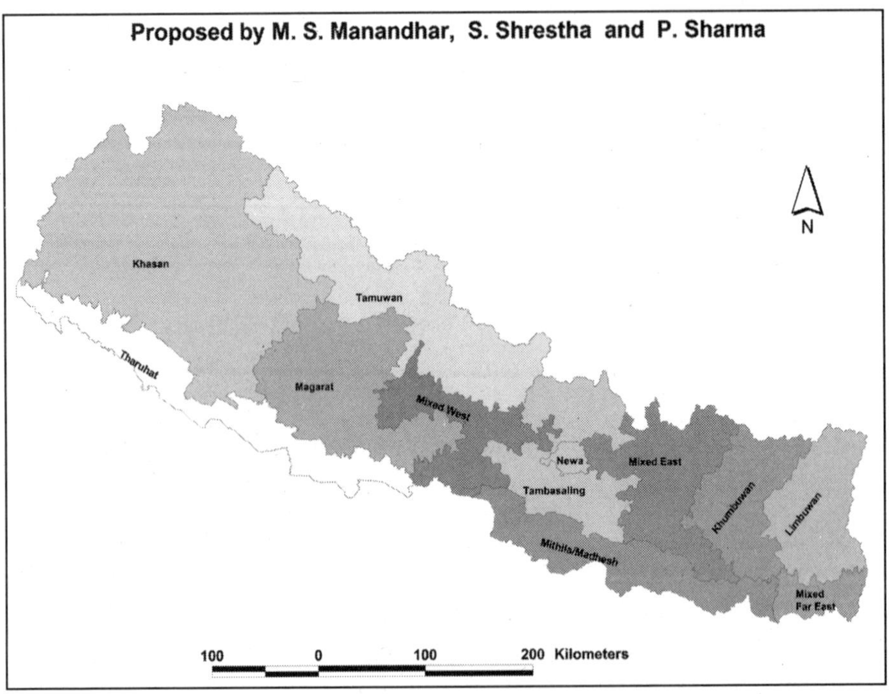

Figure 5: Map proposed by Manhandar *et al.*, 2008

BIBLIOGRAPHY

Acharya, B. 2006. 'Nepalko Bhukti (Chhetriya) Samrachana', in Sitaram Tamang (ed.), *Nepalko Sandarvama Rayako Punasamrachana* (Third Edition), Kathmandu, Samana Prakashan, pp. 400-413.

Acharya, N. 2005. *Rajyako Loktantrikaran: Nepalka Lagi Loktantrik Rajya Samrachana*, Kathmandu, Sabtsar Prakashan.

Acharya, R. C. 2007. 'A Model for Political Restructuring and Electoral System of Federal Nepal: Building on the Strength of Ethnic Diversities and Regional Complementarities'. Unpublished paper presented in the 'Interaction Program', 10 May 2007, Kathmandu, organized by the National Peace Campaign.

Adhikari, R. 2006. *Sambidhan Sabha, Loktantra ra Punarsamrachana*, Pokhara, New Helo Stationary and Books House.

Baral, B. 2004. *Yesto Hunu Pardachha Rajyako Samrachana, Kathmandu, Swayatta Shasan Sarokar Manch Nepal.*

Baral, L. R. 2008. *Sanghiya Pranaliko Auchitya ra Bisheshata*, Shichhyak, Baishakh, 2065, pp. 43-48.

Bhandari, B. P. 2007. *Sambidhan Sabha, Loktantra, Samabeshikaran ra Sanghiya Rajya Pranali*, Kathmandu, Pairabi Prakashan.

Bohara, A. K. 2008. 'Comparative Federal Structure: A Workable Political-economy Approach for a New Nepal' (http://www.nepalnews.com), 10 August 2008. Some versions of this article appeared in the *Kathmandu Post* in April 2007 (http://kantipuronline.com/kolnews.php?&nid=105529), and *Nepali Times*: 'Equal opportunity poverty: Identity politics obscures all-round deprivation', by Dr. Alok K. Bohara and Mani Nepal (http://www.nepalitimes.com/issue/338/Nation/13285).

CBS. 2007. *Nepal in Figures*. Central Bureau of Statistics, National Planning Commission Secretariat, Government of Nepal, Kathmandu.

Chapagain, P. S. 2008. 'Land Labour and Agricultural Change in Upper Manang Valley: Understanding Meanings and Processes'. Unpublished Ph.D. dissertation, Tribhuvan University, Kirtipur, Kathmandu, Nepal.

Dahal, D. R. 2007. 'Rastra Punanirman: Samajik-Saskritik Paridrishya', in *Nepalma Sanghiya Shasan Pranali, Chunauti ra Abasarharu*, Kathmandu, National Peace Campaign, pp. 99-138.

Ghimire Parashuram. (n.d.). *Rajyako Punasamrachanako lagi Sambidhan Sabha*, Kathmandu, na (Political Scientist).

Gurung, H. 2000. '*25 jilla ko naya Nepal. Bikendrikaran ko jukti, jilla sudhrinikaran*', Himal, 16-29 fagun (2058 BS), pp. 38-41.

Gurung, H. (ed.). 2005. *Nepal: Atlas and Statistics*, Himal Books, Kathmandu.

Gurung, H. 2006. 'Decentralization and Regional Development: A Perspective', in D.N. Dhungel and A. M. Amatya (eds.), *Nepal: Conflict Resolution and Sustainable Peace, Decentralization and Regional Development*, Kathmandu, Institute for Integrated Development Studies, vol. 1: 7-50.

Gurung, K. B. 2006. 'Adibasi Janajati Adhikar ra Rajyako Punasamrachana', in Sitaram Tamang (ed.), *Nepalko Sandarvama Rayako Punasamrachana* (Third Edition), Kathmandu, Samana Prakashan, pp. 85-108.

Gyawali, C. K. 2007. *Sanghiya Shahan Pranali: Rayako Punarsamrachana,* Kathmandu, Ms Amita Gyawali ra Janasewa Printers Pvt. Ltd.

Jha, A. N. 2006. 'Sanghiya Shasan Byabasthako Adharma Rajyako Punasamrachana', in Sitaram Tamang (ed.), *Nepalko Sandarvama Rayako Punasamrachana* (Third Edition), Kathmandu, Samana Prakashan, pp. 247-358.

Khanal, K. 2007. 'Sanghiya Rajyasamrachana: Manyata ra Abhyas', in *Nepalma Sanghiya Shasan Pranali, Chunauti ra Abasarharu*, Kathmandu, National Peace Campaign, pp. 193-222.

Khanal, N. R. 2002. 'Land Use and Land Cover Dynamics in the Himalaya: A Case Study of the Madi Watershed, Western Development Region', Nepal, unpublished Ph.D. dissertation, Tribhuvan University, Kirtipur, Kathmandu, Nepal.

Lal, B. C. (n.d.). *Sanghiya Swashasan Tira*, Janakpurdham, Sudhama Prakashan.

Lawati, M. 2006. 'Rajyako Punasamrachanako Sawal ra Samabeshi Prajatantra', in Sitaram Tamang (ed.), *Nepalko Sandarvama Rayako Punasamrachana* (Third Edition), Kathmandu, Samana Prakashan, pp. 229-248.

LRMP. 1986. *Land Utilization Report*, Kenting Earth Science Canada and Department of Topography, HMG/Nepal, Kathmandu, Nepal

Mabuhang, B. K. 2007. 'Rajya Punanirmanka Lagi Sanghiya Shasan Pranali: Jatiya Jansangkhik Dristikon', in *Nepalma Sanghiya Shasan Pranali, Chunauti ra Abasarharu*, Kathmandu, National Peace Campaign, pp. 59-98.

Manandhar, M. S., Shrestha Shobha and Sharma Pushpa. 2008. *Sanghiya Rajyako Banot*, Kantipur, 27 Poush, 2064.

MoF. 2007. *Economic Survey: Fiscal Year 2006/07*, Ministry of Finance, Kathmandu.

Neupane, G. 2000. *Nepalko Jatiya Prasna: Samajik Banot ra Sajhedariko Sambhabawana*, Kathmandu, Centre for Development Studies.

Pokharel, S. 2006. *Sambidhan Sabha, Rajyako Punarsamrachana ra Samabeshi Loktantra*, Kathmandu, Nepal Trade Union Mahasangh (Gifont).

Shaha, R. 2001. *Modern Nepal: A Political History 1769-1955*, New Delhi, Ajay Kumar Jain for Manohar Publishers and Distributors.

Sharma S. P. 2007. 'Sanghiya Shasan Pranalika Arthik Pakchhya: Chunauti ra Abasar', in *Nepalma Sanghiya Shasan Pranali, Chunauti ra Abasarharu*, Kathmandu, National Peace Campaign, pp. 175-192.

Sharma, P. 2008. *Unraveling the Mosaic: Spatial aspects of ethnicity in Nepal*, Himal Books, Kathmandu.

Sharma, P., Khanal, N. R. and Chaudhary, S. 2009. *Towards A Federal Nepal: An Assessment of Proposed Models*, Kathmandu, Himal Books for Social Science Baha.

Sharma, P. 2007. 'Sanghiya loktantrik ganatantra nepalko pradeshik swarup', Mulayankan, Srawan. A slightly revised version in 'Sanghiya rajyaka adhar: Bhugol ra yojanaparak drishti', in *Nepalma Sanghiya Shasan Pranali, Chunauti ra Abasarharu*, Kathmandu, National Peace Campaign (2007), pp. 139-174.

Shrestha, R. 2006. 'Rajyako Punasamrachana ra Samabeshi Loktantraka Adharharu', in Sitaram Tamang (ed.), *Nepalko Sandarvama Rayako Punasamrachana* (Third Edition), Kathmandu, Samana Prakashan, pp. 9-56.

Singh, P. B. 2006. *Sambidhan Sabha, Rajyako Punasamrachana ra Samabeshi Loktantra: Yek Prastab*, Kathmandu, Prem Bahadur Singh (Member of Parliament-Politician)

Tamang, P. R. 2006. 'Loktantra ra Rajyako Punasamrachana: Abashyakata, Abadharana ra Prarup', in Sitaram Tamang (ed.), *Nepalko Sandarvama Rayako Punasamrachana* (Third Edition), Kathmandu, Samana Prakashan, pp. 57-84.

Tamang, K. Y. 2006. 'Nepal Rajyako Punasamrachana ra Loktantrako Samabeshi Dhacha', in Sitaram Tamang (ed.), *Nepalko Sandarvama Rayako Punasamrachana* (Third Edition), Kathmandu, Samana Prakashan, pp. 137-168.

Thapa, P. 2006. 'Bartaman Rajyako Punasamrachana: Samabeshi Loktantra ra Bausaskritik Pahichanharuko Manyata', in Sitaram Tamang (ed.), *Nepalko Sandarvama Rayako Punasamrachana* (Third Edition), Kathmandu, Samana Prakashan, pp. 109-125.

Unpublished documents from different ethnic group organizations

Chumlung Kirant Yakthung Kendriya Karyalaya. 2007. Limbuwanko Prastabit Masyoda Khaka: Chhalphal tatha thap sujhavko laagi, Kathmandu.

Gurung (Tamu) Rastriya Parisad. 2008. Rajyako Punarsamrachana bhitra Tamuwan Swayatta Chhetra: AAjako Tato Bahas, Kathmandu.

Tharu Kalyankarini Sabha, Kendriya Karyalaya. 2008. Sanghatmak Nepalma Tharuhat-Tarai: Antarkriyatmak Bichar Gosthi, Awadharana Patra, Kathmandu.

Nepal Chepang Sangh. 2007. Brihat Chepang Rastriya Rajnitik Golmech Sammelan: Ghosana Patra, Chitwan.

Nepal Loktantrik Sherpa Sangh. 2008. Himali Sherpa Swayatta Pradesh: Awadharana Patra. Kathmandu.

Nepal Magar Sangh, Kendriya Samiti. 2007. Rajyako Sanghiya Punasamrachana ra Magarant Swayattata, Kathmandu.

Yayokhkhaa Kirant Rai Kendriya Karyasamiti. 2007. Kirant Swayatta Rajya (Prades) Ko Awadharana, Dharan.

Notes

1. For census and administrative purposes the country is divided into three geographical or ecological regions—Terai, hills and mountains which correspond more or less to the five physiographic regions previously mentioned.

1.2.2.

Tamuwan in the Gurung Imagination

Krishna Hachhethu[1]

Gurungs, who refer to themselves as Tamu in their own language, constitute 2.39 per cent of the Nepalese population; most of them are concentrated in the mountains and hills of the Western Development Region. Like other ethnic groups, they have been affected by the centuries-long state-designed project of a homogonous model of national integration, with only 1.49 per cent of the country's population reporting the Gurung language as their mother tongue.

But since the 1990s, like the Tamangs, Rais, Magars, amongst others, they have been laying claim to their own territory, Tamuwan, within a federal Nepalese state. Not all Gurungs conceive of Tamuwan in the same way, whether they are leaders of political parties, activists, members of civil society and so on, as will be demonstrated in this text.

This paper is mainly based on primary data (collected during fieldwork) and also partly on secondary information borrowed from selected literature on Gurungs. In-depth interviews were conducted with 28 persons from different walks of life, from Kaski district, and in particular local leaders of different political parties, Gurung activists and members of civil society. In addition, by means of a questionnaire distributed to Gurung communities, a quantitative study was conducted in all three sample villages/cities (Ghandruk VDC,[2] Deaurali VDC and Pokhara municipality) in Kaski district during the period April-May 2007. A total of 225 questionnaires were collected.

Identity Construction

The Gurungs have a composite identity. Apart from being Nepalese citizens, they make up a group belonging to *adivashi janajatis* (indigenous nationalities); they are a marshal ethnic group, a trait they share with other groups such as the Rai, Limbu and Magar groups; and they are Tamu, a specificity of their own, which distinguishes them from other indigenous nationalities.

The Gurungs' original religion is Bonism; there are, however, Buddhist Gurungs and Hindu Gurungs. The effect of associating or assimilating Gurungs to the Hindu religion is also reflected in the hierarchal division among Gurungs.

Indeed, Gurungs are split by an internal division—which is the object of much discussion—with some belonging to the said 'superior' *car jat* (four clans) and others to the 'inferior' *sora jat* (sixteen clans). There is a distinct division between *car jat* and *sora jat* in both their social practices and economic status. Marriage between *car jat* and *sora jat* is theoretically forbidden, though it is in fact becoming more widespread. *Car jat* still refuse to share food or drink with *sora jat*. (Macfarlane 1976: 19). Contrary to *car jat, sora jat* eat buffalo (Sharma, 1982: 240). In general, *car jat* possess more land than *sora jat*. The *sora jat* often work in *car jat* fields (Pignède 1993: 179). *Car jat* are Buddhist while *sora jat* profess Bonism (Khadga Bahadur Gurung, 13 May 2007). Religion-based divisions between these clans are further accentuated by the fact that *sora jat* are predominant in Pe Lhun Sangh, an organization advocating the promotion of Bonism. On the other hand, *car jat* control Boudha Argahaun Sadan, an organization whose aim is to spread Buddhism among Gurungs.

The identity of a community is generally measured by its distinctness from others in terms of its history (origin), area of habitation (homeland), language, religion and culture. The following parts provide an overview of these features in the Gurung community.

Origin

The identity of the Gurungs is the subject of debate. The various sources relating to the history of Gurung ethnicity differ. One such source is the *vamsavali* which state that the Gurungs were originally a Hindu clan and they refer to the origin of *car jat* Gurungs and *sora jat* Gurungs—the former being descendants of those born of wives belonging to the royal family and to a priest family, while the latter are descendants of those born of spouses belonging to a slave family (Sharma 1982: 232-239). Macfarlane has a different interpretation. He considers that *car jat* come from pre-Aryan India, while *sora jat* descend from nomadic Tibetan tribes (Macfarlane 1976: 19).

As for British army officers who travelled Nepal extensively in the early nineteenth century, they considered the Gurungs to be a Mongoloid group that lives traditionally in the hills of Gandaki province, professes Bonism or Buddhism and speaks a Tibeto-Burman language (Gurung 2004: 4-5).

Another source relating to Gurung ethnicity is the empirical study carried out by foreign anthropologists. They, like British army officers, identify the Gurungs as being a distinct group of Mongoloid stock who have retained their original language and religion despite having been influenced by the Hinduization process which first started in the sixteenth century. Generally speaking, foreign anthropologists' research focuses on both the religious differences depending on the Gurungs' geographical distribution —for instance, those living in the northern Himalayas profess Bonism and Buddhism, while those having migrated to the southern slopes are mostly Hinduized—and on the occupational changes the Gurungs have undergone, from pastoralist to farmer to soldier.

And finally, Gurung ethnicity has been influenced by the rising ethnic movement in the post-1990 period. A large amount of Gurung literature, produced by native Gurungs (Tamu 1997, Gurung 2003), breaks away from the previous trend of assimilating them to Hindu religion and society. The Gurungs claim to be *adivasis janajatis* with their traditional homeland/kingdom in Gandak province. Both Gurung activists and academics raise a common voice against the internal distinction between *car jat* and *sora jat* which is well reflected in the writings of Pignède, "The Brahmins and the men of the south have divided our people into castes, but previously we were all equal and we wish to become so again". (Pignède 1993: 155). A leading Gurung scholar has also asserted that the artificial internal hierarchy (between *car jat* and *sora jat*) does not match the physical structure of the Gurungs, and their language and religion (Gurung 2004: 31).

The last three sources agree that Gurungs certainly originate from a group of mongoloid stock from the north. Culturally, they are very akin to the Tibetan population. Their language belongs to the Tibeto-Burman group. Still, the main problem in seeking the truth about the origins of the Gurungs is the lack of any credible history of this community. This task is rendered more difficult by the fact that the Gurung language does not have its own script (Gurung 2004: 8). The history of the Tamus is preserved in the myths and legends of the Tamu pye (Bonism). Hence, it is almost impossible to reconstruct with any certainty the history of the Gurungs (Pignède 1993:14), because the oral history and legends do not provide a single, coherent story. Macfarlane has collected five different stories, based on legends, about the origins of the Gurungs (Macfarlane 1997: 200-202).

- Story one: The main route (of Gurung migration) was down to Burma, then westwards through Assam and eastern Nepal to their present settlement, where they have been living for over seven hundred years.
- Story two: The Gurungs were nomadic shepherds who came down over the high pastures of Tibet, through the Kingdom of Mustang to settle on the southern slopes of the Annapurna range.
- Story three: *Car jat* Gurungs came from the south, from northern India, and *sora jat* from the north, Tibet.
- Story four: During the Kirat period, they (Gurungs and some other *janajatis*) lived in the foothills and large inner valleys of Nepal. They appear to have fled to the safety of the green mountain summits after the Kirat rule was overthrown in the first century A.D. They were driven further north with the invasion by the Indo-Aryans, who infiltrated Nepal in greater numbers during the period of the Muslim attacks on India from the fifteenth century onwards.
- Story five: The Gurungs originated in Mongolia and western China about eight or nine thousands years ago. They ventured down through China, reaching the Yarlu Valley of the Lhoka region of eastern Tibet in about 1000 B.C. Here by 1000 B.C. they were known as Tamus. Some Tamus settled in the northern Bagmati region, having passed through the Kerung and Kuti, and they became Tamangs. Others crossed over into Mustang. When they moved on to Manang in

about 500 A.D., they crossed the Annapurna Range onto the southern slopes to a large valley called Kohla. From about 1000 A.D. onwards, the Gurungs settled in a number of villages along the slopes of the Himalayas. In about the thirteenth century the Rajputs started to move up from northern India and began to encounter the hill peoples, entering Gurung territory in the sixteenth century.

Homeland

Despite conflicting stories about the origins of the Gurungs, both written and unwritten histories about them provide a common answer to the question: is there a place on present-day Nepalese territory that could be acknowledged as the Gurungs' traditional homeland? Whatever the author, this is always to be found in the same area: west of Budhigandi, east of Kaligandaki, south of the Himalayas, and north of Gandaki (Sharma 1982: 230); their traditional territory extends from Gorkha district east through Lamjung and Kaski to Syangja, throughout the Gandaki Zone (Bista 1980:75); Gurung country lies in the high valleys of the southern slopes of the Annapurna Range and Himalchuli in the central region of high hills. To the south it is bordered by a series of small plains and large flat valley bottoms like that of Pokhara. The chain of hills separating the Kali Gandaki Valley from that of the Modi forms the western border, whilst the Buri Gandaki Valley constitutes the eastern border (Pignède 1993: 4). The concentration of the Gurung population settled in the Western Development Region—60 per cent in 16 districts and nearly half in five hill districts of the Gandaki Zone—also hints at the whereabouts of their traditional homeland.

Table 1: Gurung population in Gandaki zone

Districts	Population	Percentage
Kaski	69,038	12.70
Syanja	31,567	5.83
Tanahu	39,478	7.25
Lamjung	56,140	10.33
Gorkha	64,240	11.82
Total in Gandaki Zone	260,463	47.91
In other districts of Western Development Region	66,982	12.32
Total in 16 districts of Western Development Region	327,445	60.23
Total Gurung population in Nepal	543,571	100.00

Source: Harka Gurung *et al.* 2006. *Nepal Atlas of Ethnic and Caste Groups.* Kathmandu, NFDIN.

Before the Rajputs seized power in the sixteenth century, the Gurungs were governed by the Ghale dynasty whose descendants now form a Gurung clan. There might not have been a single political territory ruled by one Gurung king. There might have been many Gurung rulers, each of them exercising political and administrative authority within their own boundaries. The prevalence of a chiefdom instead of a kingdom on Gurung territory can be explained by a legend about the creation of Ghandruk, a Gurung-dominated village, 750 years ago. According to Khadga Bahadur Gurung (8 April 2007), the legend says:

> On his way to hunt, a shepherd belonging to Tilmo (a clan of *sora jat* Gurungs) planted seeds of barley in a place now known as Ghandruk, and found them to be already growing on his way home. He then settled there and married a Thimche (another clan of *sora jat* Gurung). In the course of time, the population grew, local disputes broke out and therefore a search was made to find someone who could settle their problems and rule over them. They finally invited a person from Mustang known as Gungar (ancestor of the person interviewed) and he then ruled over them as their chief.

Language

The Gurungs have their own mother tongue. In 2001, those speaking the Gurung language numbered 338,925 (62.35% of the total Gurung population, CBS 2002: 72, 83). In 1991, they amounted to 50 per cent. Harka Gurung gives three reasons for the difference between the number of Gurungs recorded and the number of Gurungs speaking their mother tongue: 1) the censuses were conducted in the Nepali language, hence there was no doubt some bias on the part of the enumerators and respondents; 2) some people belonging to other ethnic groups such as the Tamang, Thakali, Dura and Bhote groups like to register themselves as Gurung, yet they have a different mother tongue; 3) Gurungs who migrate from their homeland forget their mother tongue (2004: 76). Gurungs from the western hills—a traditional Tamu homeland—for the most part retain their mother tongue.

Table 2: Gurung-speaking population in the Gandaki zone

Districts	Population	Percentage
Kaski	56,223	16.59
Syanja	29,032	8.58
Tanahu	33,010	9.74
Lamjung	55,176	16.28
Gorkha	48.772	14.39
Total in Gandaki Zone	222,213	65.56
In other districts of Western Development Region	37,405	11.04
Total in 16 districts of Western Development Region	259,618	76.60
Total Gurung-speaking population in Nepal	338,925	100.00

Source: Harka Gurung *et al.* 2006. *Nepal Atlas of Language Groups.* Kathmandu, NFDIN.

Table 3: Census and survey data on Gurung-speaking population of Kaski district

	Kaski	Ghandruk	Deurali	Pokhara
1. Gurung population (2001 census) (No.)	69,038	2,384	683	33,795
2. Gurung as mother tongue (2001 census) %	81	100	92	75
3. Gurung as mother tongue (2007 survey) %	81	87	78	77

Both census and survey data show that Gurung populations living in rural areas retain their mother tongue more than those living in the city of Pokhara. To what extent does this figure reflect the actual situation regarding the Gurung language? In fact, there are fewer Gurungs (25%) who understand and speak their mother tongue than those claiming Gurung to be their mother tongue (Table 4). Similarly, among the 81 per cent of respondents who reported Gurung to be their mother tongue, only 21 per cent said that they can actually read and write in it. This is true in the case of both rural (Ghandruk and Deurali) and urban (Pokhara) sample areas. Depending on who the interlocutor is, communication in the Gurung language is also dwindling at a considerable rate: 62 per cent with parents, 49 per cent with siblings, 46 per cent with spouses, and 27 per cent with children. This pattern is the same for Ghandruk, Deurali and Pokhara. Perhaps this is why, in this region, a substantial number of the Gurungs questioned (48%) favoured Nepali as the only official language, and why there is a majority of 52 per cent of respondents who are in favour of a bi/multilingual policy. Data clearly reveals a trend in curbing the use of the Gurung language in subsequent generations; whereas identity politics prompts the group to make other claims.

Table 4: Understanding of and communication in mother tongue among Gurungs from Kaski district

In percentage

		Ghandruk	Deurali	Pokhara	Total
1.	Gurung as mother tongue (2007 survey)	87	78	77	81
	Understanding of mother tongue				
2.	Understand and speak Gurung language	67	46	55	56
3.	Can read and write in Gurung language	29	8	26	21
	Communication in mother tongue				
4.	Communication with parents	67	68	50	62
5.	Communication with siblings	52	60	32	49
6.	Communication with spouse	46	61	26	46
7.	Communication with children	21	47	7	27

Culture and Religion

Culture and religion are factors that provide a basis for identity politics. The Gurungs' traditional religion is shamanism or Bonism, but they were subsequently drawn to Buddhism, while some adopted the Hindu religion. A Japanese anthropologist, Kawakita, gives the following geographical distribution: Lama (Buddhist) Gurungs in the northern Himalayas, Bone Gurungs in the high hills and mountains, and Hinduized Gurungs in the low hills (cited in Gurung 2004: 5). As the census included Bone religion in Buddhism, most Gurungs registered themselves as Buddhists.

Table 5: Religious status of Gurungs from Kaski district

	Kaski	Ghandruk	Deurali	Pokhara
1. Gurung population (2001 census) (no.)	69,038	2,384	683	33,795
2. Buddhist population (2001 census) * (%)	87	92	98	90
3. Buddhist population (2007 survey) (%)	74	80	72	68

* This figure includes other populations who profess Buddhism

The size of the Buddhist population recorded in the 2001 census is higher than that in the survey because census data included other populations who also profess Buddhism. The survey data for Kaski district show that though a clear majority of Gurungs are Buddhist—more than a quarter of them—a substantial number are Hindu. Gurungs have been influenced by the Hindu religion since the beginning of the fifteenth century, when Gurung territory came under the influence of Hindu invaders from India. The process of the Gurungs' Hinduization was expedited by their association with Hindu kings during the unification and expansion of the Nepalese territory in the late eighteenth century. The Kali Bahadur regiment (in the Nepalese Army) was solely recruited from among the best Gurung soldiers. In the Kali Prasad régiment, Gurungs were the majority (Pignède 1993: 18). Besides, their integration into the British and Indian armies since the early twentieth century also contributed to their Hinduization. In order to enrol in the British or Indian army—the most sought after job for a Gurung—a person had to register themselves as Hindu (Khadga Bahadur Gurung, 8 April 2007). The process of the Gurungs' Nepalization reached its climax during the Panchayat period (1960-1990). Regarding cultural pressure against the Gurung language and traditions, Macfarlane observed: "Over the twenty-five years since I first visited (in 1972) the Gurungs, the spread of Nepali-medium teaching, the effects of the radio, the growing dominance of the towns, have all eroded the language and culture [...] There is a loss of confidence in the value of Gurung traditions and culture: a revolution of rising expectations, a growing frustration and disillusionment, especially among the young" (Macfarlane 1997: 187). The influence of the Hindu way of life among Gurungs is evident in many aspects. They celebrate Dashain as an important festival. Even in places where Gurungs are the dominant settlers, there are several Dalit and Brahmin households, as well as other Hindu ethnic groups.

Despite the increasing influence of the Hindu way of life, Gurungs are known to be a group who respect and retain their traditions. Even Hinduized Gurungs use a Lama and a Bone priest in addition to a Brahmin when performing their rituals. With regard to the Gurungs' attribute of adopting both indigenous and exogenous values, Macfarlane observed, "They appear to incorporate elements of both Tibeto-Burman and Indo-Aryan origin. This is especially evident in their religion and ritual, which is a bewildering mixture of Hinduism, Buddhism and beliefs very similar to the old bon religion of Tibet" (1976: 12). Whether Buddhist or Hindu, Gurungs have their own distinct cultural identity. Dor Bahadur Bista wrote: "Gurungs are clearly Mongoloid in their features; their language is predominately Tibeto-

Burman; very few Gurungs intermarry with non-Gurungs; Rodhi is the unique institute of their customary governance; Sorathis is the Gurungs' own dance; Lhosar is their main festival and the traditional dress of Gurung men includes a short blouse tied across the front and a short skirt of several yards of white cotton material wrapped around the waist and held by a wide belt" (1980: 75-77).

Lahure is an identity that Gurungs share with some other ethnic groups, such as the Rais or Limbus. Their choice of employment in order of preference is: the British army, the Indian army and the Nepalese army. When referring to the socialization of Gurungs during childhood, Pignède observed,

> From a young age the small boy has got used to the idea that one day he will be a soldier like his father. There was no small boy aged two or three years who did not know how to make *salaam*, that is to make a British military salute, when an adult asks him to do it. A young boy is proud to wear a soldier's belt or the green wool beret of the Gurkha gunners of the Indian army. When his father or a man from the village comes back on leave, the child will listen enraptured to the stories of military life which are recounted to him. He hears talk of planes, boats, cars and large towns full of well dressed people who earn a lot of money. The man on leave brings back books and weekly papers full of photos representing that life of which he knows nothing. (1993: 221)

This is in keeping with the Gurungs' tradition of migration whenever better economic opportunities present themselves. The history of their migration follows the change in occupation from pastoralist to farmer to soldier and on to other walks of life. The data collected reveal changes in occupation over the last three generations, with a decline in agriculture amounting to a drop from 77 per cent (grandfather) to 44 per cent (father) to 36 per cent (respondents). Such a rapid shift explains the change in the Gurungs' migration pattern. Roughly speaking, up until the early 1970s Gurungs migrated mostly to enrol in the army, after which they retired and moved back to their village. Then, from the mid-1970s, there was a wave of migration to Indian cities (Bombay, Delhi, etc.), to the Middle East, followed by retirement or a move down to Pokhara (Macfarlane 1997:189). In 1957, there were only 33 Gurung households in Pokhara, but in the 2011 census they will no doubt prove to be the largest population (Baral 2003: 47). Gurung migration to this urban area coincides with the rapid urbanization of Pokhara and the change in its demography. "Until the mid-1960s, Bahun, Chhetri and other Hindu castes dominated the population of the town. Gurungs and Magars were almost negligible in number. But the migration of Gurungs and Magars since the 1960s, especially those employed in the British and Indian armies, has changed the ethnic composition of the town. In the 1990s, there was a slight reduction in the proportionate number of Chhetris and Newars, but an increase among the Magars, Tamangs, other groups and Tarai people. In other words, there has been an increase in the number of minority groups in Pokhara" (Adhikari 2007: 32). Table 6 shows the Gurung migration pattern in sample cites of Kaski district.

Only 53 per cent of respondents said that they were actually living in their current dwelling-place. The number of Gurungs who have migrated recently is as high as 74 per cent in the case of Pokhara, with the lowest figure for migration standing at 28 per cent for Ghandruk VDC. Nearly half of all respondents reported that they had benefited from family members' foreign employment. Of 117 recorded migrations for employment, 96 per cent were to foreign countries. Among these, 37 per cent of migrants journey to Arab countries, 36 per cent to South East/Far East Asian countries, 16 per cent to the UK, 4 per cent to other European countries, and 7 per cent to India. 87 per cent of respondents acknowledged that they had benefited from the remittance economy which contributes to increasing family income to varying degrees; to a considerable degree (46%) and to some degree (54%).

Table 6: Migration of Gurungs from Kaski district

In percentage

		Ghandruk	Deurali	Pokhara	Total
1.	Living since birth	72	57	26	53
2.	Family members having migrated for employment	40	50	54	48
3.	Migration within the country	-	13	3	4
4.	Migration outside the country	100	92	95	96
5.	Money received from family member working outside	85	84	92	87
6.	Rise in family income from remittance (to a considerable degree)	27	64	44	46
7.	Rise in family income from remittance (to some degree)	71	35	56	54

In sum, Gurung/Tamu identity is associated with five distinct characteristics:

- A group belonging to indigenous nationalities which prefer ethnic identity to national identity (details in following section).
- A highly mobilized Mongoloid tribe in terms of migration.
- A Tibeto-Burman tribe which largely retains its traditions.
- A group which has become prosperous thanks to the remittance economy.
- A group that is increasingly conscious of its own ethnicity.

Inclusion and Exclusion of Gurungs

At macro level, indigenous nationalities are generally classed under the heading 'excluded groups'. To be excluded means living in one, or in a combination, of these three underprivileged conditions: (1) inferior social status, (2) deprived of any development and (3) under-representation in the political sphere. The Gurungs are one of the 59 officially recognized indigenous nationalities of Nepal. Do they constitute an excluded group?

Contrary to the generalization whereby indigenous nationalities including the Gurungs are among the excluded groups, 80 per cent of Gurungs questioned do not perceive themselves as excluded; 70 per cent are of the same opinion regarding Bahuns/Chhetris. Those who are considered to be excluded groups are 'other *janajatis*' (by 52% of respondents) and Dalits (by 83% of respondents).

Similarly, more than 80 per cent of rural Gurungs do not consider themselves to hold an inferior position vis-à-vis Bahuns/Chhetris. Nevertheless, Pokhara seems to be influenced by the general idea that affirms the ascendancy of Bahuns/Chhetris in the social order, as attested to by 59 per cent of urban respondents.

However, 64 per cent of Gurungs from Kaski district who were interviewed do not think that Gurungs dominate Bahuns/Chhetris in politics. Why therefore do these Gurungs from Kaski district perceive themselves as an included group with a higher social status? Probably because a majority (55%) consider that Gurungs are richer than Bahuns/Chhetris and because only a fourth of them consider that they are poor (Table 7).

Table 7: Self-perception of inclusion and exclusion

In percentage

		Ghandruk	Deurali	Pokhara	Total
	Included groups				
1.	Bahun/Chhetri	33	79	93	70
2.	Gurung	82	76	81	80
3.	Other *janajatis*	31	47	63	48
4.	Dalit	17	11	22	17
	Comparison with Bahun/Chhetri				
5.	Social status of Bahun/Chhetri is higher than that of Gurung	10	27	59	32
6.	Gurungs are richer than Bahuns/Chhetris	85	34	49	55
7.	In politics, Gurungs have an advantage over Bahuns/Chhetris	70	29	14	36
	Economic conditions of the household				
8.	Very good economic conditions	16	4	25	15
9.	Good economic conditions	43	78	64	61
10.	Poor economic conditions	41	18	12	24

The position of Gurungs from Ghandruk VDC represents a particular case. For two thirds of them, the Bahun/Chhetri group is an excluded one. They disagree with the general statement that Bahuns/Chhetris enjoy a higher social status than Gurungs. On the contrary, they affirm the ascendancy of Gurungs over Bahuns/Chhetris in both political and economic spheres. This broadly reflects the distribution of economic and political power among different caste/ethnic groups at local level in Ghandruk village.

Thus, we can conclude from this picture that, at macro level, development is not generally taken as a factor for measuring inclusion and exclusion. Obviously, urban dwellers, irrespective of caste and ethnicity, enjoy greater privileges than rural dwellers, whether road developments, hospitals, educational institutions or other services. Yet at micro level, there is a correlation between political influence, economic conditions, social status, and development projects (Table 8).

Why do Gurungs therefore perceive themselves as less privileged than Bahuns/Chhetris in the political power structure? Does this perception match the reality?

Table 8: Availability of services and facilities

In percentage

		Inside house or nearby	Within 1 km	Within 2-3 km	Within 4-5 km	More than 5 km
Drinking water	Ghandruk	81	15	2	-	1
	Deurali	43	55	1	-	-
	Pokhara	100	-	-	-	-
	Total	75	24	-	-	1
Post office	Ghandruk	-	32	11	9	48
	Deurali	-	24	58	-	-
	Pokhara	4	32	46	17	-
	Total	1	29	43	8	17
Telephone	Ghandruk	7	37	9	9	36
	Deurali	99	1	-	-	-
	Pokhara	94	3	3	-	-
	Total	64	15	4	4	13
Bus Depot	Ghandruk	1	-	-	4	95
	Deurali	73	27	-	-	-
	Pokhara	48	33	6	13	-
	Total	39	19	2	5	34
Health Post	Ghandruk	3	35	19	15	27
	Deurali	-	24	76	-	-
	Pokhara	35	46	16	3	-
	Total	12	35	37	6	10
Hospital	Ghandruk	-	-	1	-	99
	Deurali	-	-	-	-	100
	Pokhara	10	43	23	23	-
	Total	3	13	8	7	69
Primary School	Ghandruk	6	53	30	5	5
	Deurali	22	78	-	-	-
	Pokhara	41	57	1	-	-
	Total	22	63	11	2	2
Lower Secondary School	Ghandruk	5	47	25	7	16
	Deurali	-	24	74	-	1
	Pokhara	41	58	1	-	-
	Total	14	43	34	3	6
Secondary School	Ghandruk	4	26	12	7	51
	Deurali	-	24	74	-	1
	Pokhara	41	58	1	-	-
	Total	14	35	29	3	19
Campus	Ghandruk	-	4	-	-	96
	Deurali	-	23	75	-	1
	Pokhara	36	42	22	-	-
	Total	11	22	31	-	35

Gurungs are generally seen as a group less politically aware. This is more or less confirmed by both Gurung and non-Gurung respondents of all three sample cites. Two representative voices, one expressed by a Gurung and another by a Bahun, are that of Kisam Gurung (Vice-President of Pelhu Sangh, Ghandruk village) who said, "I am more interested in social services than politics" (21 April 2007) and that of Tek Nath Ghimere (President of the Village Unit of the NC, Deurali VDC) who declared, "Gurungs are less aware as people. One Bahun can easily tackle ten Gurungs. A Brahmin with only the SLC can win over a Gurung with a university degree. If some Gurungs do exceptionally take an interest in politics they are no doubt motivated by a desire to gain status, not by wealth. Otherwise, they don't like to take on political leadership. They always say that they would back or support a leader from another community, Bahun in most cases. Soon after completing their school education, the Gurungs' one-item agenda is to enrol in the British army and to become rich" (30 April 2007). In referring to this trend, Ms. Karma Tamu (Chairperson of Tamu Dhin, Kaski) made the remark that "We gain wealth but at the cost of political power" (8 June 2007). These observations are confirmed by the Gurungs' poor participation in formal party organizations.

Table 9: Gurungs in party organizations in Kaski district

	Total	Representation of Gurungs	
District level	No.	No.	%
District Population	280,527	69,038	18.1
Nepali Congress	48	6	12.5
CPN-UML	37	5	13.5
RPP	24	7	29.2
Municipal level			
Population of Pokhara city	156,312	33,795	21.3
Nepali Congress	33	1	3.0
CPN-UML	34	2	5.9
RPP	12	0	0.0

Except at district level within a small conservative party, RPP (Rastriya Prajatantra Party, in which they are well represented (29.2%), Gurungs are under-represented in the two major political parties, whether it be at district level or at municipal level (Table 9).

However, quite another picture emerges when scrutinising Gurungs' political power in terms of their representation at different levels of elected bodies.

In every case, at district, municipal or village level, their representation is much higher than their proportion within the total population (Table 10). Ghandruk is an extreme case since 46 per cent of this VDC's population are Gurung, yet they won 91 per cent of VDC seats in both the 1992 and 1997 local elections.

Generally speaking, those who control the party organization also dominate elected bodies. Yet Gurungs in Kaski district show that the distribution of power between the organizational and elected wings of a political party is not necessarily balanced since they are under-represented in party organizations, but they benefit from a higher level of representation at all levels of elected bodies. Two factors may explain this situation.

Table 10: Gurungs in Elected Bodies in Kaski District

	Total	Representation of Gurungs	
District level	No.	No.	%
Population	280,527	69,038	18.1
Parliamentary Representation (1991-2008)	13	6	46.1
DDC Representation (1992)	15	7	46.6
DDC Representation (1997)	15	3	20.0
Municipal level			
Population	156,312	33,795	21.3
Representation in Pokhara municipality (1992)	20	6	30.0
Representation in Pokhara municipality (1997)	20	6	30.0
Village level			
Population in Ghandurk VDC	5,138	2,384	46.4
Representation in Ghandruk VDC (1992)	11	10	90.9
Representation in Ghandruk VDC (1997)	11	10	90.9
Population of Deurali VDC	2,882	683	23.7
Representation in Deurali VDC (1992)	11	6	54.5
Representation in Deurali VDC (1997)	11	4	36.4

One is the relationship between political and economic power. Gurungs are rich, hence they have the economic power to gain political power. Damodar Ghimire (a local UML leader from Deurali VDC) admitted that they consciously selected Bhoj Bahadur Gurung as the successful candidate for the post of Vice-President in the 1992 local election, and again as President in the 1997 local elections despite there being many other qualified party workers from Bahun and Newar communities who had long served the party, the reason being that he is rich and can afford the election costs. Furthermore, although Brahmins represent the largest section of this village population, Bhoj Bahadur Gurung was elected because Gurungs in general have a tendency to cast a block vote, unlike voters from other communities whose split their vote in accordance with their preference for a party (28 May 2007). A study also explains why both the largest parties, NC and UML, give preference to Gurung candidates; this is because this community makes up a substantial proportion of the population in the district and they have a tendency to vote for candidates from their own community (Baral 1999: 127). These observations match the findings of this study.

Voting for a particular party proved to be slightly more common in the 1999 parliamentary elections than in the 1997 local elections. However, the trend has shifted over the last two general elections, the 1999 parliamentary elections and the 2008 CA elections, since the proportion of respondents casting their vote for a party dropped by 6 per cent in Ghandruk, 15 per cent in Deurali and 20 per cent in Pokhara. This obviously means that Gurung voters give priority to the kinship and ethnicity factor, with their number increasing by 10 per cent in Ghandruk, by 13 per cent in Deurali and by 20 per cent in Pokhara. Part of the reason behind this decline in party-based votes is the general disillusionment with political parties. In response to the question "which party do you like most?" 65 per cent said they like none of the existing parties, including the CPN (Maoist).

Table 11: Gurungs' voting pattern

In percentage

		1997 Local Election	1999 Parliamentary Election		2008 Constituent Assembly Elections	
Ghandruk	Party	45	43	- 2	37	- 6
	Kinship and Ethnicity	54	56	+ 2	64	+ 10
Deurali	Party	53	55	+ 2	40	- 15
	Kinship and Ethnicity	48	45	- 3	59	+ 13
Pokhara	Party	51	61	+ 10	41	- 20
	Kinship and Ethnicity	48	39	- 9	59	+ 20

Movements

Gurung ethnicity has been undergoing a revival since the fall of the partyless Panchayat system in 1990. The history of Gurung activism dates back to 1954 with the creation of Gurung Kalyan Sangh. Some Gurung organizations were set up in the 1980s as a reaction to the relaxation of rules in the partyless Panchyat system following the 1980 referendum. Nevertheless, their activities only became visible and effective after creating an open political environment with the restoration of democracy in Nepal in 1990. During the initial phase in the post-1990 period, Gurung activism was mainly confined to the revival of a uniform code of the Tamu community's customary practice. Three major resolutions passed by a nationwide Gurung conference held in the second week of March 1992 were:

1. Gurung history was written and distorted by Brahmans.
2. There are no inferior and superior clan groups in Gurung society.
3. Traditional Gurung priests are the Pa-chyu and the Khilibri; lamas are a more recent addition (see Macfarlane 1997: 185).

In retrospect, this constituted an assertion of their rejection of the centuries-long assimilation process to the hill Hindu high-caste culture. Until the 1990s, asserting cultural identity was predominant in Tamu activism. This is reflected in

Harka Gurung's appeal to the Tamus about how to include Gurung identity in the 2001 census. For purposes of the census, he suggested that they identify themselves as (1) Gurung in the section on caste/ethnicity, (2) Buddhist in the religion section and (3) Gurung in the mother tongue section (Gurung H. 2004: 76-77). To the surprise of observers of recent Gurung activism which has been dominated by demands for inclusion through federalism, proportional representation and reservation, the Gurung movement has been largely apolitical until recently.

Five major Gurung organizations are: Boudha Argahaun Sadan, Tamu Dhin, Tamu Chonj Dhin, Tamu Pe Lhun Sangh and Tamu Baudha Sewa Samittee. The headquarters of the first four organizations are in Pokhara, the last one, Tamu Baudha Sewa Samittee, being in Kathmandu. These are largely non-political organizations. The single most objective of Boudha Argahaun Sadan is to promote Buddhism among Gurungs. The main focus of Tamu Pe Lhun Sangh is to bring the Gurungs back to their primitive religion, Bonism. The scope of Tamu Dhin and Tamu Chonj Dhin is broader since they take an interest in other aspects of Gurung society, i.e. language, culture, festivals, education etc. As the leaderships of most of these five pioneering, active organizations have directly or indirectly focused on the patrimonial practice of the former partyless Panchayat system, they have consciously left aside party politics from their organizations' main concerns and activities. Regarding another dimension of the leadership of Gurung organizations, Ganesh Man Gurung stated that the above-mentioned organizations were formed by a handful of intellectuals and ethnically conscious 'early risers' who had migrated from villages to the city, and Pokhara in particular. Their leadership is in the hands of Gurung intellectuals and ex-Gurkha (British and Indian) army officers (2003: 35-36). The Panchayat guaranteed its stronghold among these two groups until its downfall in 1990. Most ex-Gurkha army officers show less interest in party politics.

Yet recently, Gurung movements have become radicalized and politicized in a response to the CPN (Maoist)'s successful attempts at blending class ideology with ethnic identity policies. Tamu Mukti Morcha (TMM) is one of several of its frontier organizations, which has been quite active in rallying Gurungs to the party's agenda and goal. Some of TMM's manifest objectives are (2007):

- Establishment of Tamuwan as a separate province within the proposed federal structure.
- Ethnic autonomy and right to self determination of Tamuwan.
- Preservation of Tamu's history, culture and tradition.
- Promotion of the Tamu language.
- Recognition of Tamu Lohasar as a national festival.
- Tamu ownership of natural resources on its territory.

The victory of the CPN (Maoist) in three out of four seats in Kaski district in the April 2008 CA election testifies to the successful mobilization around Gurung ethnicity. This was, in some manner, predicted by a UML leader from Kaski district. While explaining the strategic goal behind forming Loktantrik Tamu Sangh by the UML, Kaski district, he said, "Since Gurungs are rich, the communist slogan of economic equality could not appeal to them. Nor are they attracted to our other agenda of gender equality and women's property rights. They are sensitive to the one single most important issue: their lack of governance" (12 April 2007). This is confirmed by Man Bahadur Gurung, a former Deputy Mayor and former leader of the UML, as he explained why he moved over to the CPN (Maoist): "the UML had no political incentive for the Gurungs. Neither did the party advocate ethnic autonomy and proportional representation. The municipal government, whether under NC control over the period 1992-1997 or dominated by the UML in the period 1997-2002, had no policy or plans for promoting Gurung interests" (21 May 2007). The absence of any ethnic content in local governance was also acknowledged by the former Vice-President of Kaski DDC[3] (Shree Nath Baral, 15 April 2007). This was true in the case of village governance in Deurali and Ghandruk VDCs. The NC had not caught on to the emerging trend of politicizing ethnicity. Its district president admitted: "our party has not yet sent any guidelines on the integration of *janajatis* in politics. We therefore have no such policy, nor are we contemplating any" (Yagya Bahadur Thapa, 19 May 2007).

The changing ethnic landscape of Nepalese politics has significantly impacted Gurung activism which has evolved to adopt the most radical content and agendas. Demands listed according to 20 items, prepared by Tamu Dhin and submitted to a meeting of Gurung CA members, represent the voices of other Gurung organizations. These demands are registered under four categories (May 2008):

Culture:

- Recognition of Tamu as a national language and inclusion of this language in the education syllabus.
- Celebration of Tamu festivals, such as Lhosar and Tuhute, as national festivals.
- Establishment of a separate government department for the preservation of Tamu dress, culture and tradition.
- To declare as martyrs those who sacrificed their lives for the Tamu cause.
- Free schooling up to higher education.

Economy:

- Provision to ensure the rights of those working in foreign countries.
- Constitutional provision to use knowledge and skills of ex-army men.
- Nationalization of natural resources.
- Project for the utilization of national resources on Tamu territory.
- To use income collected from mountaineering expeditions in Annapurna and Machapuchhe ranges to develop these areas.
- To ensure employment for young Tamus in Tamuwan province.
- Declaration of Gandak territory as tourist destination area and special programme for developing tourism in this area.

Representation:

- To ensure representation from the Tamu province in the list of 26 CA members nominated by the government.

- Meaningful integration of Tamus and other backward communities in governance.
- Implementation of agreements made between NEFIN and the government.

State restructuring:

- Formation of federal provinces on the basis of caste/ethnicity, language and geography.
- Declaration of Gandak province as autonomous Tamuwan with right to self declaration.
- Establishment of basic rights of Tamus to the post of Chief Minister, Mayor, Chair of DDC, CDO in Tamu autonomous Gandak province.
- Creation of a ministry for indigenous nationalities.
- Implementation of provisions for a secular state.

To what extent is Gurung activism making ground at grass roots level? Is it still confined to an elite-based movement or has it spread downwards to village level. Data from the survey inform these questions. An overwhelming majority of 71 per cent of respondents were aware of the existence of Tamu organizations. However, they consider that only half of them are really functional. There is much rural-urban variation. Only 12 per cent of respondents from Deurali VDC and 27 per cent of respondents from Ghandruk VDC participate in the activities of Tamu organizations. Respondents from Pokhara municipality acknowledge the existence and activities of Tamu organizations at a higher level, obviously because the

Table 12: Awareness of, association with and participation in Tamu organizations

In percentage

	Ghandruk	Deurali	Pokhara	Total
Awareness				
Awareness of existence of Tamu organizations	59.3	60.8	97.0	71.0
Participate in activities of Tamu organizations	27.4	12.3	77.2	36.5
Association				
Membership of Tamu organizations	12.3	2.7	22.7	12.2
Nature of organization: Tamu activism	70	100.0	93.3	85.2
Participation in movements organized by				
Gurung organizations	6.5	21.6	43.8	22.8
Political parties	12.3	26.0	18.8	18.8
CPN (Maoist) during insurgency	8.6	20.3	5.8	11.6
NGOs	22.1	30.6	32.4	28.1
CBOs (Community Based Organizations)	74.4	64.4	68.1	69.2

headquarters of most Tamu organizations are in Pokhara, and this city is the centre for any demonstrations by Gurung movements. This also explains the reason behind rural-urban differences in membership of Tamu organizations (23% in Pokhara, versus 3 in Deurali and 12 in Ghandruk VDC) (Table 12). Among respondents who said they are members of formal Tamu organizations, most of them are associated with those that are fully devoted to Gurung activism.

Participation in public affairs is not merely confined to ethnic organizations. There are many other platforms that allow for the public's participation. Gurung respondents in the three locations take a greater part in CBO programmes and activities (69%). Their participation in movements initiated by political parties and in insurgencies launched by the CPN (Maoist) is low, 19 and 11 per cent respectively. Respondents' participation in NGOs is relatively higher (28%) than in Gurung organizations (23%), except in Pokhara.

Besides Tamu organizations, there are many organizations which have directly or indirectly served to promote the Gurung cause. NGOs, INGOs, political parties and local government might have manifest and latent programmes aiming to serve the interests of Gurungs. To the question: "to what extent do organizations work for the Gurung cause?", 83 per cent think that this cause is absent from the TMM programme, while only 45 per cent think it is absent from INGO programmes.

Table 13: Contributions of organizations to the Gurung cause

In percentage

	Ghandruk	Deurali	Pokhara	Total
Tamu Organization	20	9	43	23
NGO/Civil society	21	7	18	15
VDC/Municipality	14	4	4	7
Political Party	4	0	2	2
Tamu Mukti Morcha (Maoist)	0	0	4	1
INGO	13	67	25	41

Concerning activities undertaken by organizations and contributing to the Gurung cause, in the three locations (though ranked differently), Tamu organizations, INGOs and NGOs come well ahead of VDC Municipality, political parties and TMM (Table 13).

The lay Gurung is obviously attracted to several agendas circulated by Tamu activists and Tamu organizations. But their preference is closely linked to those issues which have a direct bearing on increasing their standard of living. When asked to choose the issue they like most, two thirds of respondents preferred 'reservation of jobs for Tamus in education and employment'. Other issues, which fall in the category of collective incentives, received little preference. For instance, only 16 per cent of respondents listed the 'promotion of the Gurung language and culture' as their top priority. Similarly, 15 per cent opted for increasing Gurung representation in the state apparatus. Those who ranked the 'secular state' their top priority only amount to 4 per cent.

Table 14: Top priority issues

In percentage	Ghandruk	Deurali	Pokhara	Total
Promotion of the Gurung language and culture	21	15	10	16
Reservation of jobs for Tamus in education and employment	67	63	64	65
Increase in Gurung representation in state apparatus	9	15	20	15
Secular state	3	4	7	4

Tamu organizations' potential to rally Gurungs to the ethnic cause is relatively high since many respondents expressed more trust in formal and informal Tamu institutions than in other organizations (Table 15).

Table 15: Trust in public figures and organizations

In percentage

	Ghandruk	Deurali	Pokhara	Total
Tamu organizations	60	42	65	57
Tamu Mukhiya	58	68	63	63
Member of Parliament (elected in 1999)	8	13	15	11
Leader of local party	13	41	24	27
NGO/Civil society	41	33	43	40
Chair of VDC or Mayor of Municipality	46	63	30	47
Ward Chair	42	70	29	48

Tamu organizations and Tamu mukhiya enjoy 'a lot of trust' on the part of the majority of Gurung respondents in our survey; 57 per cent and 63 per cent respectively. The common pattern for all respondents of the three sample areas is that they have more trust in political figures working in elected bodies at grass roots level (47-48%) than politicians working at a higher level and in the organizational wing of a political party, i.e. Member of Parliament and leader of local party organization.

Impact of Gurung Movements

The impact that decades-long Gurung movements have had on the Gurungs is manifest in more than one aspect, such as less discrimination between *car jat* and *sora jat*, assertion of the Gurung identity, support for ethnic aspects of state restructuring, etc.

One of the major objectives of Tamu organizations is to abolish the 'artificial' division between *car jat* and *sora jat* which, for 45 per cent of respondents in the survey, still exists. Several Tamu activists from Ghandruk VDC and from Deurali VDC are of the opinion that a sort of hierarchical stratification between *car jat* and *sora jat* still exists, which is attested by the ongoing system of banning marriage between these clans. Nevertheless, public opinion on prohibiting inter-clan marriage

Table 16: State of division between car jat *and* sora jat

In percentage

	Ghandruk	Deurali	Pokhara	Total
1. *Car jat* Gurung is different from *sora jat*	59	41	32	45
2. Less discrimination between the two	53	75	61	63
3. Marriage between these two is prohibited	67	20	41	43
4. *Car jat* are richer than *sora jat*	14	18	12	15
5. *Car jat* have taken the lead over *sora jat* in politics	9	13	11	11

differs, whether it concerns Gandruk, Deurali or Pokhara, and a mean of 43 percent are of the opinion that the old practice has already been abandoned. There has been a substantial reduction in internal differences between *car jat* and *sora jat* regarding

Table 17: Asserting the Gurung identity

In percentage

	Ghandruk	Deurali	Pokhara	Total
Preferred identity				
1. Nepali	16	20	27	21
2. Gurung	29	65	39	44
3. Mixed	62	15	33	35
4. Very proud to be Gurung	63	63	66	64
5. Very proud to be Nepali	58	43	62	54
Relations with neighbours from other communities				
6. Very good	32	35	46	37
7. Reasonably good	67	63	52	61
Cooperation between ...				
8. Gurungs and Brahmins/Chhetris	97	90	84	91
9. Gurungs and other *janajatis*	100	76	88	87
10. Gurungs and Dalits	99	91	78	91
11. Gurungs and Madheshis	100	100	60	74
Trust in others				
12. Gurungs	100	96	97	98
13. Brahmins/Chhetris	70	39	52	53
14. Newars	48	27	38	35
15. Magars	95	71	96	89
16. Other *janajatis*	84	67	77	76
17. Madheshis	58	18	19	32
18. Dalits	69	47	50	56

their economic status and political power. Only 15 per cent of respondents believe the statement that *car jat* are richer than *sora jat*. And only 11 per cent find differences between these two groups in terms of their wielding political power, with *car jat* having more power at a higher level than *sora jat*. The younger generation is believed to break the old feudal tradition of dividing the Gurungs into 'superior' *car jat* and 'inferior' *sora jat*.

At the same time, assertions regarding the Gurung identity as promoted by Gurung movements are gathering momentum. For Gurungs, ethnic identity and national identity overlap to a large extent, yet 44 per cent of those questioned put their Gurung identity before their Nepali identity. And, as we have noted, the number of those who reported Gurung as their mother tongue increased from 50 per cent in the 1991 census to 62 per cent in 2001. Likewise, a tendency towards ethnic-based voting has been on the rise among the Gurungs in subsequent elections.

Unlike the sudden explosion of Madheshi identity in a way that proved hostile to Paharis, the Gurungs' assertion of their identity seems to be of a non-confrontational nature. Relations between the Gurungs in our survey and their neighbours from another community are relatively good (Table 17). Inter-community relations are established by a spirit of mutual cooperation at the highest level. Yet we once again find this opposition between Paharis and Madeshis in so far that they have much more trust in other *janajatis* like themselves—mainly Magars—than in Madeshis.

The impact of Gurung activism is also reflected in the increasing tendency for the average Gurung to turn their attention to the state restructuring agenda even though a large proportion of Gurungs are not yet fully aware of it.

Table 18: Understanding of and in favour of state restructuring agenda

In percentage

	Have not heard	Heard but do not Understand	understand and in favour	Understand but not in favour
Cultural rights	52	15	31	2
Reservation system	55	15	27	2
Ethnic autonomy	57	13	25	4
Inclusive democracy	41	28	29	1
Proportional representation	58	13	27	2
Tamu autonomous province	50	14	24	12

Those who have not heard of expressions such as cultural rights, reservation, ethnic autonomy, inclusive democracy, proportional representation, and Tamu autonomous province constitute almost half of the total number of respondents. Around 16 per cent do not understand the meaning of these expressions though they have indeed come across them. Approximately one third of respondents understand the notion of state restructuring agenda. All of them agree with the content except for the claim upon the Tamu autonomous province about which public opinion is divided.

The Gurungs' attachment to the state restructuring project is so strong that a majority of 53 per cent of respondents equate democracy with state restructuring. For them, state restructuring could mean many things. An overwhelming majority of 97 per cent recommend 'special provisions for backward communities'. Support for a secular state is as high as 91 per cent. Those in favour of a bi/multilingual policy constitute 52 per cent. However, the substantial number (48%) opting to retain Nepali as the only official language, suggests a rift between activists and the lay Gurung, which is reasserted regarding their opinion on federalism. Only 46 per cent of the total number of respondents subscribe to the idea of transforming Nepal into a federal state.

Table 19: Restructuring Agendas

In percentage

1.	Democracy is understood as restructuring the state	53
2.	Special provision for backward communities	97
3.	Nepal should be declared a secular state	91
4.	There should be a bi/multilingual policy	52
5.	Nepal should be federal	46

Gurung activists do not want to compromise on their demand to constitute Tamuwan based on ethnicity claims. The qualitative information collected through in-depth interviews shows the prevalence of three different opinions on the proposed Tamuwan province under Nepal's federal scheme. Those who still have hesitations about federalism argue that it would create ethnic conflict (Amar Tamu, 20 April 2007). Shree Nath Baral's view is representative of the voices of many people belonging to hill caste groups. He is in favour of limited federalism, vesting a large part of power at the Centre (15 April 2007). Gurung activists working in Tamu organizations or in the CPN (Maoist) raise a common voice in favour of establishing an ethnic-based Tamuwan province with provisions for ethnic autonomy and rights to self-determination, and of ensuring the basic rights of Gurungs in Tamuwan province (Ram Kaji Gurung, 22 May 2007; Man Bahadur Gurung, 21 May 2007; Karma Tamu, 8 June 2007).

Conclusion

The idea put forward by activists of creating Tamuwan differs somewhat from the overall Gurung population's general understanding. For a Gurung, Tamuwan is a village/hamlet associated with geography, kinship, a shared culture, language, and history. It is a place governed by Gurung customary laws where the Tamu way of life prevails. However, for Gurung activists and organizations, Tamuwan province is set to become a political unit with an extended territory that integrates several scattered Gurung hamlets along with some other areas predominantly occupied by non-Gurung settlers. Gurungs would prevail in the social, economic and political power structure of this prospective province. The claim over Tamuwan province is a culminating point in the surge in Gurung ethnicity. The revival of Gurung ethnicity in the post-1990 period aims, on the one hand, at moving away from the past record of their assimilation to hill high castes in terms of their language and religion and, on the other hand, at sharing national resources and at proper representation in the country's political power structure. The Gurung movement is therefore a combination of ethnic revival and self-assertion with the purpose of gaining power. The Gurung people's receptivity to the agendas of Gurung activism is revealed by the findings of this study. For instance, Gurungs set more store by ethnic identity than by national identity; the number of people who speak Gurung as their mother tongue increased from 50 per cent in 1991 to 62 per cent in 2001; now around two-thirds of Gurungs identify themselves with Buddhism; they have more trust in formal and informal Tamu institutions than in other non-ethnic organizations (such as NGOs, political party, etc.); and they are increasingly becoming ethnic voters rather than party voters. If a surge in ethnicity is one of the requirements for creating an ethnic-based province, the Tamuwan dream is likely to become reality.

In retrospect, establishing a separate province for Gurungs is a very difficult task. Gurung activists claim that the western hills, the Gandaki zone in particular, is their traditional homeland. This claim is not far from the truth since nearly half of the total Gurung population and two thirds of the Gurung-speaking population reside in five districts of the Gandaki zone (Kaski, Syanja, Tanahu, Lamjung and Gorkha). Yet in their traditional homeland, an area between Kali Gandaki and Budhi Gandaki which includes three more districts (Parbat, Mustang and Manang) in a neighbouring area, Dhaulagiri zone, Gurungs are now the minority, numbering only 24 per cent and the majority group residing at present in traditional Tamuwan is the hill Bahun/Chhetri group that constitutes 54 per cent of the population distributed over this area. In a cluster of the eight districts mentioned above (five districts of Gandaki zone and three districts of Dhaulagiri zone), the Gurung-speaking population constitutes only 15 per cent, whereas people who speak Nepali as their mother tongue are the overwhelming majority, representing 72 per cent. Similarly, Buddhists residing in this area make up only 20 per cent, whereas Hindus are dominant (78%). Perhaps this is the reason why Gurung activists and organizations are championing their basic rights in the proposed Tamuwan province. To what extent would it be practical to constitute a province in the name of and in favour of one particular ethnic group (Gurung) in an area where another ethnic group (Bahun/Chhetri) represents more than twice the population of the group concerned (Gurung)? Indeed, the present territorial division does not offer an ideal situation for creating Tamuwan in the Gandaki zone; instead, it calls for the reallocation of territory while excluding predominantly non-Gurung areas and including other districts of Gurung settlement, such as Manang and Mustang. Creating a new territory in such a way facilitates the prospect of creating Tamuwan province in which the Gurung population would be the majority or at least the largest group.

Until now, asserting the Gurung identity vis-à-vis other social groups has been largely non-confrontational, contrary to what many people assume, i.e. that designing ethnic based federalism would lead to ethnic conflicts. The ultimate decision, however, lies with the Constituent Assembly which, of course, considers ethnicity as one of the factors but not the only factor, for creating provinces in a new federal Nepal.

REFERENCES

Adhikari, J. 2007. 'Urbanization and Change in Pokhara, Nepal', in Hiroshi Ishi *et al.* (eds.), *Nepalis Inside and Outside Nepal*, New Delhi, Manohar, pp. 17-57.

Baral, T. 2003. 'Samajik Paribartan tafa Pokharaka Gurung Samudaya' ('Towards Social Change of Pokhara's Gurung'), *Journal of Political Science*, vol. V & VI.

Baral, U. 1999. 'Political Participation: A Study of Local Elections 1997 in Pokhara Submetropolis', *Journal of Political Science*, vol. 2, no. 1.

Bista, D. B. 1980. *People of Nepal* (Fifth edition), Kathmandu, Ratna Pushtak Bhandar.

CBS. 2002. *Population Census of Nepal*, Kathmandu, CBS.

Gurung, G. M. 2003. 'Sansthagat Bikas ra Antarsambhanda' ('Institutional Development and Interrelationship'), *Smarika* (National Convention of Gurungs).

Gurung, H., Yogendra G. and Chhabi L. C. 2006. *Nepal Atlas of Language Group*, Kathmandu, NFDIN.

– 2006. *Nepal Atlas of Ethnic and Caste Groups*, Kathmandu, NFDIN.

Gurung, H. 2004. *Janajati Sherofero* ('Around Janajati'), Kathmandu, NEFIN.

Macfarlane, A. 1976. *Resources and Population: A Study of the Gurungs of Nepal*, Cambridge, Cambridge University Press.

Macfarlane, A. 1997. 'Identity and Change among the Gurungs (Tamu-mai) of Central Nepal', in David Gellner *et al.* (eds.), *Nationalism and Ethnicity in a Hindu Kingdom*, Amsterdam, Harwood Academic Publisher, pp. 185-204.

Pignède, B. 1993. *The Gurungs* (English edition), Kathmandu, Ratna Pushtak Bhandar.

Sharma, J. L. 1982. *Hamro Samaj Yak Adhyan* ('Our Society: A Study'), Kathmandu, Sajha Prakashan.

Tamu, D. J. 1997. *Tamu jatiko Parichaya* ('An Introduction to Tamu Ethnicity'), Kathmandu, Bhadra Kumari Ghale.

Text of Constitution and Political Report of Tamu Rashtriya Mukti Morcha. 2006.

Text of Tamu Dhin's 20 Point Demands. 2008 (May).

Notes

1. The author thanks the Social Inclusion Research Fund for providing him with a grant to carry out a study on the 'Interface between State and Ethnicity in Nepal'. This paper is part of that study and its initial outline was presented at a seminar on 'Territorial Changes and Territorial Restructuring in the Himalayas', organized by CNRS on 17-19 December 2007.
2. VDC: Village Development Committee.
3. DDC: District Development Committee

1.2.3.

Cultural Territoriality in the Context of Modern Nepal: Some Examples from the Adivasi/Janajati and Madhesi Groups

Dilli Ram Dahal

Introduction

Since the emergence of the idea of nation-state in the late nineteenth century (Nugent 1994), the concept of state building, self-ruled cultural territory and national consciousness has proved to be a paradox in many countries of the world. This was already the case of Nepal as soon as it emerged as a unified state in 1769. It has become even more pronounced since the onset of multi-party democracy in 1990, now that the state is depicted as i) unitary in nature (Hachhethu n.d.), and with ii) only a few groups (particularly the Brahmins and Chhetris) using much of the nation's resources and historically wielding political power (Bhattachan 1995, Gurung 2002, Lawati 2005). In recent years, the leaders of various ethnic/caste groups of Nepal, in resisting the power of the state, have clung to a myriad of oppositional local cultural identities in the name of Adivasi/Janajati, Madhesi, Khambhu, Kirati, and so on, claiming their separate cultural identity and territory, as if this cultural identity were primordial. Gurung (2003) also notes the ethnic/cultural events from 1779 to 1964 concerning various Adivasi/Janajati groups who competed for space within the state. Their demands today focus on the creation of 'autonomous regions', or separate cultural territories based on ethnic/caste principles, or a federal structure of government based on ethnic/caste cultural identity. In a word, the concept of 'cultural territory' with reference to particular ethnic/caste group is now a burning issue as far as the state is concerned, despite the fact that Nepal came into being as a multi-cultural nation in 1769.

The question of ethnic/caste based cultural territory was first proposed by the Nepal Sadbhawana Party (NSP) in 1999. They proposed five autonomous regions: Eastern Mountain and Hill, Western Mountain and Hill, Eastern Tarai, Western Tarai and Central (NSP 1999). Nepal Rastriya Jana Party (NRJP) stands for ethnic federalism and proposes 12 provinces: Khasan, Jadan, Magarat, Tamuan, Tamba Saling, Nepal, Khumbhuwan, Limbuwan, Kochila, Maithila, Bhojpuri and Awadhi (Tamang 2001: 416). Others demanding ethnic-based states with a right to self determination are Limbuwan Mukti Morcha and Khumbhuwan Mukti Morcha.

The cultural territorial issue has now spilt over into state politics with various cultural groups within Nepal claiming their traditional habitat as their own territorial state. Many NGOs/INGOs working in Nepal have adopted the same language, talking of the 'legitimate right of people'. This has become a 'potent area of political contestation'.

The public and academic discourse on this issue has intensified since the inclusion of the Maoists in the government peace process in 2006 and the promulgations of the Interim Constitution of Nepal in January 2007. However, the most important, yet unexamined, aspects of these ethnic/caste-based territorial claims which need to be taken into consideration are: i) the making of Nepal as a nation and culture of various groups of Nepalese people with their 200-year-old origins, history and population size, and ii) the migration patterns of various groups of people within the country and outside over the last 50 years. My argument here is that if our notions are guided by the ethnic/caste-based principles of cultural territory, we must attempt to understand the entire Nepalese culture in a historical and modern context. It is futile to discuss the concept of culturally based territory and its viability in the context of modern Nepal without understanding the formation of the Nepalese state over the last 240 years.

In this article, I argue that the concept of ethnic/caste-based cultural territory becomes pertinent only when we understand these diverse cultural groups of people living either in one place or spread over the whole country. My discussions and analysis on ethnic/caste-based cultural territoriality are largely confined to two broad cultural groups of people: the Adivasi/Janajati and Madhesi. This is simply because they are the most important cultural groups at this juncture in Nepalese democratic politics, when the concept of cultural territoriality is becoming a more and more redundant category in the changing social, economic and political Nepalese landscape, given the rapid process of urbanization and globalization.

This paper is organized into three broad sections: i) an overview of Nepalese culture highlighting the history and migration patterns of the various groups in question, ii) the basis and limits of ethnic/caste-based cultural territory in modern Nepal, and iii) discussions and conclusions.

1. A brief overview of Nepalese culture

Here, the political history of Nepal needs to be discussed briefly in order to understand the broad concept of the cultural territory of the various groups living there today. Though the history of Nepal is depicted as being more than 2,000 years old, little is known about the culture of the people and their territorial boundaries. There are three reasons for this: i) a lack of written history about the various groups of people living in a given area, ii) a lack of archaeological evidence to justify their existence and their cultures in a particular area, and, iii) a lack of anthropological and linguistic surveys to trace their history (either written or oral or other kinds of evidence) to identify their language and culture. For lack of scientific evidence, it is very difficult to substantiate even today who the real 'indigenous groups' or native vs. outsiders are in the context of modern Nepal (Bista 1990, Dahal 1993, Pradhan 1994).

Prior to the 'unification' of Nepal by King Prithivinarayan Shah in 1768 and its subsequent consolidation (Stiller 1973, Regmi 1976), the present-day western part of Nepal was divided into many kingdoms and principalities called *Baise* (22) and *Chaubise* (24). In the east, Nepal was divided into three major kingdoms: Bijaypur, Chaudandi and Makwanpur. Even the Kathmandu Valley was divided into three different kingdoms (Kathmandu, Patan and Bhaktapur) ruled by the Malla Rajas. Prior to the Sugauli Treaty (1816), the territory of Nepal was forever shifting and was only properly demarcated after this treaty. In other words, it took almost 75 years for the Shah rulers to territorialize Nepal as one nation (from the conquest of Nuwakot in 1741 to the Sugaluli Treaty in 1816). Even the four present-day districts of the western Tarai—Banke, Bardiya, Kailali and Kanchanpur—commonly known as 'Naya Muluk' (New Territory) were awarded to Prime Minister Jung Bahadur Rana as Birta by the British government when he helped the British to quell the 'Sepoy mutiny' in British India in 1857.

This lengthy, peaceful move towards the Hindu culture by various groups of people in Nepal was forcefully imposed as late as 1854, when the Old Legal Code Muluki Ain was implemented. This Code categorized the people of Nepal into four distinct orders, without considering the diversity of Nepalese society and culture:

- Tagadhari (castes wearing a sacred thread)
- Matwali (liquor consuming castes)
- Pani na calne choi chito halnu paparne (castes from whom water is not accepted but whose touch does not require purification by water)
- Pani na calne choi chito halnu parne (untouchable castes)

According to Sharma (1977), this code embodied certain distinct features: commensality, supremacy of Hindu values and religious orthodoxy, and caste as the basis of social mobility. This particular Hindu model of ruling the state according to a hierarchical structure created a certain uneasiness among the non-Hindu groups of Nepal. Likewise, the Khas language, which eventually became the official language of Nepal, Nepali, and lingua franca of most Nepalese people over the years created tension among various groups of people whose mother tongue was not Nepali.

It is a sorry state of affairs that up until 1990 there was little official information about how many ethnic/caste groups lived in Nepal. The 1991 census, following the establishment of democracy in 1990, provided information on 60 such groups. The 2001 Census revealed that there are 100 distinct ethnic/caste groups, 92 languages and 9 religious groups in Nepal. In other words, a larger societal compositional trend is essential to our understanding of the ethnic autonomy and cultural territory of a particular group.

A rapid overview of the cultural groups of Nepal will help us to place the current claims in their context. Broadly, five major cultural groups can be identified in Nepal (Dahal 1995): the caste-origin Hindu groups; the Newars; the Adibasi/ Janajati or nationalities; the Muslims; and the others (Marwaris, Bangalis, Sikhs, and so on).

1.1. The caste-origin Hindu group is further divided into:

- The caste-origin Hill Hindus: the 2001 census recorded only nine groups in the caste-origin Hill Hindu groups (including five Hill origin Dalit groups). The social structure of caste-origin Hill Hindu groups is simple; there are not four Varna (lacking Vaisya) in this category and they only have one language (Nepali as mother tongue) and one racial group (Caucasoid). Excluding the Hill Dalits, this group alone comprises 30 per cent of the total population of Nepal (Brahmins, Chetris and Thakuris) (Central Bureau of Statistics: CBS, 2002). The population of Hill Dalit groups is 7.0 per cent of the total population (2001 census).
- The caste-origin Tarai Hindus: the social structure of the caste-origin Tarai Hindu groups is more complex; there are four Varna groups with a distinct hierarchical structure within them. The 2001 census recorded 43 distinct caste-origin Tarai Hindu groups (including 11 Tarai Dalit groups). These various cultural groups belong to four distinct language groups: Maithili, Bajika, Bhojpuri and Awadhi but belong to a single racial category, the Caucasoid. Without the Tarai Dalits, the population of these various groups is 15.1 per cent of the total population. The total population size of the 11 Tarai Dalit groups is 3.8 per cent.

The Dalit population (both Hill and Tarai Dalits) comprises 11.6 per cent of the total Nepalese population (2001 census).

1.2. The Newar cultural group presents the most complicated social structure among all groups in Nepal, truly reflecting the model of four Hindu Varna categories and is divided clearly into two distinct religious (Hindu and Buddhist) and racial (Caucasoid and Mongoloid) groups. King Jayasthiti Malla (1380-1394 A.D) had classified the population of the Kathmandu Valley into 64 castes with different functional and occupational categories. There are still said to be about forty Buddhist and Hindu castes easily identifiable within the Newars, and they share a single common denominator, Newari as their mother tongue. Today, even if a large number of Newars find this classification problematical, the Newars have been scheduled by the Adivasi Organizations under the Adivasi/Janajati category and their population makes up 5.6 per cent of the total population (2001 census), including the Newar Dalits.

1.3. The Adivasi/Janajati category is described in great detail in the following section.

1.4. The Muslim group is divided into three types: the Tarai Muslims, the Hill Muslims or Churautes and the Kashmiri Muslims living in the Kathmandu Valley (Bista 1967). In fact, the Churautes are a small group and the Kashmiri Muslims have not been recorded separately, so the major Muslim group under discussion is the Tarai Muslim group. The population size of Tarai Muslims is 4.3 per cent of the total population.

1.5. Other groups such as Marwaris (including Jains), Sikhs, and Bangalis, live in small numbers in different parts of Nepal, but mostly in the Nepal Tarai and Kathmandu. Each of these groups has its own language, culture and religion with a distinct hierarchical structure.

At the interface between Hindu and Buddhist civilizations (Fisher 1978), made up of two distinct racial groups—Caucasoid (Indo–Aryan) and Mongoloid[1]—and including various cultural groups, Nepalese culture is an ethnic melting pot, "an ethnic turn-table" as Tony Hagen (1971) puts it.

2. Basis and limits of cultural territorialism

Now the basis and limits of cultural territorialism or of autonomous cultural regions as demanded by various ethnic/caste groups of Nepal will be discussed briefly. The examples here are cited from the Adivasi/Janajati and Madhesi groups. The basis and limits of an ethnic/caste based territorial structure are analyzed considering three factors: cultural traits (population size and distribution, language and religion), history and migration patterns.

2.1. The Adivasi/Janajati group

Today, various cultural groups in the broad Adivasi/Janajati cultural category are demanding their own autonomous region within the federal structure of government such as: Tharuhat (by the Tharus); Kochila (by the Meches, Koches and Rajbansis); Tamuwan (by the Gurungs); Magarat (by the Magars); Limbuwan (by the Limbus); Tambasaling (by the Tamangs); and Khumbuwan (by the Rais). Within the Adivasi/ Janajati category, the Newars have also been demanding their own separate autonomous cultural region (Newa Rajya), and leaders of this group have reiterated their demands time and time again for the Newari language to be adopted as the official language.

Demands for a caste-based autonomous cultural region are made through two channels: the Federation of Indigenous Nationalities, and Leaders of the respective cultural group.

Before discussing the viable separate cultural territory among various cultural groups within the Adivasi/Janajati category, it would be worthwhile providing some knowledge about these groups in the context of modern Nepal.

According to Pari Thapa, a Member of Parliament from the Samukta Janamorcha Party, the term Adivasi/Janajati was only coined in 1986 at a meeting of Janajati people. Nevertheless, some of these groups have been living in Nepal for centuries and call themselves the "First Settlers" or the "Sons of the Soil".

Others, however, have a more recent past; they came to Nepal either from India or Tibet over the last 150-400 years. Considering their relatively low social, economic and political status, the Government of Nepal formed a task force in 1996 for the identification and bettering of these various groups. According to the newly enacted National Foundation for Development of Indigenous Nationalities Act, 2002 (Adivasi-Janajati Utthan Rastriya Prathisthan, 2058 B.S. or 2002), the government provided a definition of Adivasi/Janajati and identified 59 groups within this broad cultural category. The Foundation also identified them in different ecological regions: 18 groups from the Mountains, 23 from the Hills, 7 from the Inner Tarai and 11 from the Tarai. The listed number of Adivasi/Janajti groups has been further recorded as 81 by a Technical Committee recently set up by the Government of Nepal (2009). The definition of Adivasi/Janajati is as follows:

> Indigenous nationalities' means a tribe or community as mentioned in the schedule, having its own language and traditional rites and customs, distinct cultural identity, distinct social structure and written or unwritten history" (National Foundation for Development of Indigenous Nationalities: NFDIN 2003).

But Adivasi/Janajati per se is not a homogeneous group and their heterogeneity extends to language, religion and culture. Their heterogeneity can be better explained through two broad regional groups: the Hill Adivasi/Janajati (consisting of 49 groups) and the Tarai Adivasi/Janajati (consisting of 10 groups).

Of the 59 listed Janajati groups, CBS has provided data on the population of only 44 groups (including the Newars and Mundas; in fact, the Munda group was not listed as Janajati) and therefore the size of the population of 16 (44 + 16 = 60, and Munda is added here as a separate Janajati category) other groups (12 from the Mountains, 3 from the Hills and 1 from the Tarai) is not yet known. This is simply because many of these groups could not be identified as a separate cultural group in the regions where they have been listed.

The population of all the identified Janajati groups in the 2001 census was 8,473,429 including 2,030,851 Tarai Janajatis (24 per cent of the total Janajati population). If this figure is added to that of the unidentified Janajati group population (the unidentified Janajati numbered 5,259), the total size of the population of Adivasi/Janajati groups in Nepal, according to the 2001 census, amounts to 8,478,688 or 36.6 per cent of the total population. Among them, the Magars, Tamangs, Newars, Rais, Gurungs and Limbus are the largest groups in terms of population size. All together, these six groups comprise 65.5 per cent (total pop. 4,221,387) of the total Hill Janajati population. Similarly, the Tharus alone comprise 58.8 per cent of the Tarai Janajati population. In other words, 63.9 per cent of the total Janajati population include these seven groups: Magars, Tharus, Tamangs, Newars, Rais, Gurungs and Limbus. On the other hand, there are 23 very small Janajati groups whose population is less than 0.1 per cent of the total population of Nepal. Some of these groups are: Kusunda (164), Yehlmo (579), Raute (658) and Munda (660). These groups may find themselves on the verge of extinction if their culture is not suitably protected.

Under this broad Adivasi/Janajati banner, only two groups that are claiming traditional cultural territories are under review here. They are the Limbus of Limbuwan (Hill Janajati) and the Tharus of Tharuhat (Tarai Janajati).

2.1.1. Limbuwan

The Limbus of today are thought to be the original settlers of Limbuwan. According to Chemjong (1967), Limbus are the descendents of the ancient Kirantis (Chemjong 1967). But what is perceived as the cultural groups called Kiranti is rather vague and blurred in the context of Nepal (Chatterji 1952; Chemjong 1967; Dahal 1975). This traditional Kiranti land was divided into three regions during the Rana rule for administrative purposes: Wallo Kirant (or Near Kirant, or an area where mostly Sunuwars, Rais, Sherpas have settled today), Manjh Kirant (between the area of Dudh Koshi and the Arun River, and an area where Rais and other groups have settled) and Pallo Kirant or Far Kirant (east of the Arun River or Limbuwan, where mainly Limbus and other groups have settled).While referring to the Hodgson collection, Bairaji Kaila (2006) notes the four boundaries of the original Limbuwan as follows: east of the Arun River up to Balumsomgola, west up to the Mechi River, from the south of Tokpegola and the north of Chihurama. It is hard to demarcate this area today as Limbuwan or the traditional Kiranti land as the land of the Limbus.

The first historical record concerning the Limbus dates from around the eighteenth century, by which time a number of small chiefdoms in Limbuwan were governed by the kingdom of Bijaypur ruled by the Sen dynasty of Makwanpur (Caplan 1970: 14). But with no document to substantiate this, Chemjong (1967) argues that there were a number of Limbuwan chiefdoms which were not subservient to Hindu kings. So far, little is known about how many Limbu chiefdoms there were within Limbuwan, and the kind of economic and political relations one Limbu chiefdom maintained with other Limbu chiefdoms.

Present-day Limbuwan includes six districts of east Nepal which are: Sankhuwasabha, Terathum, Dhankuta (Koshi zone), Taplejung, Pachthar and Ilam (Mechi zone). The post-2005 ethnic/caste uprising in Nepal further motivated the Limbus to extend their traditional territory to Jhapa, Morang, Sunsari and Siraha districts of the Nepalese Tarai, where a large number of Hill Brahmins, Chhetris and Tarai groups such as the Yadavs, Dhimals, and Rajbanshis live today. In fact, the two districts of eastern Nepal, Tarai-Jhapa and Morang, have already been recognized as 'Kochila' or the land of Rajbanshis and Tajpuriayas. On the other hand, some of the Madhesi leaders in the Tarai have started claiming all 20 districts of the Tarai (including Jhapa and Morang) as the Madhesi people's homeland or territory, and thus the presence of Hill people in the Tarai is allegedly no longer tolerated.

Taking into consideration the present ethnic/caste composition of the traditional Limbuwan area, the four major ethnic/caste groups with the highest number living in Limbuwan today are as follows (Table 1).

No doubt, of the total 359,379 Limbu population in Nepal, 333,633 Limbus (92.8%) speak Limbu as their mother tongue and more than 90 per cent of Limbus are followers of the Kirat religion (CBS 2001). But in the so-called 'Limbuwan', the Limbu population is the highest in three districts only: Taplejung and Panchthar (Mechi zone) and Terathum (Koshi Zone). In terms of population size, there are more Brahmins/Chhetris (26.7%), followed by the Limbus (23.5%) and Rais (16.3%). The Limbus are not even the fourth largest group in Sankhwasabha district. Though the Hill Brahmins and Chhetris are recorded separately, they rank either second or third in all these traditional Limbuwan districts. In addition, the Sherpas rank fourth in the size of population in Taplejung, the Magars rank fourth in Dhankuta and the Tamangs rank fourth in terms of population size in both Terathum and Sankhuasabha districts. Rai is another significant population group in Limbuwan today. But the Rais have already claimed a separate territory for themselves as 'Khumbuwan' adjoining the districts of Limbuwan or the west of the Arun River. Though the Rais and Limbus look physically similar, they were never historically united to fight against the aggression of the Gorkhali rulers. Culturally, the Rais are not a homogenous group and are divided into more than 20 distinct sub-groups with different languages and cultures. Many Rais do not even want to label themselves Rai today and call themselves a separate cultural group such as Kulung, Thulung, Sampang, Lhorung, Bantawa and so on.

Table 1: Population size of four major ethnic/caste groups living in the traditional Limbuwan, 2001 census

Districts	Total Pop.	Limbu	Rai	Brahmin Hill	Chhetri	Tamang	Magar	Sherpa
Sankhu wasbha	159,203	7,584 (4.8%)	37,725 (23.7%)	10,476 (6.6%)	30,931 (19.4%)	15,048 (9.5%)	–	–
Terathum	113,111	40,020 (35.4%)	–	17,771 (15.7%)	21,506 (19%)	6,548 (5.8%)	–	–
Dhankuta	166,479	22,849 (13.7%)	38,257 (23%)	(10,647)	33,921 (204%)	– –	16,165 (9.7%)	– –
Taplejung	134,698	56,324 (41.8%)	–	13,974 (10.4%)	15,982 (11.9%)	–	–	12,585 (9.3%)
Panchthar	202,056	81,403 (40.3%)	28,157 (13.9%)	25,304 (12.5%)	21,520 (10.7%)	–	–	–
Ilam	282,806	40,524 (14.3%)	68,901 (24.4%)	42,805 (15.1%)	38,320 (13.5%)	–	–	–
Total	1,058,353	248,704 (23.5%)	173,040 (16.3%)	120,977 (11.4%)	162,180 (15.3%)	–	–	–

Source: CBS, 2002

In addition, a large number of Limbus who bettered their lives by joining the British and the Indian armies are gradually emigrating from their traditional homeland, preferably to urban areas. The CBS (2001) recorded a significant number of Limbus in municipalities such as Damak, Dharan and Kathmandu.

The languages spoken as the mother tongue in Limbuwan are as follows (Table 2).

In terms of their mother tongue, 45.2 per cent of the population speak Nepali as their mother tongue, followed by Limbu (22.5%) and Rai (15.7%).

The number of people with different religious faiths in Limbuwan is given in Table 3.

In terms of religion, 44.1 per cent of the population is Hindu, followed by Kirati (39.1%), Buddhist (15.8%) and others.

To sum up, Limbuwan is inhabited by several cultural groups among which Limbus are not the majority. The issue here is: how are we to demarcate a traditional Limbuwan area as an Autonomous Cultural Region or territory in the name of the Limbus in the midst of Rais, Hill Brahmins, Chhetris, Magars, Tamangs and others?

Table 2: Major languages (mother tongue) of people in the traditional Limbuwan, 2001 census

Districts	Total Pop.	Language (No. of speakers and %)			
		Nepali	Limbu	Rai	Others
Sankhuwasbha	159,203	73,970	7,864	38,071	39,298
Terathum	113,111	59,838	37,407	2,095	13,771
Dhankuta	166,479	74,705	22,893	39,443	29,438
Taplejung	134,698	49,833	53,691	6,214	24,960
Panchthar	202,056	67,556	79,949	26,976	27,575
Ilam	282,806	152,804	35,868	53,340	40,794
Total	1,058,353	478,706 (45.2%)	237,672 (22.5%)	166,139 (15.7%)	175,836 (16.6%)

Source: CBS, 2002

Table 3: Number of people with different religious faiths in Limbuwan, 2001 census

Districts	Total Pop.	Religion (No. of followers and %)			
		Hindu	Buddhist	Kirati	Others
Sankhuwasbha	159,203	74,745 (46,9%)	39,451 (24,8%)	43,853 (27,5%)	1,154 (0,7%)
Terathum	113,111	57,994 (51,3%)	12,836 (11,3%)	41,367 (36,6%)	914 (0,8%)
Dhankuta	166,479	82,442 (49,5%)	22,875 (13,7%)	59,843 (35,9%)	1,319 (0,8%)
Taplejung	134,698	49,186 (36,5%)	25,159 (18,7%)	58,556 (43,5%)	1,797 (1,3%)
Panchthar	202,056	69,112 (34,2%)	22,894 (11,3%)	108,104 (53,5%)	1,946 (1%)
Ilam	282,806	133,718 (47,3%)	44,354 (15,7%)	102,302 (36,2%)	2,432 (0,8%)
Total	1,058,353	467,197 (44.1%)	167,569 (15.8%)	414,025 (39.1%)	9,562 (0.9%)

Source: CBS, 2002

2.1.2: Tharuhat

Another example here is the Tharu community that claims that their traditional habitat should be recognized as Tharuhat. The Tharus of Nepal have claimed two types of cultural identity in recent years: the Adivasi/Janajati group of the Tarai and Madhese in the name of the pan-Madhese cultural identity.

It is worthwhile noting here that the Tharus are the only pan-Tarai group that spread from the far-west (Kanchanpur district) to the far-east (Morang district) along the Tarai belt and are a very large group (11% among the total Tarai population as a whole and almost 20% among the so-called Madhesi population). In the Adivasi/ Janajati category, they are the second largest group in terms of population size (1,533,879 or 6.8% of Nepal's population).

The history of the Tharus living in Nepal is not very clear. The oral history of the Tharus indicates that they came to settle in the Nepal Tarai from Rajputana, India, a long time ago. The way Tharu women dress, adorn themselves and treat their husbands reveals their Rajput ancestry (Rajaure 1977). Some even claim that they are descendents of the Sakya dynasty, a progeny of Lord Buddha (Singh 1999). The other interesting fact about the Tharus is the Tharu language which comes under the family of Indo–Aryan languages (Subba 1977). As the Tharu population spreads from east to west, their culture and language differ from the eastern to the western region of Nepal. The Tharu group encompasses: Rana Tharus (Kanchanpur and Kailali districts), Dangaura Tharus (Dang and Surkhet districts), Saptariya Tharus (Saptari district), Chitauniya Tharus (Chitwan district) and Morangiya Tharus (Morang district). The Tharu language itself is heavily influenced by the local culture. For example, the Tharu language in the far-west closely resembles the Awadhi language, whereas the eastern Tharu language is more influenced by the Maithili language. In addition, in terms of culture, the Dangaura Tharus raise pigs and eat pork, whereas the Saptariya Tharus do not (Dahal and Mishra 1993).

Despite their distribution over the Tarai region as a whole, Tharus are the biggest in number in only four districts of the Tarai: Sunsari in the eastern Tarai, Dang in the mid- western Tarai and Bardiya and Kailali in the far-western Tarai. In a word, it is very difficult to demarcate a territory for the Tharus which can be marked as Tharuhat Autonomous Cultural Region within the Tarai.

2.2. The Madhesi Cultural Groups

Conceptually as well as empirically, the term "Madhesi" is vague as it connotes different meanings to different researchers and people. In general, the term is understood as: a) inhabitants of the Tarai region of Nepal, showing a geographical identity of a population as a whole (Gaige 1975, Dahal 1978, Jha 1993); b) a term used by Hill people (*Pahade*) to address the Tarai people or the Plains people (*Madhese)* to mark their physical and cultural identity (such as those with a dark-brown complexion or those wearing *dhotis* (Dahal1978); and c) a term meaning historically the people of Madhya Desha, showing a close affinity and connection with people living in northern India, particularly in Bihar and Uttar Pradesh (Bista

1967, Chowdhary 2007). The use/misuse of this term has been further complicated because there are few historical or any other kind of written document to show the historicity of the Tarai people in general and of the Madhese group in particular.

Using generally accepted knowledge about the Tarai that the hill people were not the original inhabitants (no doubt many of these hill people were absentee landlords during the Rana regime) and considering the overall present settlement of people in the Tarai, the so-called 'Madhesi people' of the Nepalese Tarai today can be broadly categorized into four distinct cultural groups: i) a group of people who have been living in the Nepal Tarai for centuries, and who prefer to call themselves 'indigenous' (though many groups have no written history at all) of the Tarai. Such groups are: Tharu, Dhimal, Gangain, Jhangad, Danuwar, Koche, Meche, Rajbansi and a few others; ii) The Tarai caste groups of Hindu caste origin, a broad cultural group with a social structure whereby the Maithil Brahmin group places itself at the top and untouchables such as the Doms at the bottom; iii) the Indian-origin Nepali merchant groups such as the Marwari, Bengali and Sikhs and; iv) the Muslims.

Now, considering their origin, language, culture and history, all these four groups make up a diverse society, representing hardly one closed Madhese cultural group or the primordial group as such, except for their geographical identity as plain dwellers. A summary of some of their distinct socio-economic and cultural traits given in Table 4 amply testifies to this.

Historically, little is known about the human settlement of the Nepalese Tarai. Up until 1960, the whole region of the Nepalese Tarai was called *kala pani* (black waters) or 'death valley' by Hill people. Settlement in the area was avoided by Hill people because of the black waters, particularly the high incidence of malaria. Up until 1960, 60 per cent of the land in the Tarai was covered by forests, and the central and eastern Tarai was almost completely forested (Gurung 1974).

As to the question of who is and who is not an indigenous group in the Tarai, the history of the Rajbansi, Koche and Meche shows that they are linked to the Koch kingdom in India. Their history in the eastern Tarai goes back 300-400 years (Prasai 1988, Shrestha 2007); they not only wear the sacred thread but also observe Hindu festivals and rituals and celebrate the Ramayana and Mahabharata, the greatest Hindu epics in their plays and songs. The other interesting feature is the language of the Rajbansi and Tharu groups which fall within the Indo-Aryan language family (Subba 1977). The Satar people came to settle in the eastern Tarai districts from Santhal Pragna, in India, to work as farm labourers around 100-150 years ago (Dahal 1994). So the question as to who are the oldest settlers in the Nepalese Tarai is hard to prove even for ethnic groups such as the Tharus, Rajbansis and Satars.

The caste structure of the Tarai Hindus is more complex and diverse compared to the Parbatiya Hindus. In the Tarai Hindu caste model, there are well over 20 groups in the Vaisya category alone, each with a distinct hierarchy, whereas there is no Vaisya category in the Parbate model. The Tarai Hindu caste groups are primarily predominant in Dhanusa, Mahottari, Siraha, Sarlahi, Saptari, Bara, Parsa, Rautahat and Kapilbastu districts. Again from a linguistic viewpoint, the area of the Hindu

Table 4: Some distinct socioeconomic and cultural traits of the Madhesi (Tarai) people

Distinct Traits	*Ethnic/Tribal Group*	*Caste origin—Hindu groups*	*Indian-origin Nepali Merchant groups*	*Muslim*
Tentative history of settlement in the Tarai	Claims to have settled in the Tarai for centuries; indigenous groups?	Settlement history over the last 100-200 years.	For the last 150 years.	For the last 100-200 years.*
Physical feature	Primarily Mongoloid	Primarily Caucasoid	Primary Caucasoid	Primarily Caucasoid
Language	In general, each group has its own language. Except for the Tharus, and Rajbansis who have a Indo-Aryan language, the others belong to Tibeto-Burman language group.	Indo-Aryan language group. Three dominant language groups: Maithili, Bhojpuri and Awadhi; also Bajika. Hinduism of the great tra dition; always a Brahmin	Indo-Aryan language group; Marwari, Sikh and Bengali.	Urdu-Hindi.
Religion	Animism, local Hinduism or folk religion. They have their own priest for rituals and other religious activities.	priest for a ritual except for 'untouchables' who call upon their own priest.	Hinduism and Sikhism.	Islam.
Social Structure	Clan organization; no hierarchy within the group, bride price or bride service; sexually more re-laxed than the Hindus; age at marriage higher than for Hindu groups; remarriage permitted for both sexes; women have a	Caste organization; distinct hierarchy among various groups; dowry; sexually less permissive society; remarriage of women not permitted; women have relatively lower status and less autonomy	Same as the caste-origin Hindu group.	Clan-like organization; no social hierarchy, women have l ower status than men, divorce is easy.

	relatively higher status and more autonomy than Hindu women.	than the ethnic/tribal group.		
Economy	Relatively poor group, except for a few Tharu and Rajbanshi landlords; bonded labour is present among the Tharus; primarily agriculturists.	Relatively better in terms of economy except for the 'untouchables' who are poorest; untouchables are mostly landless; economy: a griculture, service, trade, occupational workers and labourers.	Rich group in the Tarai, many of them are merchants; little agriculture and service as such by groups.	Relatively poor group in the Tarai; owns little land, most of them are specialised workers.
Politics	Except for Tharus and Rajbansis, few other groups engage in politics; political culture has yet to emerge.	Most sensitive group in the political sphere except for the untouchables; politics in the Tarai is the realm of a few high-caste members of this group.	Not involved in current Nepalese political debate.	Few are engaged in politics; political culture is gradually emerging.

* The settlement history of the Kashmiri Muslims in the Kathmandu Valley goes back as far as the sixteenth and seventeenth centuries (Bista 1967).

caste groups can be distinctly demarcated into three geographical belts within the Tarai: the Maithili belt (eastern Tarai districts such as Dhanusa, Saptari, Siraha, Sarlahi, and Rautahat), the Bhojpuri belt (Bara and Parsa) and the Awadhi belt (the districts such as Kapilbastu, Banke and Bardiya, west of the Narayani River).

The merchant group, such as the Marwari, has its own complex social structure including the Brahmin, Rajput, Vaisysa and Dalit categories. Their arrival in the Nepalese Tarai is relatively recent and coincided with the development of local market centres.

As the road transportation network improved throughout the kingdom, there was increased mobility and settlement among the Sikh populations in Nepal. The history of the Muslims in the Nepalese Tarai goes hand in hand with the other Hindu caste groups of the Tarai, who arrived in Nepal as labourers and construction workers; they came mostly from Bihar and Uttar Pradesh, India, with some arriving from Bangladesh after 1971 (Dahal 1978).

'Madhesi culture' as a whole is the most heterogeneous in terms of hierarchy and social structure compared to the 'hill people's culture'. Based on the 1991 census data, Subedi (2002) noted 14 districts of the Tarai (excluding Jhapa, Morang, Chitwan, Rupendehi, Kanchanpur, Bardiya) which are the most heterogeneous in terms of ethnic/caste composition. Based on the 2001 census, Dahal (2003) also noted similar findings. The total percentage of the four most represented groups (largest in number) in some of the Tarai districts hardly accounts for more than 40 per cent, for example in Sarlahi (36%), Dhanusa (37.3%), Morang (37.2%), Parsa (38.2%) and Bara (40.1%), suggesting that they are the most heterogeneous districts in terms of ethnic/caste composition. In fact, among the Hindu Tarai caste groups Yadavs are the majority in some eastern and central Tarai districts only, such as Saptari, Sirha, Dhanusa, Mahottari and Sarlahi. Yet, despite there being so much caste diversity within the region, Yadavs are the major political actors in the Tarai.

One other reason why it is difficult to declare Madhesh an Autonomous Cultural Region is the migration of hill people to the Tarai over the last 50 years.

For the Tarai region as a whole, there was no concept of land shortage until the mid 1970s. With the abolition of all forms of land grant systems and contractual arrangements that had contributed to the expansion of cultivated land, the state itself undertook the task of further expanding cultivated land in the region. The implementation of the Rapti Valley Development Programme in 1956 was an inaugural effort in this direction. In 1964, the Resettlement Company was established as a separate organization for the systematic implementation of resettlement programmes. By 1977, the Company had implemented ten major programmes in different districts of the Tarai region: Jhapa, Chitwan, Nawalparasi, Bardiya, Kailali, and Kanchanpur (Elder *et al.* 1976).

In the initial stages, the purpose of launching the programme was to resettle the landless from the hill region, victims of floods and other natural disasters. Later on, the programme was used to settle Nepalese people who were forced to return from Burma and Assam because of nationalistic movements in these areas. Retired army personnel were also accommodated, especially along the frontier zones with India, as a defence measure to protect the territory in the open border area. These developments were accompanied by malaria eradication as well as the construction of highways and roads to improve transportation in the region. All these events made the Tarai an attractive destination for the landless and marginal farmers of the hill region. During the 1960s, 92.2 per cent of the country's total inter-regional migrants selected the Tarai as their destination. Corresponding figures in the 1980s and 1990s were 77.9 per cent and 82.9 per cent, respectively. However, the flow of migrants to the Tarai gradually declined and only 76.8 per cent of inter-regional migration was to the Tarai according to the 2001 census (K.C. 2003: 133-4). This is because the hill peoples have found other better alternative areas for temporary or permanent labour migration (such as the Middle-East countries, Kathmandu and elsewhere). In addition, the Land Reform Programme introduced in 1964 forced a large number of absentee Hill landlords to settle in various parts of the Tarai. Finally, the Tarai became a prominent region with the growing number of Indian migrants who came to do business and to run industries, as it was easy to trade goods and services from the Tarai to India (Gaige 1975).

The influx of the Hill population has been a significant demographic event over the past few decades. Several towns have developed where hill meets plain, though with a majority of Hill people. These include, from west to east, towns such as Mahendranagar, Tribhuvan Nagar, Butwal, Bharatpur, Hetauda, Dharan, and Damak. The western districts of Kanchanpur, Kailali, Bardiya, Banke, and Dang Deokhuri are 30 to 50 per cent Tharu and the Hill people make up the second largest group. Rupendehi is made up of 60 to 70 per cent Hill people and half of the population of Nawalparasi district consists of Hill people. In Chitwan and Jhapa districts, 70 to 80 per cent of the population is of Hill origin.

In short, the percentage of Hill people in the Tarai, which was close to 33 per cent as recently as in the 1991 census, increased to almost 37 per cent in the 2001 census.

In a word, today the Tarai has become very mixed in terms of ethnic/caste composition. The greater ethnic diversity in the Tarai has meant that the term 'Madhesi' has become increasingly contested in recent years (Dahal 2007).

Another complicated issue is the definition of Madhesi and Madhesh as a cultural region within the Tarai. There is great confusion between Madhesi as a cultural group and Madhesh as a geographical region or a separate cultural territory in the Tarai. It is still not known whether Madhesh represents all 20 districts of the Tarai or only a few Tarai districts. Tharu leaders and intellectuals have already declared that they do not like to call themselves 'Madhesi' (Chowdhary 2007). Similarly, the Rajbansis, Koches and Meches have already proposed their traditional habitat (Jhapa and Morang districts) as the Koch Autonomous Region. The other dominant group living in the Tarai is the Muslims. The highest number of Muslims is to be found in five districts of the Tarai: Rautahat, Bara, Parsa (central Tarai), Kapilbastu (western Tarai) and Banke (mid-western Tarai). Muslims do not compromise with other groups in terms of their religious values. Up until today, little is known about how the Hindu Madhesi group of the Tarai and the Muslims manage to live together in harmony within the single Hindu Madhesi cultural identity.

If one considers the recent Adivasi/Janajati slogan "rights to self-determination of people", the issue becomes further complicated when conceiving a territory which might be declared 'Madhesh' because of the settlement of Tarai Adivasi Janajati groups such as the Tharus, Rajbashis, Dhimals and others. Moreover, if nine districts of the Tarai such as Siraha, Saptari, Mahottari, Dhanusa, Sarlahi, Bara, Parsa, Rautahat and Kapilbastu are considered to be 'Madhesh', is the large Muslim population in these districts willing to be governed as if part of the Hindu Madhesi culture? Similarly, the large hill population in the different districts of the Tarai over the last 50 years is another major challenge to declaring 'Madhesh' a single cultural territory.

In short, it is hard to demarcate a separate cultural territory which can be branded 'Madhesh' within the Tarai in the name of pan-Madhesi cultural identity.

3. Discussion and conclusion

In considering the changing demographic factors and the process of modernization and globalization in Nepal, the section below briefly discusses why the ethnic/caste-based model of a separate cultural territory or of an autonomous region is not a viable model for Nepal.

3.1. Changing ethnic/caste landscape and the issue of cultural territory

Amidst the diversity of the various groups, the 1991 and 2001 censuses of Nepal clearly show that ten groups (Hill Chhetri and Brahmin, Magar, Tharu, Tamang, Newar, Kami, Yadav, Muslim and Rai Kiranti) represent close to 70 per cent of the total population. The population of Hill Brahmins and Chhetris is spread over the country in such a way that their presence can be felt in every Village Development Committee, district, and in the nation as a whole. The literacy rate of these groups is very high, constituting 62.2 per cent of highly educated people, with 66.2 per cent of them making up the government elite (2001 census). In brief, whether the system remains a separate cultural territory or ethnic/caste-based model of federal structure, there will forever be far more hill Brahmins and Chhetris qualifying as teachers, administrators, leaders of political parties, politicians, planners or advisors than any other groups unless strict measures are adopted to include the latter in the overall system. Thus, high-caste hegemony will definitely persist in politics and economics in any form of future government.

The other complexity in the ethnic/caste-based territory concerns the 32 smallest ethnic/caste groups whose population altogether accounts for just 0.71 per cent of Nepal's total population. There are ten distinct groups whose population size is less than 0.1 per cent of the population. Little is known about how these marginalized ethnic/caste groups, who are from various linguistic, ethnic, cultural and linguistic backgrounds, will participate in the ethnic-based sub-cultural category of federalism. Many of these groups came into existence mutually respecting each other's culture. Even in the culturally based territorial region, there is always the risk that the small elite of a particular cultural group (such as the Thakalis) will appropriate the political sphere while the others will be deprived of what they deserve. So far, no scholar has clearly analyzed inequality and conflict within the Adivasi/Janajati or Madhesi group. The internal conflict and the underdevelopment of the Madhesi group as a whole are only partly related to domination by the hill population; the major problems linked to the Madhesi groups' underdevelopment stem from the structural constraints within their system such as landlordism, untouchability, dowry, oppression and exclusion of women, dacoity, and so on (Ramesor Sha Teli, Kantipur, 13 March 2007).

No doubt, true participation and inclusion of all groups will strengthen the democratic process and development of the nation. The discontent of those who are excluded will turn into resentment and cause fresh conflicts. In the Tarai itself, beyond the established political parties in the region, new faces of leadership with new political associations are emerging for the sake of Madhese solidarity such as the Madhesi Cobra, Madhesi Tiger, Madhesi Janatantrik Morcha and so on. The question here is how to unite these different factional groups as one force behind the common goal of the Madhesi people.

To sum up, Nepal needs to take the path from segmentation to integration, fully united within diversity and pluralism.

3.2. Migration and Urbanization

The ethnic/caste-based concept of cultural territoriality loses some of its force if we take into consideration the larger process of migration and urbanization that has taken place in Nepal over the last 50 years.

Migration has played a key role in settling people from one area to another over the last 50 years. They have generally moved for reasons of livelihood, employment and access to other better facilities (education, health, transportation and communication services). The substantial rate of migration by hill peoples to the Nepalese Tarai over the last 50 years is one of the major demographic factors changing the social landscape of the Tarai. In recent years, international migration of young people has also risen, resulting in an increase in income, which is used to purchase land and houses in the urban areas of Nepal. Urban growth has been very rapid in Nepal over recent years. The internal conflict due to the Maoist insurgency over the last decade also forced many people to migrate, particularly to urban areas of Nepal, for their own safety and for economic reasons. The trend to migrate to urban areas in Nepal accounted for 16.3 per cent of migrants in 1981; this increased to 17.2 per cent in 1991, then further increased to 24 per cent in 2001 (K.C. 2003). The total urban population stood at 3.2 million. By 2001 the proportion living in urban areas had reached 13.9 per cent, while the number of towns increased from 10, during the 1950s, to 58. There has been obvious constant growth since the 1950s in the proportion of the population living in urban areas. It is expected that by the year 2027, the urban population in Nepal will be close to 11.35 million or 31.3 per cent of the total population (Subedi 2006).

The process of urbanization is considered vital for the development of the nation (Joshi 2006). As there are better facilities in urban areas, more and more people have chosen to live in such areas. For example, Tamangs constitute 8.5 per cent of the population in the Kathmandu district (Newar 29.6%, Brahmin 20.5%, Chhetri 18.5%). In Damak, of a total population of 35,009, there are Hill Brahmin (23%), Newar (6.3%), Rai (6.5%), Limbu (4.9%), and other groups (59,3%). In Biratnagar, of a total population of 166,674, there are Hill Brahmin (17.5%), Chhetri (9.6%), Muslim (9.2%), Marwari (3.9%), Tarai Brahmin (3.8%), and other groups (56%). In Dharan, of a total population of 95,322, there are Rai (19.2%), Newar (14.2%), Limbu (10.8%), Chhetri (10.1%) and other groups (45.7%). In Rajbiraj, of a total population of 30,353, there are Yadav (14.2%), Hill Brahmin (11.2%), Muslim (9.8), Baniya (7.9%) and other groups (56.9%).

Even in Rajbiraj municipality, district headquarters of Siraha district and the heart of Madhesi culture and population, the proportion of Hill Brahmins is the second highest. Likewise, the Rai and Limbu population in Dharan and Damak municipalities (eastern Tarai towns) is significant. Young educated people tend to move to urban areas for multiple reasons. In other words, urbanization leads to the development of people and it is the means of ensuring spatial and cultural integration for better planning. The basic assumption that people will remain static or stay in one place or that culture does not change is a short-sighted way of viewing the future of Nepal and of the Nepalese people. This means that the concept of traditional territory will gradually lose its meaning in the process of urbanization over the years to come.

As Subedi has rightly pointed out (2002: 112-117), the "Nepalese social landscape is the diversification where several ethnic and caste groups live side by side in the district with or without an identifiable cluster of settlement."

While for various reasons the process of migration and urbanization increases in Nepal, the ethnic/caste-based model of the political structure is forever weakening. If money does not go to the poor, the poor move to where the money is (Linder 1994). Employment will play a major role in shaping the future of the nation and the political structure of the Nepalese government.

Culture is a means, not an end in itself. Cultural identity is always constructed over time and space; it does not remain static, and changes in culture take place due to the economic, social and political conditions of the local area, region, country, and of the process of globalization as a whole. It does not remain eternal or primordial in nature. In other words, ethnicity is not a primordial attachment that stems from the given social system, but something that can be created and recreated by the elite to suit certain economic and political circumstances in the country.

To conclude, it is somewhat difficult to demarcate one single cultural zone or territory for a particular cultural group of people considering the following factors i) size of population of various groups with their language and religion spread over different regions and over Nepal as a whole; ii) internal and international migration of people and the globalization process over the last 50 years, and iii) demographic and cultural dynamics of various groups of people over the last 50 years.

From the very day of its modern existence, Nepal has remained a multicultural nation, though the concept of multiculturalism has remained an artificial one. It should not be forgotten here that multiculturalism is both a liability as well as an asset for the nation. Over the years, the central political dialectic has consisted in a set of conflictual relations between the country's many native communities and the expanding state. When the state's goal is cultural homogenization, differences emerge because resistance is natural. The changing ethnic/caste landscape and the migration and urbanization patterns that have emerged in Nepal over the last 50 years clearly indicate that a separate cultural territory for a particular ethnic/caste group such as the Tharuhat, Madhesh, Limbuwan and so on is not a viable model in the context of modern Nepal. My only concern is that the mythic imagination of a cultural territory does not seem 'real' at present as it was 238 years ago.

REFERENCES

Bhattachan, K. B. 1995. 'Ethno politics and Ethno development: An Emerging Paradigm in Nepal', in Dhruba Kumar (ed.), *State, Leadership and Politics in Nepal*, Kirtipur, CNAS.

Bista, Dor B. 1967. *People of Nepal*, Kathmandu, Ministry of Information and Broadcasting.

Caplan, L. 1970. *Land and Social Change in East Nepal*, London, Routlege and Kegan Paul.

CBS. 2002. *Population Census 2001*, Kathmandu, Central Bureau of Statistics.

Chatterji, S. K. 1952. *Kirat Janakriti, the Indo-Mongoloids: Their Contributions to the History and Culture of India, Journal Royal Asiatic Society of Bengal Letters*, vol. XVI, no. 2.

Chemjong, I. S. 1966. *History and Culture of Kirat People*, Phidim, Limbuwan.

Chowdahry, M. 2007. *Nepalko Tarai and Yaska Bhumiputraharu* ('Tarai of Nepal and its Sons of the Soils'), Kathmandu, Chowdhary Publications.

Dahal, C. 1994. *Satars of Jhapa District*. Jhapa.

Dahal, Dilli R. 1975. 'Kirat Reviewed in the Nepali Context', *Panchayat Darpan*, vol. 5, no. 15: 85-96.

– 1978. *Indian Ethnic Groups in the Nepal Tarai*, Kirtipur, CNAS (Mimeographed Report).

– 1993. *Nepalka Adibasiharu? Samajik Samrachana ra Arthik Sthiti ra yes bare Utheka Kehi Prsnaharu* ('Indigenous Groups of Nepal? Their Social Structure and Economic Condition and Some Questions Raised on this Issue'). A paper presented at the seminar on 'Indigenous Peoples of Nepal' organized by CNAS and the World Indigenous Peoples Year, Kathmandu, National Committee.

– 1995. 'Ethnic Cauldron, Demography and Minority Politics: The Case of Nepal', in Dhruba Kumar (ed.), *State, Leadership and Politics in Nepal*, Kirtipur, CNAS, T.U.

– 2003. 'Social Composition of the Population: Caste/ethnicity and Religion in Nepal', *Population Monograph of Nepal*, vol. 1, Kathmandu, Central Bureau of Statistics.

– 2007. 'The 'Madhesi' People: Issues and Challenges of Democracy in the Nepal Tarai', in Gellner, David and K. Hacchethu (eds.), *Democracy in South Asia*, New Delhi, Sage Publications.

Dahal, Dilli R. and S. Mishra. 1993. 'Hierarchy, Social Structure and Inter-group Relations: Case Studies of some VDCs of Sirha and Saptati Districts'. A report submitted to Churiya Forestry Project, Kathmandu, GTZ.

Elder, J. *et al.* 1976. *Planned Resettlement in Nepal Tarai*, Kirtipur, CNAS, T.U.

Fisher, J. 1978. *Himalayan Anthropology: The Indo-Tibet Interface*, The Hague, Mouton.

Gaige, F. H. 1975. *Regionalism and National Unity in Nepal*, New Delhi, Vikas Publishing House.

Gurung, H. 1974. 'The Population Aspect of Development', *Population and Development in Nepal*, Kirtipur, CEDA.

– 2002. 'Fragile Districts: Futile Decentralisation', Readings on Governance and Decentralisation, Kathmandu, Institute of Governance and Development.

– 2003. *Trident and Thunderbolt: Cultural Dynamics in Nepalese Politics*, The Mahesh Chandra Regmi Lecture 2003, Lalitpur, Social Science Baha, 24 April 2003.

Hacchethu, K. (n.d.). 'Nepal: Inclusion, Federalism and Autonomy'. Manuscript.

Hagen, T. 1971. *Nepal: the Kingdom of the Himalaya*, New Delhi, Oxford Publications.

Jha, H. 1993. *The Terai Community and Nation Integration in Nepal*, Kathmandu, Center for Economic and Technical Studies and FES.

Joshi, J. 2006. *Regional Strategies for Sustainable Development*, Kathmandu.

Kaila, B. 2006. 'Hodgson Pandulipima Limbu Bhasa Sambandhi Sanchipta Bibarad' ('A short description of the Limbu Language based on Hodgson collections'), *Baha Journal*, vol. 3, no. 3: 105-135.

K.C. Balkumar. 2003. "Internal Migration in Nepal", *Population Monograph of Nepal*, vol. 2, Kathmandu, CBS.

Lawati, M. 2005. *Towards a Democratic Nepal: Inclusive Political Institutions for a Multicultural Society*, Kathmandu.

Linder, W. 1994. *Swiss Democracy, Possible Solutions to Conflict in Multicultural Societies*, Macmillan, Palgrave.

NFDIN (Nepal Foundation For Development of Indigenous Nationalities). 2003. 'National Committee for Development of Nationalities', Kathmandu, NFDIN.

Nugent, D. 1994. 'Building the State and Making the Nation: The Bases and Limits of State Centralization in 'Modern Peru'', *American Anthropologist,* 96 (2): 333-369.

Pradhan, R. 1994. 'Natives by any other names', *Himal*, June-July.

Prasai, G. K. 1988. 'Colorful Tribes of the Tarai Belit of Nepal', *Ankho (Eye)*, Shantinagar, Jhapa.

Raujaria, D. 1977. 'The Tharus of Dang-Deokhuri', M.A. thesis in Anthropology, Submitted to CNAS, T.U.

Regmi, M. C. 1976. *Land Ownership in Nepal*, Berkeley, University of California.

Sharma, P. R. 1977. 'Caste, Social Mobility and Sanskritization in a Tribal Hindu Society: A Study of Nepal's Old Legal Code', in Iijima S. (ed.), *Changing Aspects of Modern Nepal*, Tokyo, ISLCAA.

Shrestha, K. K. 2007. 'Rrajbashi Samajko Samrachana (Social structure of the Rajbanshi Community)'. Ph.D. dissertation in Anthropology submitted to the Dean's Office of Humanities and Social Sciences, Tribhuvan University, Kirtipur.

Singh, R. P. 1999. 'Personal Communication', Kathmandu.

Stiller, L. F. 1973. *The Rise of the House of Gorkha: A Study of the Unification of Nepal, 1768-1816*, New Delhi, Manjushri Publication.

Subba, S. 1977. 'Studies in Bilingualism in Nepal', 3 volumes, Kathmandu, Institute of Nepal and Asian Studies, Tribhuvan University.

Subedi, B. P. 2002. 'Concentration or Diversification: Geography of Ethnic Diversification in Nepal', *Population and Development in Nepal*, vol. 10: 109-126.

Subedi, B. P. 2006. 'Spatial Dimension', *Nepal Population Perspective Plan*, Kathmandu, New Era.

Tamang, P. 2001. (2058, B.S.) 'Kendrikit Sasanata ra Sayata Sasanko Sawal' ('Question of Centralized System and Autonomous System'), in Pratush Onta *et al.* (eds.), *Chapama Janajati* ('Nationalities in Media'), Kathmandu, Ekta Books.

Teli, R. S. 2007. *Nepal Taraika Madhesiharu* ('Madhesi of Nepal Tarai'), Kantipur, March 13.

Notes

1. Besides the Caucasoid and Mongoloid groups, there are two small groups, Satar (Austro-Asiatic) and Jhangar (Dravidian).

I. 3. VILLAGE-COMMUNITY ROOTS OF TERRITORY-MAKING

1.3.1.

From the 'Borders of the Country' to the 'Special District': the Vicissitudes of Kham-Magar History

Anne de Sales

This chapter is an attempt to trace the outlines of the history of one population in particular, the Kham-Magar, with a view to shedding light on the various kinds of territorial frames in which the group has been set, from their first appearance in the literature down to the present time, following the choice by the Maoist leaders to make it their iconic stronghold. The further back into the past we go, the rougher and more hypothetical are the outlines of the history of this ethnic minority. Historical accounts are provided on the one hand by nineteenth century British visitors and, on the other, by a few Nepalese historians who themselves use various sources ranging from the Puranas to recent chronicles. Historical accounts as told by the inhabitants of the area are mostly legendary but they tell us something about the territories that the people see as theirs, and even if they do not necessarily correspond to a political reality, are likely at least to express principles underlying their constitution. Other accounts, such as the narratives related by a traditional village headman, provide valuable information concerning the administrative functions that he used to fulfil until the Panchayat reforms of the 1960s. The headman's travelling between the village and various politico-administrative centres extends the range of the territorial sketch beyond the village and opens it onto the wider national frame. Looking back into history also sheds light on more recent developments. As a matter of fact, as the title of this study indicates, the initial situation has been reversed, albeit temporarily. This population on the margins has been given a central part to play in the political life of the country. Its specific territorial setting may have been a determinant factor in this clear case of cultural politics.

I. The Kham-Magar country

The Kham-Magars inhabit the upper valleys to the north of Rolpa, the east of Rukum and the north west of Baglung, an area also known as the Four Thousand Hills (*chaar hajaar parbat*)—the valleys along the Bheri river, the Sani Bheri, the Uttar

Ganga and their tributaries (Fig. 1). They are grouped into 14 Village Development Committees comprising around 50,000 people—an approximate number since the census lumped them together with the Magars, the largest ethnic group in Nepal, of which they represent only 3 per cent. The expression Kham-Magar dates from David Watters' first study of the Tibeto-Burman language called Kham, in 1973.[1] His analysis casts doubt on the notion that the four clans of Kham speakers (Budha, Gharti, Rokha and Pun) and the clans who speak Magar proper (Rana, Thapa, Ale) might have a common origin. In the literature predating Watter's publications, the Kham-Magars can be distinguished from the Magars only by their clan names combined with their location. They are largely absent from the nineteenth century British writings on Nepal since they were not recruited into the Gurkha regiments, or at least engaged to a much lesser extent than the Magars or the Gurungs.

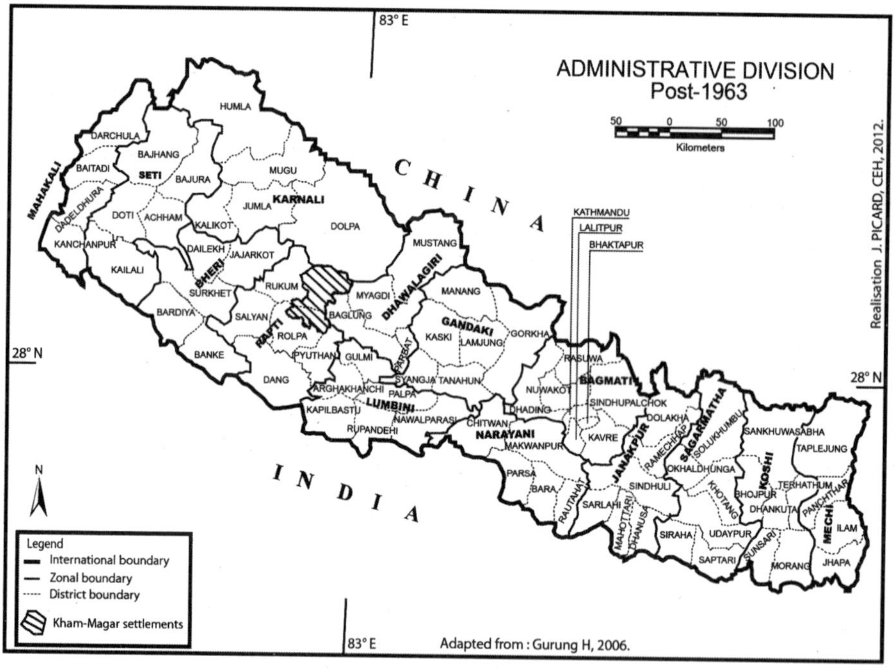

Figure 1: Post-1963 administrative map (Gurung H., 2006)

There are about a hundred settlements of Kham speakers, with a dozen numbering up to 300 houses with smaller satellite hamlets. These compact villages are built on slopes at an altitude varying between 2,000 and 2,500 m. The flat-roofed houses are gradually being replaced by sloping roofs, but the architecture remains characteristic of the north-west of Nepal, with the living area built above the stables. High-caste Hindus did not settle in this high and remote region, and apart from the two service castes, the Blacksmiths and the Tailor-Musicians, and a few families of Gurungs, only Kham-Magars inhabit these villages that are united by a strong sentiment of belonging to one 'country' (*desh*).

The Kham-Magar villages have a mixed economy based on agriculture and animal husbandry. Their transhumant lifestyle draws a web of trails that extend beyond their territory into adjacent districts: in the summer the shepherds take their large flocks of sheep and goats to the north, up to Dolpo. Each village (or, in the past, each clan) owns a large tract of grazing land that is blocked by high mountains with only a few passes to the northern side of the Himalayas; in the winter families travel down to the foothills of the Mahabharat range to the south, keeping to the ridges with their flocks. This transhumant lifestyle is combined with trading since the animals carry back home rice and other necessities that have been exchanged in the bazaars against wool or home-woven blankets. This pattern, which once used to dictate the pace of village life, is on the wane, and only the northern villages still have large flocks of sheep. Most families have sold their animals and reinvested the cash in mules—a lucrative business now that shops have opened in the villages and more goods need to be transported from the southern markets. Villagers' movements are more and more restricted to accompanying the cattle from the permanent villages to the summer residences. Or, if they travel, this is much further away from home, to distant countries where they emigrate in search of work. I shall return to this gradual change in the living space of the villagers, but let us start at the beginning, when the people who may have been among the ancestors of the Kham-Magars were frontier people. These investigations are made with a view to identifying the historical and mythical space within which this population has developed.

II. The 'frontier place'

At the very beginning of the nineteenth century Nepal was known as the Dominions of the House of Gorkha, in which Gorkha itself was a small kingdom in the centre of the territory (Fig. 2). The king and his descendants had extended their conquest well beyond what are now the national borders, to include Kumaon and Garwhal in the west and Sikkim in the east.Francis Hamilton, surgeon to the British Governor of India and polymath, stayed fourteen months in Kathmandu (1802-1803) before spending two years on the border of Nepal and India. His first task was to commission several maps of the country for the East India Company and provide a description of the Kingdom. It may not be irrelevant to recall at this point that ten years later the British would declare war on Nepal and reduce by a third the size of this ambitious nation in the making. In his introduction to the famous *Account* (Hamilton 1990 [1819]) we learn that Hamilton's informants came from various walks of life and belonged to various castes: a Buddhist lama, a slave of the king of Gorkha who happened to be a gifted and well-travelled botanist, two notables from the Kiranti group in the east and several Brahmans from Kumaon in the west. These learned travellers were likely to provide complementary information on the various parts of the country. However, the map that resulted from these various surveys reveals a disproportionate blank in the centre—the area that is the focus of the present study: to the east of the Karnali basin, more precisely east of the river Bheri and to the west of the Kali Gandaki. Hamilton admits that his informants did not know much about this area apart from the fact that the principality of Rugum (sic) had a common border with the kingdom of Parbat (Hamilton 1990 [1819]: 276). He also mentions that towards the northwest the area is called Seshant, "a frontier place" mainly populated by Gurungs or "Barbarians rejecting the doctrine of the Brahmans", whereas the area to the south is called Khasant, the country of the Khas, who are primarily Hindu and speak Nepali (Hamilton 1990 [1819]: 274-75). Local accounts will provide more insights about this mysterious country of the Sesh.

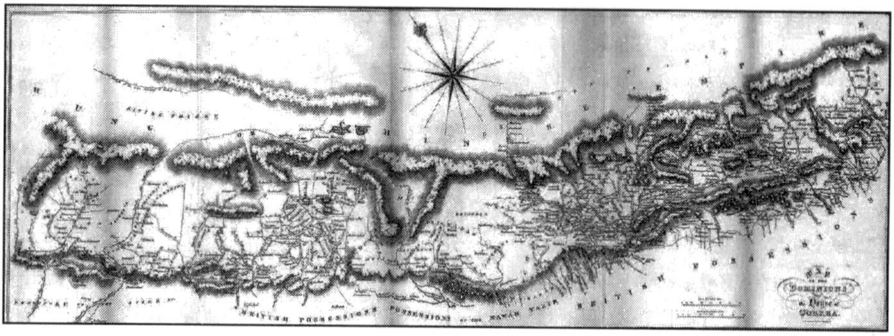

Figure 2: Map of the Dominions of the house of Gorkha, Hamilton, 1990 (1819)

The 83rd meridian that cuts through the middle of the area under scrutiny constitutes an important bioclimatic frontier. Further west, away from the influence of the monsoon, the climate becomes drier, and Mediterranean species become more abundant. Recent research[2] has drawn attention to this natural frontier, a line that is not marked on the topography but that nevertheless sets the western hills apart from the rest of the country: a number of features concerning the adaptation to the environment are visible in the habitat and in farming techniques and tools (see above and note 2). Other cultural and religious features are understood to be the remnants of the Malla Empire that dominated the western region from the twelfth to the fourteenth century. Founded by the Khas populations who long ago entered Nepal from the west, this decentralised empire included the Tibetan kingdoms of Guge and Purang as well as the Indian provinces of Kumaon and Garhwal. The collapse of the empire saw the rise of a confederation known as the 'Twenty-two kingdoms' (Baisi) to the west of the 83rd meridian, followed in the fifteenth century by the emergence of the 'Twenty-four kingdoms' (Chaubise) to its east. This frontier between the two confederations makes Rolpa and Rukum the two easternmost states of the 'Twenty-two' (the case of Pyuthan is a subject of debate among historians), and the Kham-Magars the only hill population speaking a Tibeto-Burman language in the western hills, a region otherwise dominated by caste people.[3] Several clues of a linguistic, mythico-ritual and historical character will shed more light on the Kham-Magar country.

David Watters noted scattered groups of Kham speakers to the west of what nowadays appears as the homeland of the Kham-Magars (Jajarkot, Dailekh, Kalikot, Accham and Doti, Fig.1), suggesting that the Kham language was "one of the first

Tibeto-Burman languages encountered by the Khas-speaking Aryans on their eastward migration out of northwestern India into the Himalayan foothills…" He remarked that "a full 25 per cent of the vocabulary can be traced to Indic sources, much of it predating modern Nepali" (Watters 2004: 1-2). Families or clans of Blacksmiths and Tailor-Musicians have been affiliated to Kham-Magar clans for a long time, and are said to have migrated together with them. Since loanwords in Kham specifically concern the material culture of the artisans, these two groups may have been at the origin of the presence of old Nepali words in the Kham language.

Shamanic ritual songs would tend to corroborate Watter's linguistic observations concerning the provenance of this population (or groups of it). Villagers from the northern valleys (Taka, Sera, Lukum) take their animals grazing to Dhorpatan where they possess communal pastures. When a villager falls ill, his or her soul is supposed to leave the body of its owner and follow the same route as the flocks, up to the high pass of Jaljala, which is located on the eastern border of the modern Dhorpatan hunting reserve (see below). Beyond this point, where the river of forgetfulness flows (sings the shaman), the soul will forget its owner and never return; it has left the Kham-Magar country for the other world. Knowing that there is no equivalent point towards the west, one may wonder whether this mythical border is not also a landmark that used to divide western Nepal from the rest of the country and the two confederations of kingdoms, the Twenty-two and the Twenty-four.

The Kham-Magar country is also delimited by a northern frontier. The influence of the Khas extended over the Magar petty states in the south from the east of the Bheri to the Kali Gandaki (an area corresponding to mediaeval Magarant) while the states in the north, namely the modern districts of Humla, Rukum, Manang, Mustang, Lamjung and Kaski, were under the domination of 'Tibetan kings' belonging to the so-called Jad population.[4] Chronicles mention that Rukum was finally taken by the Mallas from a Jad king.[5] Dhor is located on the route that used to link Sinja and Jumla, the winter and summer capitals of the Malla kings, to Pokhara via Beni—an important Northern route for communication and military conquest. It just so happens that local folklore mentions a 'Tibetan king' (*bhote raajaa*) or 'black king' (*kaala raajaa*) to whom the Magars used to pay taxes and who used to rule over Dhor and Jaljala. This black king is also said to have left behind him many of his servants or slaves, whose descendants are considered to have been the source of one branch of a Kham-Magar clan, the Gharti. The vanquished king was made a powerful spirit, with whom the shaman has to negotiate the souls of his patients and whose name, Karbir Masan, recalls the taxes or tribute (Nep. *kar*) that were paid to him.

The use of a plough of a Tibetan type in certain villages may be another testimony to a Tibetan authority over the Kham-Magar population, or at least some part of it, in the past.[6] Other instances recall that the Kham-Magar territory adjoins an area under the influence of Tibetan or Tibetanised populations: as mentioned above, the transhumance routes bring the shepherds and their flocks to Dolpo in the summer; Kham-Magar villages are also the first settlements encountered by Tibetan traders who come down from Tarakot with the loads of woollen belts they used to exchange for maize. At a more general level, and even though the history of their migrations is far from being straightforward, the Kham-Magars look to the north as their place of origin, and the birthplace of their first shaman, Ramma Puran Tsan. It is remarkable that this myth of a northern origin of the Kham-Magar is not accompanied by any Buddhist influence, suggesting that if the Jads were of Tibetan stock, they were not Buddhist.

The historical, legendary and linguistic clues gathered so far present a group of highlanders, speaking a Tibeto-Burman language, being pushed towards the eastern and the northern frontiers of the Malla Empire. Although the topographic relief does not impose an isolation comparable to the enclaves of Tibetan populations further to the north, the Kham-Magar territory nevertheless appears as a sort of ethnic enclave on the southern border of the medieval Seshant, the 'frontier place', and under the conflicting authorities of Khas and Jad kings. This characteristic of being governed by others persisted in the period following the collapse of the Khas Empire when the Thakuri principalities[7] that emerged in its place were fighting each other, leading to an ongoing modification of their borders.

III. The "remaining states"

A local account by Syam Shahi, a descendant of the king of Musikot, sheds further light on the area.[8] Musikot is currently the headquarters of the modern district of Rukum.

> A Rajput prince from Riwa in India (Madya Pradesh) came to take refuge in Jumla. The king of Jumla said: 'Oh! *jatkomanche ayo*! Here are caste people! Let's give him a daughter!' With his daughter, he gave the newcomer the states of Parbat as a dowry, including Baglung, Myagdi, Argha Kanchi and Gulmi with the injunction that the newcomers follow the local customs of the Khas, specifically the cults to the god Masta and to the ancestral deity Braha. The power of the new king of Parbat grew bigger and bigger to the point that the Jumli king, feared that he would be surpassed by his son-in-law. The two powerful kingdoms fought over Mustang and the states of Rukum, Rolpa and Jajarkot belonging to the Twenty-two, and Sallyan and Dang belonging to the Twenty-four. In the absence of a clear victory on either side, the Magar population of these districts was left to decide on its own administration. Since then, they were called *sesh raajya*, meaning 'the remaining states'.[9] There were several *sesh*, but nowadays only Black Sesh *kalo sesh* and White Sesh *gora sesh* remain, confined to the district of Rolpa.
>
> Once left to themselves, the Magar fought each other and came to the conclusion that only a king from outside could bring peace. They turned to the king of Parbat who refused to provide them with one of his seven sons as their future king. So a Magar delegation led by Lal Bujakad went at night and kidnapped the tallest son, who also happened to be deaf and mute. They stuffed him inside a sack and carried him back home. The king of the Sesh was known thereafter as the *jholiko raja*, the 'bag king'.
>
> Sixteen generations after the event, the Sesh area was divided among three brothers: Rukum went to the eldest, Jajarkot to the second and Sallyan to the third. Eventually, the three states were conquered by Rana Bahadur Shah.

The account takes up recurrent features of the history of the region: caste people leaving India and taking refuge in Nepal among the Khas, who are not Hindu in quite the same way as the Rajput; the giving of a daughter, along with land, to the new settlers and the potential conflicts between allies which follow, something of a pattern in narratives concerning the history of Nepalese settlements;[10] the endemic rivalry among local clans leading to their calling a Hindu king from outside to bring peace. The story, as it was told, suggests a clear association between the status of the areas that are 'leftovers' of an indecisive war between more powerful states, and the status of the Magar population left to itself.

What Syam Shahi refers to as the Sesh states (Rukum, Jajarkot and Sallyan) does not correspond to Seshant, or 'frontier place', which Hamilton located to the northwest of Beni, with the exception of Dhorpatan. The Sesh states coincide with the old district of Sallyan, one of the 35 districts established after the Conquest (Fig. 3). This district included the following modern districts: Sallyan, Jajarkot, Rukum and half of Rolpa, the other half being included in the old district of Pyuthan. It seems that two administrative circumscriptions (*thum*) in Rolpa and Pyuthan were called White Sesh and Black Sesh. With the administrative reorganisation of the country in 1962, these two ancient *thum* came under the sole jurisdiction of Rolpa.

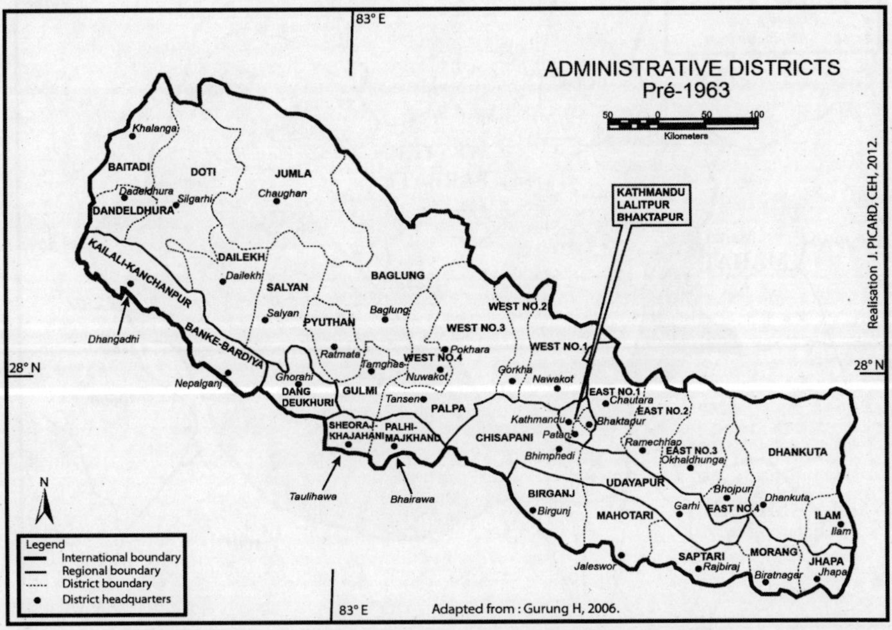

Figure 3: Pre-1963 administrative map (Gurung H., 2006)

Now, David Watters (2004) applies the name 'sheshi' to one of the five Kham-Magar speaking territories (Fig. 4) that is located precisely in the southernmost part of the Kham speaking territory, bordering Sallyan. Unfortunately Watters does not suggest any etymology of the term.[11] However it is used by 'caste people' in the districts adjacent to Rolpa and Rukum (Sallyan, Jajarkot, Pyuthan and Dangdeokhuri) when they see the Kham-Magar shepherds coming down from the hills to the plain with their sheep in winter. The expression therefore refers to the highlanders entrenched on their hilltops at the 'end of the country' or on the 'frontier area', as Hamilton suggested. We could even speak of the 'dead end' of the country, if we combined the various meanings of the term *sesh* according to Turner (note 9). However, it should be clear that this expression is not used by the Kham-Magars about themselves; it is an exonym that conveys an image that others have of them. They themselves may have the feeling of being neglected in the 'corner of the country'[12] as illustrated by examples cited below, but this should not overshadow the much more robust feeling of identity and independence that they have nurtured so far, partly because, it may be stressed, they have been left to themselves for so long.

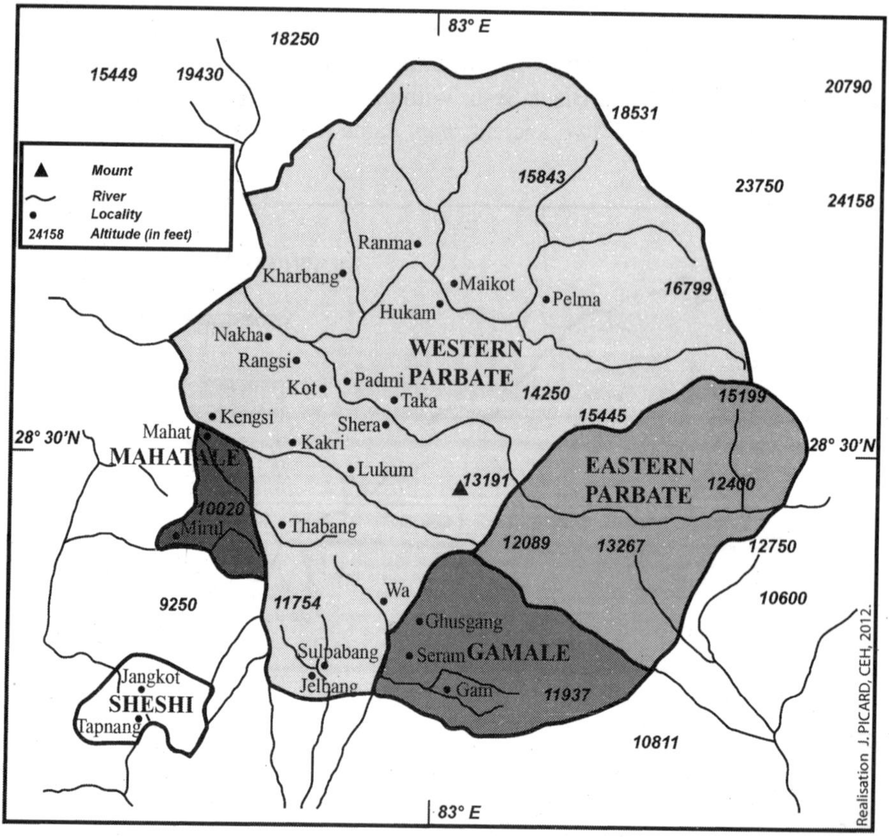

Figure 4: Kham-speaking territories showing five major dialect groupings (Watters n.d.)

In the perspective of this reflection on the notion of territory, a 'remaining state' that does not have a clear affiliation is almost a non-territory, at least from the point of view of the rulers. In contrast to the fluctuating borders of the principalities to which the Kham-Magars may have been affiliated in the past, the Kham-Magars themselves have a more defined territory, at the centre of a network of relationships with other centres.

IV. A ritual territory, administrative networks

The ritual term for 'village' illustrates the local anchoring of the population in its territory: while in the everyday register of Kham the word for village is *nankhar*,[13] in shamanic songs the term used is *nam*, 'world' or 'earth', and when associated with a quasi-homonym, *nem*, 'sky', it means 'universe'. A village includes the area around the settlement, its 'wilderness' (Nep. *ban*), and all the communal lands that are the main traditional source of wealth, also through the taxes that the outsiders have to pay for grazing their animals there. The fact that a hundred dialects can be identified on the Kham-speaking territory—almost one per settlement—adds to the idea that the village constitutes a world in itself. Several other features enhance this characteristic: each big village has its own ritual calendar, its own corpus of ritual songs and even specific festivals, as many variations on the same themes.

These 'village worlds' are not completely independent from one another. They used to be linked by a ritual dance. This dance was performed by male dancers, the *nachane,* and female dancers, the *jaulani*, under the guidance of a guru.[14] The troupes would travel from village to village during the winter months. Each performance was associated with the expulsion of demons and was concluded by a sacrifice to the goddess Kali. The point is that dancers from outside the village would come to expel the village demons, a rule that implied a sort of ritual exchange between the 'village worlds'. Although it is not practised anymore, this ritual suggests that the Kham-Magar territory was, if not enclosed (there was no ritual encirclement of the territory, for example), then at least somehow checked or re-defined by the religious dancers and given a ritual existence. In the same way, shamanic journeys delimit a territory: either in the search of a soul or in order to expel diseases, the shaman's song takes him to a great number of villages and places that he describes or sometime just lists, thereby drawing a ritual map of the Kham-Magar territory. This territory is not unconnected to the rest of the country: in one song in particular, the shaman 'went' to Rukumkot, the capital of the ancient kingdom (de Sales 1994), before driving evil spirits out to the south.

The functions of the village headmen as tax-collectors and their responsibilities regarding the maintenance of law and order took them regularly to the district headquarters, weaving a web of relationships that extended the village beyond its territory *strictu sensu*, in the same way as shamanic journeys or the transhumance routes linked the Kham Magar territory to the adjacent territories. Sri Bahadur Pun is the last headman (*mukhiya*) of the village of Lukum. There was no school in the 1930s and he was sent to work in the house of a literate Brahman in Rukumkot, in the hope to learn there how to read and write. He succeeded his father as headman at the early age of 17, in 1948, and kept this position for almost 15 years until the Panchayat reform of 1962. He was then elected Pradhan Panch. The Kham-Magar villages were divided at the time between the jurisdictions of three of the old districts inherited from the conquest: Sallyan, Baglung and Pyuthan.[15] Sri Bahadur refers to

Salyan as Salyan *gaunDa*. This term, that designates military headquarters in hill districts, recalls that there was no definite demarcation between the civil and military administration under the Ranas. These garrison towns, headed by a Bara Hakim, incorporated a number of *thum* or county. Sallyan comprised six, including Rukum, each of which had been ruled in the past by a small 'king'.[16] It seems that Sri Bahadur spent quite a lot of time at Salyan headquarters. Apart from his responsibilities as a tax collector, he was involved in numerous cases, not only on his own account but also as a mediator in others' trials, many of which were triggered by boundary conflicts.

V. Stress on boundaries

Sri Bahadur recalled an interesting development in the way in which land was registered and taxes paid under the Ranas. Initially, each piece of land had to be registered under the name of its owner, the taxes being calculated according to the number of pieces (*kittaa*, 'item') rather than according to their size. The amount due for two small fields would be higher than for one big field corresponding to the same surface. The description and the registration of the property used to be based on topographic features: "from this river to the top of the mountain, this is my *dartaa* ('entry in the register')". Should the owner of a field wish to divide his property, he would drive a stone into the ground, and this would be sufficient to indicate the boundary between the two shares (Photographs 1 and 2). No specific measurement was made. Towards the end of the Rana regime, in 2006 BS (1949), a new way of registering the land was introduced: each piece had to be measured in hands (*haat*) along each of the 'four sides', *caar killaa*.[17] Apparently this generated quarrels over the ownership of land: the more precise the definitions, the more likely a source of dispute they became. A later development was the cadastral survey or *naapi* system, locally known as the 'chain survey', because the unit of measure was no longer the Nepali hand but the six Imperial inches marked on a metal chain. Yet more accurate measurement was needed.

Photographs 1 and 2: Boundary stones

Marie Lecomte-Tilouine mentions an alternative method of registration of the land according to which the piece of land was delimited in the document by the names of the owners of the fields surrounding it.[18] In this case, too, no specific measurement seemed to be needed. The author provides a villager's sketch of a place that illustrates the Nepalese conception of space whereby one point is defined by the four directions: the place is represented in the centre of a square of which the four sides, matching the four directions, are named. If the 'four sides' express a "fundamental structure of the conception of space" in Nepal and more generally in the Hindu world, what needs to be stressed here is that while the centre is defined, the boundaries are not.

The stress on the perimeter and its precise measurement in the identification of a piece of land illustrates the changing conception of space and ownership that gradually took place in Nepal after the beginning of the nineteenth century. This evolution resulted from several factors among which demographic pressure on arable land is the most obvious. The area under cultivation was continually expanding under the Ranas and the dramatic acceleration in population growth in the 1940s must have made the situation more critical. However, it was mentioned earlier that the Anglo-Nepalese war (1814-1816) had put an end to the Gorkhali expansion and deprived Nepal of one third of its territory. From then on "the country's borders, which had already lost their former fluidity [...] became of increased psychological importance" (Whelpton 2005: 56). There is, no doubt, a difference between farmers' and rulers' conceptions of ownership. Richard Burghart made it clear that prior to the confrontation with the British there was no necessary overlapping of the territory of the possessions of the Gorkhali king with the territory covered by his religious authority, neither was there an exclusive affiliation of his subjects to only one polity (Burghart 1984). Farmers, by contrast, would always know the precise delimitation of the plot of land they cultivate. However, I would suggest that the geo-political factor had repercussions at local level and that the conflicts over the ownership of land that were reported by Sri Bahadur Pun need to be situated within the general evolution that Nepal was undergoing at the time. The stress on boundaries reflected the will to increase control over the country. The Panchayat reforms would later develop this trend in conformity with the process of building a modern Nepalese nation.

VI. Towards national integration: a paradox

The many works by Nepalese scholars who studied these reforms in the 1970s and the 1980s develop the ideological rationale behind the creation in 1962 of 14 zones and 75 districts and, in 1972, of four development regions: the need to reduce regional disparity for socio-economic unification, consolidation of national integration, emphasis on development, administrative convenience and so forth. As stated above, the original 35 districts were to a certain extent the result of the historical conquest and had not been created with a view to achieving uniformity in size or administrative convenience. In actual practice, Nepal was a loose union of almost autonomous districts over which central government often failed to exercise full and effective political control (Shresta 1981: 184); hence the 75 smaller districts, more compact and easier to administrate under the jurisdiction of a prefect or Chief District Officer.

Rukum and Rolpa, and therefore the Kham-Magar villages under their jurisdiction, are grouped together with Salyan, Pyuthan and Dang to form the Rapti Zone. This Zone covers roughly the old Salyan district or the Sesh states as Syam Shahi saw it, and became infamous in the course of the Maoist insurrection (1996-2006). The recent history of this region has been scrutinized in the hope of understanding why the Maoist rooted their movement precisely in this area and particularly among the Kham-Magar population. This is not the place to retrace this history.[19] However, two events that took place respectively in the 1970s and the 1980s will illustrate the population's reactions to the intervention of the state in their territory.

The first concerns the relocation of the district headquarters from Rukumkot to Musikot at the beginning of the 1970s. Eastern Rukum, and the Kham-Magar villages in particular, initially fell under the jurisdiction of Baglung headquarters, quite far away across difficult terrain. The residents of eastern Rukum had been asking for years that their area be affiliated to Rukum district, to which they felt historically attached. Rukumkot, the old capital, also presented the great advantage of being much closer and more accessible. Their demand was finally met at the beginning of the 1970s: eastern Rukum was incorporated into Rukum District, with Rukumkot as the District Headquarters. However, this victory was short-lived: in 1973 powerful families from Western Rukum managed to have the headquarters of the district transferred to Musikot, the rival capital of the district.[20] Eastern Rukum residents perceived this transfer as an act of marginalisation, with their area being bypassed by development projects and employment opportunities. This marginalisation was bitterly felt and resulted in violent demonstrations against the state. I would suggest that in the context of the Panchayat ethos, which focussed on economic development, being left at the margin of the nation had acquired more acute significance. The effort towards national integration made the Kham-Magars appear by contrast more 'backward', in the conflated dimensions of time and space.

The second event concerns an area that has been mentioned above: the vast plain of Dhorpatan where villagers from the neighbouring valleys graze their cattle. Even though they have been fighting among themselves for generations over this pastureland, the Kham-Magars all agree that Dhorpatan is their ancestral land. In 1983, following surveys conducted by UNESCO and the WWF the place was transformed into the Royal Dhorpatan Hunting Reserve. The original agreement, according to which the Kham-Magar community would receive a fee in exchange for hosting the hunters of the Himalayan Blue Mountain Sheep, was not respected, and the Kham Magars opposed the private tour operators, furious about this encroachment on their territory and its jurisdiction.[21].

VII. The 'special district'

Now, in its attempt to mobilise the countryside the CPN (Maoist) focussed on this area by using a long-standing communist presence in a few chosen places.

Communist activists from Pyuthan involved the Kham-Magar village of Thabang in Rolpa, on the border with Rukum, as early as 1954. After the inception of the People's War in 1996, partly through sheer coercion and partly through a complex mechanism of factionalism, the Maoist influence gradually spread throughout the Kham-Magar villages, even though its success varied from one place to another. This remote region, several days' walk from any road and densely forested, offered a perfect retreat for the Maoist fighters. Although the leaders of the movement are Brahmans, several of the high-ranking commanders and commissars of the Maoists' armed front were Kham-Magars from Rolpa (such as Pasang — *alias* Nanda Kisor Pun — and Ananta,— *alias* Barsaman Pun) and Rukum.

In April 2002 the Kham-Magar country was declared a 'special district', *vishes jillaa* or the '76th district' of Nepal. It was constituted from seven VDCs in eastern Rukum and five VDCs in northern Rolpa.[22] It was given a People's Government a year later, but this was dissolved in October 2004 by the Central Committee of the Party. The reason put forward by the Party was that the special district was an experiment in autonomy at local level and it lost its relevance after the establishment of the Magarant Autonomous Region (9 January 2004). After the dissolution of the 'special district', the 12 Kham-speaking VDCs were included in the 'central base area' (*aadhaar ilaakaa*) of the Rapti Zone with Rolpa, Rukum, Salyan, Pyuthan and Dang in October 2005. These districts now form the western part of Magarant, understood (in the words of the Maoist Policy and Programme) as comprising "all the mountainous territory from the Kali Gandaki River to the Bheri river, where the Magars have been living as original inhabitants since historical time".[23]

In other words, the Kham-speaking villages constituted the core or, as it were, the embryo of the medieval Magarant that was about to be reborn under the firm tutelage of the CPN (Maoist). The irony is that the Kham speakers have always been considered as the 'poor relatives' of the Magars proper: it was not until several years after 1990 that they were given official representation in the National Magar association. Through the Maoist insurrection the Kham speakers, or at least their leaders, were given the opportunity to occupy an eminent place in national politics as well as towards their more privileged Magar cousins.

It would not be the first time that in the process of building a nation, a peripheral or marginal and underprivileged territory is transformed by the political culture into the emblem of the nation. In an earlier work I have shown how Rolpa in particular was exalted in the epic style of communist propaganda as a metonym of the country: "Rolpa is not a district, it is Nepal. It is the source of revolution, the centre of hope. Glory to Jaljala, glory to Sisne!"[24]

The Maoists turned a blank space on the map into the heart of the new nation. However, the ethnicisation of the Nepalese territory by the Maoists marks—in the words of the leaders—a necessary phase in the process of the political maturation of the people and their mobilisation towards the ultimate communist conquest that will make castes and ethnic groups irrelevant. The rebels needed to convince the Nepalese population that communism was not only an imported ideology but was practised in its most natural state among the indigenous populations of Nepal, although it had been compromised by Hinduism and the feudal caste system. They turned towards the Kham-Magar communities—as peripheral a location as there could be—in search of legitimacy for their ambitions to conquer the power at the centre. The inhabitants of the 'special district', however, are fully aware that they have been both instrumental in and instrumentalised by revolutionary cultural politics. And if, partly in spite of themselves, they set their seal of authenticity on the revolutionary movement, they may well expect more than a minor place in the history books in return.

Conclusion

In the near-absence of written documents, the various territorial frameworks within which the Kham-Magars have developed offer the most valuable material for a sketch of their history. Let us take them up in turn. Linguistic criteria delimit a first territory on the eastern frontier of the western hills and suggested that the current Kham-Magar settlements represent the last stop in the migration of this population, whose ancestors entered what is now Nepal from the west. The transhumant lifestyle extends the village territory up to several days' walk from the main settlements between the high mountains in the north and the southern plains. This traditional space, within which economic activities (sheep herding, trading) develop coincides with the ritual map that the shamans follow in their search for vitality or in their efforts to drive out evil spirits and protect the village territory. The shamanic territory, so to speak, seems rather conservative, perhaps because it preserves the memory of ancient historical divides between political entities such as the mediaeval frontier between the two confederations that emerged after the collapse of the Khas Mallas.

By contrast with this well-delimited ethnic enclave, the politico-administrative web of relationships that connect villages to the relevant administrative centres have changed over the course of history, and the Kham-Magar localities have been subjected to conflicting powers: first, between the Khas Mallas and the Jads; then between powerful Thakuri principalities (Jumla and Parbat) before their conquest in the eighteenth century; later still under the Panchayat, in the second half of the twentieth century, between ambitious politicians who are the descendants of the petty kings of the past. This recurrent historical feature associated with their territory led to a lack of integration and visibility of the Kham-Magar country at national level, something that the Maoists used to their advantage.

Although the insurrection lent a prominent place to the Kham-Magar territory, it also froze its development for a decade: villagers were not allowed to leave their occupied districts in search of work, and in times of crisis could not even travel outside their settlements; NGOs were banned. The situation is now changing, and in 2008 the village committees received far more generous budgets than in the past, while development organisations were again welcome. In spite of these efforts to keep the population in the villages, the region is facing something as important as the insurrection in terms of the consequences it has for the localities: outmigration on a massive scale. This phenomenon would deserve another study, but if the vicissitudes of the Kham-Magar history are to be understood within the construction of the Nepalese nation, the national framework will be insufficient to understand the developments that are occurring at regional and global levels.

REFERENCES

Burghart, R. 1984 'The Formation of the Concept of Nation-State in Nepal', *Journal of Asian Studies*, 44: 101-25.

Dollfus, P., M.Lecomte-Tilouine and O. Aubriot. 2001. 'Un araire dans la tête… Réflexions sur la répartition géographique de l'outil en Himalaya', *Techniques et Culture*, 37: 3-50.

Gaborieau, M. 1978. 'Le partage du pouvoir entre les lignages dans une localité du Népal central', *L'Homme*, 18: 37-67.

Gersony, R. 2003. 'Sowing the wind…History and Dynamics of the Maoist Revolt in Nepal's Rapti Hills', Report submitted to Mercy Corps International.

Gurung, H. 2006. *Nepal Atlas and Statistics*, Kathmandu, Himal Books.

Hamilton, F.B. 1990 (1819). *An Account of the Kingdom of Nepal and the Territories annexed to this dominion by the House of Gorkha*, New Delhi, Asian Educational Services.

Lecomte-Tilouine, M. 1997. 'Le paysage népalais, exégèse et appropriation du pays', in Smadja, J. (ed.), *Histoire et devenir des paysages en Himalaya*, Paris, CNRS Editions, pp. 165-192.

Lecomte-Tilouine, M. (ed.). 2009. *Bards and Mediums. History, Culture and Politics in the Central Himalayan Kingdoms*, Almora, Almora Book Depot.

Lecomte-Tilouine, M. and J.Smadja. 2003. 'Au cœur de l'ancien empire Malla, Népal occidental, un monde bouleversé par les mesures de protection des milieux', in Muxart, T.*et al.* (eds.), *Des milieux et des hommes : fragments d'histoires croisées*, pp. 161-172.

Molnar, A. 1981a. 'Flexibility and Option. A Study of the Dynamics of Women's Participation Among the Kham-Magar of Nepal', Ph.D. thesis, Wisconsin University.

Molnar, A. 1981b. 'Economic Strategies and Ecological Constraints: Case of the Kham-Magar of North West Nepal', inVon Fürer-Haimendorf, C. (ed.),*Asian Highland Societies in Anthropological Perspective*, New Delhi, Sterling Publication, pp. 20-51.

Malla, M. B. 2033bs (1977). *Sinjako Itihas Doreto* ('Guide to the history of Sinja').

Ogura, K. 2007. 'Maoists, People and the state as seen from Rolpa and Rukum', in Ishii H., D. Gellner and K. Nawa (eds.),*Political and Social Transformations in North India and Nepal*, Delhi, Manohar.

– 2008. 'Maoist People's Government, 2001-05. The power in Wartime', in Gellner, D. N. and K. Hachhethu (eds.), *Local democracy in South Asia: Microprocesses of Democratization in Nepal and its Neighbours*, Delhi, Sage.

Oppitz, M. 1991. *Onkels Tochter, Keine Sonst*, Frankfurt am Main, Suhrkamp Verlag.

Sales, A. de. 1986. 'The Nâchane of the Kham-Magar: Ethnographic notes on a group of religious dancers', in Seeland, K. (ed.),*Recent Research on Nepal*,München, Weltforum Verlag.

– 1991. *Je suis né de vos jeux de tambours. La religion chamanique des Magar du nord*, Nanterre, Société d'ethnologie.

– 1994. 'Quand le chamane fait danser les sorcières. Un voyage en pays magar',*Cahiers de Littérature Orale*, no. 35: 85-121.

– 2003.'The Kham-Magar Country, Nepal: Between ethnic claims and Maoism'in

Gellner, D. (ed.), *Resistance and the State: Nepalese Experiences*, New Delhi, Social Science Press, pp. 326-357.

Shrestha, T. N. 2038bs (1981).*Nepalese Administration. An Image*, Kathmandu, Sajha Prakashan.

Turner, R. L. 1931 (2001). *A Comparative and Etymological Dictionary of the Nepali Language*, New Delhi, Allied Publishers Limited.

Watters, D. 2004. *A dictionary of Kham: Taka dialect*, Kirtipur: Tribhuwan University.

– (n.d.). 'Kham Grammar. Cambridge Grammatical Descriptions'. Prepublication draft.

Whelpton, J. 2005. *A History of Nepal*, Cambridge, Cambridge University Press.

Notes

1. See also Watters' informative introduction to his more recent Grammar (2004). Apart from a number of articles, three monographs are available on this population: Molnar 1981a, Oppitz 1991, de Sales 1991.
2. See Lecomte-Tilouine and Smadja 2003. See also the analysis of the distribution of four types of plough in Nepal, that suggests an influence of the Khas in the west that is absent in the east (Dollfus, P. *et al.* 2001) Along with these two articles, the fruits of collaborative work between geographers and anthropologists, a collection of articles devoted to the religion and the history of Western Nepal renews our understanding of the region (Lecomte-Tilouine [ed.] 2009).
3. This border was represented by the Jaitha Lekh hill range that extends from the Dhaulagiri Massif down to Talka Peak in Dang. It is said that the last Magar king of Rukum, Dare Jaitam, married his daughter Tumbavati to the king of Gajul, Thuthasing, with Jaitha Lekh as a dowry.
4. For Mohan Bahadur Malla, the Jads are the population that would later come to be known as the Gurungs (2033bs: 26-28) and may be more appropriately seen as Tibeto-Burman groups.
5. On the basis of chronicles of Malaibam's dynasty M.B. Malla mentions a bloody battle between Adai Barman (the father of Malaibam?) and the Jad king of Rukum. A certain Ruwani Magar fought on the side of the Jad king and is said to have fled to the forests of Parbat in order to escape enslavement after the defeat (*ibid.*:41).
6. See note 2. The authors argue that the choice of a type of plough has more to do with political domination and identity than with the constraints of the environment. If we combine this study with Watters' linguistic map, it appears that the Tibetan type of plough is only present in villages where the western Parbate dialect is spoken. This would suggest that the Kham-Magar territory has been under the influence of different political entities in the course of its history.
7. At the end of Malaibam's reign, the Malla Kingdom was divided among his brothers and sons-in-law, and one of them, Pitambar, inherited the state of Rukum in 1405. From then on Rukum was ruled by Thakuri kings.
8. This legend was recorded in 1998 in Kathmandu where Shyam Shahi resides. Being related to the last king of Musikot does not make him a reliable historian but apart from the fact that he was passionate about the history of the area, his understanding of what happened represents a local consensus in Rukum district and is interesting for its own sake.
9. Nep. *sesh* or *shesh* or *sekh* : "end, conclusion, remainder in the sense of something that remained to be paid, the end of instalments like in *sekh-bhag*. It also means death like in *mera sekha pachi*'after my death'" (Turner 2001 [1931]).

10. The legends of the Kham-Magar clans confirm this pattern that was first identified by M. Gaborieau in central Nepal (Gaborieau 1978).
11. According to Watters, the term is derived from Shes Khola. Although a river by this name does not actually exist, the word *khola* 'river' is often used to refer to a region (like Thak Khola) (Watters 2004).
12. An expression used by the Tamangs of Rasuwa district to speak of themselves conveys a similar feeling: "Hami kunako manche"—"We the people from the corner (of the country)". This is also the title of a documentary on a Tamang community by Kesang Tseten (2004). However, to the best of my knowledge, the Tamangs are not called *sesh*.
13. Watters (2004) derives Kham *nankhar* from Tib. *mkhar*'fortress'. Alternatively, it may be the local, often nasalized pronunciation of Nep. *nagar*, 'city, town'.
14. See de Sales 1986.
15. Most of the Kham-Magar villages, now in the eastern part of Rukum district, fell under Sallyan jurisdiction (Sera, Padmi, Arjal, Lukum, Mahat, Sima, Kankri, Morabang), while the villages to the north (Hukam, Maikot, Taka) were under Baglung, and the villages to the south, now in Rolpa, were shared between Pyuthan and Salyan.
16. The other *thum* are Phalabang, Maichane, Musikot, Banphikot and Atbiskot.
17. According to Turner Nep. *killaa* is derived from an Arabic word meaning 'fort' or 'fortress' (*qila*). This suggests the military aspect of a border, something that is not conveyed by the other term Nep. *dhunga*, that is used in Kham for 'border stone'.
18. Lecomte-Tilouine 1997: 169-70.
19. See de Sales 2003, Gersony 2003, Ogura 2007.
20. For more details see Gersony 2003: 16-17 and Ogura 2007: 445-450.
21. Gersony 2003: 20.
22. Rukum: Ranmamaikot, Hukam, Jang, Kol, Taka Sera, Kankri, Mahat, Rolpa, Thawang, Uwa, Gam, Jelbang, Mirul.
23. 'Policy and Programme' 2003 quoted in Ogura (2008).
24. Jaljala and Sisne are the highest mountains in the two districts of Rolpa and Rukum respectively.

1.3.2.

Rai Villages as Ritual Entities, and the Making of an Ethnic Festival

Martin Gaenszle

Every spring, on the full moon of Baisakh, most Rai villages in eastern Nepal collectively celebrate rituals of fertility which include drumming and collective dancing. On this day many shamans go on pilgrimages to high mountain lakes (see Gaenszle 2008), and those who remain in the village usually play a prominent part in these communal celebrations. However, whereas this pilgrimage appears a rather uniform practice across the region, village festivals are highly localized traditions and can only be understood within the context of other rituals to which they are linked. In fact, in several Rai groups the major fertility rituals take place at a time before the full moon of Baisakh, but as I will try to show in this paper, the rituals are closely linked through their underlying concepts. These rituals are basically addressed to some form of the 'divinity of the soil', generally known in Nepali as *bhume*, and widely associated in the eastern hills with a particular notion of ancestral territory, the settlement, its lands, and its 'original' inhabitants (also see Gaenszle 2010). But the timing, the names, and the ritual performances of these festivities vary to a considerable degree among the two dozen or more Rai 'subtribes'. For example, among the Kulung Rais there are three such rituals, known as *tos* (McDougal 1979: 37, Schlemmer 2004a), while among the Thulungs the territorial rituals, which are clearly distinct here from the full moon festivities, are simply known as *bhume pujâ* (Allen 1976: 511). Therefore, careful analysis is required to discern the extent of common ground in these ideas and practices, especially since such claims are not a mere matter of scholarly interest but have recently become a matter of politics.

Since the formation in the 1990s of the Rai ethnic roof organisation, named 'Kirat Rai Yayokkha', the festival on the full moon of Baisakh is regarded as a common and rather homogenous cultural trait of the Rai community as a whole: it is known as *sâkewâ* (or *sâkelâ*). This ritual is now publicly celebrated (on the initiative of Yayokkha) in many district towns across eastern Nepal (such as Dhankuta or Khotang), and especially in urban centres from Kathmandu to Ilam and even

Kalimpong (Darjeeling) or Gangtok (Sikkim). Thus the festival has become partly a political event with a new dimension.

In this paper I will first look at the religious background of this festival in different Rai groups (Mewahang, Puma and Chintang) and analyse its transformation into a symbol of 'unity in diversity'—in particular in view of the territorial units implied. Since Rai ethnicity, which has become a strongly politicised idea and sentiment especially after the democratic movement of 1990, 'suffers' from a high linguistic diversity (as each 'subtribe' possesses its own language), the public performance of unity is an important step toward cultural as well as political recognition within the multi-ethnic nation-state.[1]The festival has become the focus of a new communal identity, especially in urban middle-class contexts. Thus the reinvented Sakewa celebration of the Rais, the former rulers of the region, can be seen as the creative reflection of fundamental changes they have dealt with in recent years.

In spite of the variation of ritual practices indicated above, I will begin by pointing out a number of features which mark Rai territorial rites as a 'polythetic' type with a certain degree of family resemblance (Needham 1975).

- These rituals generally refer to the ancestral past, i.e. the creations of ancestral beings, and a crucial part is the dramatic enactment of these creations through dance.
- Drumming and dancing by both men and women are an important part of this performance.
- The ritual cycle is closely linked to the agricultural cycle, and both are defined by a concept of time which divides the year into two halves: the rising time and the descending time (*ubhauli* and *udhauli*, see Gaborieau 1982).
- The rituals are addressed to the forces of the soil as a source of fertility and wealth. This is a disjunctive notion which posits these forces as an 'other': i.e. a world of wilderness in contrast to a world of human culture. At the same time the notion refers to a specific territory with at least roughly delineable boundaries.
- The local kinship system (of patrilineal clans and proto-clans or phratries) plays a significant part in structuring ritual events, and often the local headmen play a pivotal role, indicating that there is a strong political layer of meaning.
- The ritual is a communal event at village level. It articulates the social unit of the village in terms of ritual performances, which include a pattern of gift-offering and redistributing wealth.

In fact, as I will argue, the whole ritual complex can be seen as a ritual definition of the village unit, a unit which in terms of the spread-out settlement structure is not clear for an outside observer. This is not unique to this area but can also be found elsewhere in Nepal (e.g. Lecomte-Tilouine 1993: 263ff.) and in many parts of India. E.g. Nandini Sundar writes in reference to Bastar (central India) that "the point of [village] cohesion may be said to be the celebration of the earth festival" (Sundar 2002: 150). In the following, I will present some examples of variations among the Rai groups from the fieldwork I have been carrying out in eastern Nepal since 1984. The first example concerns the Mewahang Rais (Sankhuwa Sabha district), I studied in the 1980s and 1990s. The second and third examples are taken from the Puma Rais (Khotang district) and Chintang Rais (Dhankuta district), where I have been conducting fieldwork since 2004 in collaboration with colleagues as part of the 'Chintang and Puma Documentation Project'.[2]

Mewahang (Sankhuwa Sabha): sakhewa puja

The major Mewahang villages in Bala, Tamku, Mangtewa, Yaphu and Sisuwakhola VDC (whose total Mewahang population is about 3,500) are said to have been established by a First Settler whose deeds are usually remembered in mythical narratives. These recount how the First Settler (and his entourage) arrived in the area and took control of the territory through an act of 'domestication'.

> For example, in the village of Bala a myth is recounted in which the forefathers of the present residents are depicted as the first legitimate settlers on the territory. (The following is a summary of the myth presented in Gaenszle 2000: 288, 302.)
>
> Before the Mewahangs, there were Sampange Rai settlers on Bala territory. However, the *ca:ri* (territorial spirit) kept pestering these settlers and even killed some of them.
>
> Eventually the forefathers of the present-day Balalis, Yungthu *râjâ* and his three sons, came and took possession of the land. Though the Sampanges had erected the *sakhewalung* stone in an attempt to pacify the *ca:ri*, they were not able to celebrate the necessary rituals. Only the Mewahangs had proper knowledge of the *muddum* to pay reverence to the territorial spirit, and therefore they chased the Sampanges away and established themselves as the legitimate settlers. From then on the Mewahangs celebrated the offering for the *ca:ri* every year, giving a blood sacrifice and offerings of beer.

This myth legitimizes the status of the major clans in the village: the three sons of Yungthu are the founders of the three 'original' clans of the village, ranked in order of seniority according to their forefather's position: the descendants of the eldest son are the Temora clan, while the descendants of the second son make up the Lumluma clan, and those of the youngest are Uchongma. There are other clans, but their presence is the result of later developments (mainly immigration).

Thus the foundation myth can be seen as a 'charter' for the social order in the village. The special status of the descendants of the First Settler is also expressed in the fact that they all share a common ritual name (called *same* or *samet*): in the case of Bala, all the men of the three original clans are referred to as '*sakchami*' in ritual contexts, and all their sisters are known as '*chekhama*'. This ritual label thus defines a social unit which I will call the *samet*-group, and which can be seen as a proto-clan or phratry (see Gaenszle 2000: 117-122, Gaenszle 2002: 49).

The status of the original clans can also be detected in the settlement structure: most of these clans live in the heart of the village, around the ancestral stone; they occupy and own the best rice lands (*khet*) of the village.

Moreover, in all the villages, these original clans and in particular the seniors are granted a special privilege within rituals. This can be seen clearly in the example of Bala. The sacrifice to the *ca:ri* (territory), as laid down in the myth mentioned above, is celebrated in the month of *mâgh* (January/February). Since the offerings are presented at the ancestral stone, *sakhewalung*, the rite is also known as *sakhewa pujâ*. The senior lineage of the senior clan, i.e. Temora, finances the offering of a pig and a clay container (*ghyampâ*) of millet-beer (*jâr*). No priest is required for the rite, but according to one myth, it was previously celebrated with a drum (*dhol*) which is nowadays only used by knowledgeable shamans (*selemi*).

At the time of my research (in the late 1980s) the ritual address to the *ca:ri* in the form of the ancestral stone was delivered by the Mewahang elders. The members of the three original clans and first settlers of the village (*kipatiyâ*) assemble and receive parts of the offerings as *prasâd*. Later immigrants, like the Kulunge settlers (non-*kipatiyâ*s), however, are excluded from the rite and therefore do not receive *prasâd*.

While this rite is presently no longer celebrated on an annual basis, the focus has shifted to another ritual called *ca:ri pūjâ* which is performed in the month of Cait. This offering to the territory (*ca:ri*) and the local water tap (*ca:wa*) is organised by each local headman (*jimmâwâl*, 'tax functionary') separately, hence there are several *pūjâ*s held on the same day. Here all villagers take part, members of the original clans as well as those of immigrant clans. Traditionally, everyone had to bring a gift that they presented to the headman in charge of their area (a container of rakshi and some produce from the land). At the end, a big communal meal was hosted on the premises of their respective headman.

These rituals are described in more detail in a separate paper (Gaenszle 2010). As I argue, this ritual of *ca:ri pūjâ* can be taken as a reflection of the historical developments and changes in the political structure of the village. As the number of tax functionaries multiplied, so the ritual organisation was split up equally, altering the festival and revoking its character as a village ritual. Though the position of headman was eventually no longer limited to original clans, they still played a prominent role.

There is no dancing in any of these spring cults. However, the Mewahangs have a dance in which the origin of species is enacted. This dance, called *khamangsili*, is performed during the harvest rituals, the *nuwâgi*, which is a first fruits rite celebrated in October or November. This ritual is not referred to as *sakhewa*, but as we will see, a similar feature can be found in other traditions.

In the other Mewahang villages, similar *ca:ri pûjâ*s as in Bala take place, though the specific procedures vary (for a more detailed account see Gaenszle 2010). What is important is that all have an ancestral stone as the central focus of the rituals. All these territorial cults are dominated by the local *samet*-group, and though the ritual is no longer as exclusive as it was, a certain status distinction is still evident. It is also interesting that all villages (including Bala) have a village Devi, which is another focus at village level. However, this must be seen as a more recent development.

It is these rituals then which most clearly demarcate the unit of the village among the Mewahang Rais. It is therefore not surprising that the boundaries of the former Panchayats and present village districts (VDCs) are more or less the boundaries of these traditional territories. Though the political system has drastically changed, the old system still does affect the definition of contemporary administrative units (Fig. 1).

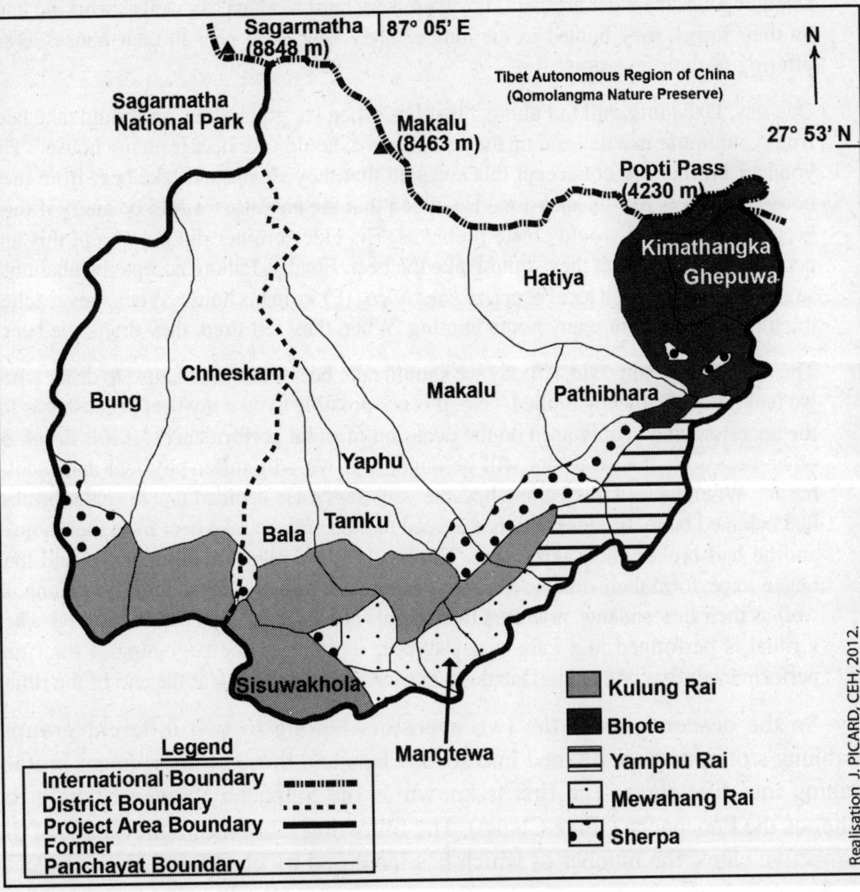

Figure 1: Mewahang villages

Puma Rai (Khotang District): phagu and sakenwa

The Puma Rais (numbering about 6,000) are in many respects close relatives of the Mewahangs. They are often regarded as a section of the Bantawa Rais, but they are linguistically as well as culturally distinct. Their villages are spread over the following VDCs: Pauwasera, Mauwabote, Diplung and Devisthan. Whereas in the Mewahang case, village districts also represent core settlements, the situation among the Pumas is far less clear-cut, as each VDC contains a number of spread settlements which are more appropriately characterized as hamlets. Most of these are Puma-dominated but there are also some Bantawas, Chetris, Kamis.

Again, in order to understand local concepts one has to look into the local mythology about ancestral migration. The Pumas derive their clans from the two

First Ancestors, two brothers called Dabalung and Palung. They settled on both sides of the valley and each carved out a territory, known as their *bobi*. A local myth explains why the two brothers quarrelled and eventually split up.

> In the past, Dabalung and Palung were brothers. Palung was the younger brother and Dabalung was the elder brother. They were very hard workers. As well as working hard on their farms, they hunted in the jungle. They kept local beer in their houses as an offering to their ancestors.
>
> One day, Dabalung said to Palung, "Brother, when we go hunting we should take beer from your house one day and on the next day we should take beer from my house". The younger brother did not accept this and said that they should not take beer from their houses as it was for the ancestors. He added that the ancestors would be angry if their beer was taken and would create problems. His elder brother did not accept this and persisted in saying that they should take the beer. Finally, Palung accepted Dabalung's suggestion and he took local beer (*rakkhe khâcâ*) (P) from his house. When they reached the jungle, they spent many hours hunting. When they felt tired, they drank the beer.
>
> The next day, Palung said, "Today we should take beer from your house to drink when we feel tired". Dabalung replied, "No, it is not possible to take my beer because it is for the ancestors and is only used on the occasion of ritual performances. If it is drunk on other occasions the ancestors will be angry and we will suffer physical and mental harm". When he said this Palung became very angry. He decided that his elder brother had behaved badly because they had agreed to take turns to take beer from their houses and he had broken their agreement. The brothers' relationship deteriorated and they began to perform their rituals separately. From then onwards Dabalung and Palung, as well as their descendants, practised their rituals in a different manner. Nowadays when a ritual is performed in a Palung house, beer is drunk at the beginning of the ritual performance. By contrast, in Dabalung houses, it is only drunk at the end of the ritual.

So the descendants of the two ancestors belong to two different groups: Dablung's offspring are divided into seven clans and those of his younger brother Palung into five clans. The first is known as the Satpacha ('Seven Clans'), the latter as the Pancpacha ('Five Clans'). The following table shows the names of the respective clans, the number of which has increased by one clan each due to clan fission.

Sâtpâchâ	**Pañcpâchâ**
Dabalung (elder brother)	Palung (younger brother)
1. Yongdohang / YongduhangĐ	1. Mithahang
2. Hadire / Hadire	2. Tungmalung / Tongmalung
3. GarajaĐ	3. Metlongthong / Metlongdong
4. ThumrahangĐ	4. Wabihang / Babihang
5. Limmachit	5. Khahong
6. Hangthangkha / Hangtangpa	
7. Dumanglung	6. Khirihang
8. Henyongcha mainly in Pauwasera,Mauwabote VDC	mainly in Diplung, Devisthan VDC

Thus the Puma settlement area consists of two territorial units (*bobi*), which are named after these clan groups: Palungbobi (which lies north of the Ruwa Khola) and Dabalungbobi (south of the Ruwa Khola). This division has no implications as far as social relationships are concerned (e.g. there is no ban on intermarriage), and seems to be largely a matter of ritual distinctions, as indicated in the myth. It is important to note that all Pumas belong to the same *samet*-group: all Puma males are *namdhungpa* and all their sisters are *namdhungma*. Thus in the Puma context the *samet*-group, other than among Mewahangs, is coextensive with the ethnic group as a whole. As in the Mewahang case, marriage within the Puma *samet*-group is possible, but whereas in the former context such a marriage tends to be avoided due to 'incestuous' connotations, in the Puma case it is the normal way to marry.

This situation may be related to the particular history of the Pumas. According to some accounts, their forefathers escaped an attack during an internecine conflict, hid in the Ruwa Khola, and eventually settled there. Perhaps it is due to this unintended or initially temporary stay that the Pumas have no communal festival. The spring festival is celebrated only at household level: each household performs the *phagu* ritual and the Sakela dance separately. This situation is unusual, and there is a myth which explains why it came about.

> Budhahang, a well-known ancestral king had brought his ancestral shrine, the *sakenwalung*, from a place in the west to Chisapani (in the Puma area). There, the stone fell to the ground and became the new place of worship. Twice every year, once in spring for *ubhauli* (the rising time) and once in autumn for *udhauli* (the descending time), people worshiped at this stone and danced the *sili* dances. But then Budhahang died, and no one was able to perform the proper rituals any more. One day a white elephant came, roamed around the place and broke off part of the stone. People were upset and decided that from now on they would make the offerings at home. This is how the ritual of *phagu puja* originated.

It is clear then that a community ritual is seen to be the norm, and when such a ritual does not exist, this calls for some explanation. Yet in recent times (in 2005) people have begun to 'rectify' the abnormal situation and have established a communal Sakenwa shrine in Chisapani. Some claim that it was there originally and had only been 'lost', but this is difficult to verify. In any case, this ritual site can be seen as an expression of Puma solidarity.

Nevertheless, this does not mean that the ritual innovation is an entirely new form of community-making. It is interesting to note that there is an important Devi, a goddess called Hongmadevi (or Kanyadevi) who has a shrine in the village of Devisthan. This deity is the only such Devi in the Puma area, and considering the situation among the Mewahangs (and elsewhere) where most villages have a Devi, this also indicates that in the Puma context, the ritual unit is more inclusive. In ritual terms the situation is clear: in view of the celebration of a *sakenwa* ritual—whether celebrated in the mythic past or only recently—the Pumas are not only a community but can also be seen as one large 'village'. This interpretation is corroborated by the ritual focus on one Devi shrine. For an outsider, the Puma settlement area is just a collection of spread settlements and does not differ much from other mid-hill areas. Only an understanding of the ritual system reveals the underlying structure of territorial units.

Chintang (Dhankuta district): *wadhangmi* festival

Likewise the Chintang community is considered to be one of the Rai groups (and usually they also regard themselves as such), but it must be stressed that they are linguistically closer to the Limbus than most Rais, as their language is classed among the eastern Kiranti subgroup (also see Gaenszle *et al.* 2005). Similarly, Chintang culture (along with Belhare and Athpare) is somewhat different from the central and western Kiranti patterns. Moreover, the Chintang group is a unique case in that all Chintangs reside in one village—likewise called Chintang: they number about 4,000 out of a total of 9,088 residents in the VDC (latter figure from the 2001 census). Thus the ethnic Chintang group is a one-village unit. Only a few households are to be found outside this VDC (mainly to the south in Ahale VDC, which was clearly set up later by emigrants).

Chintang village is a spread settlement consisting of a core area (Mul Gâu) and some 65 hamlets, often comprising only a handful of households. There are altogether 12 'genuine' Chintang clans, but what is interesting is that they do not, as is customary among the Rai groups, claim descent from one common ancestor. Instead, there are different lines which have no link except for later intermarriage. However, there is one original settler from whom the village name is derived: Chendahang. He is said to have come from the south, Barachetra, the sacred site of the confluence of the seven Koshis. He had three sons: the name of his eldest son was Arihang, the second son was Khinci and youngest son was Dalahang. These three are the eponymic ancestors of three clans which are resident today in the core village. Two more, Tuprihang and Ananda, later derived from the eldest brother's offspring. Four other clans are said to derive from the marriage of the daughter of one Sambhong to an immigrant hunter who fathered four sons—the eponymous ancestors of the four clans residing in Sambhonggau. The remaining clans have separate stories of origin. But what is of note: in spite of these varying backgrounds, all these clans have *samet namcihang tunglungmi*, i.e. all male descendants have the ritual name *namcihang* and all their sisters are *tunglungmi*. Thus, as in the Puma case, the *samet*-group is a kind of proto-clan or phratry defining the ethnic unit as a whole. And what is of particular interest to us here is that all these clans play a prominent role in the Wadhangmi festival described below.

A striking feature of the Chintang village social structure is the existence of no less than 37 other Rai clans, i.e. exogamous units that are not Chintang clans in the sense described above. Of these, 18 clans are Bantawa (with a distinct proto-clan name, i.e. *samet*), apparently settlers who came from the western bank of the Arun River. And 19 clans are simply immigrant members of other Rai subgroups that are identified by names, such as Puma, Dumi, Khaling, Camling, Kulung and Lohorung. As these are isolated households with no marriage partners from their own group, their ethnic name has become the clan name, and most have forgotten their original clan affiliation. It is not entirely clear when these groups settled in the area, but there is strong evidence that they were displaced from their homelands further westwards due to the Shah military expansion in the late eighteenth century.

There are also a few other settlers from other castes, such as Chetris, Damais, Kamis and Newars. All the settlers, Rais and non-Rais alike, worship Chintang Devi at her well-known shrine in the centre of the village (on the pilgrimage route to Barahchetra). But it should be noted that this village *devi* is unique in that she has translocal significance. Many devotees from neighbouring villages and districts, including high-caste Bahuns and Chetris, come to Chintang to worship, sacrificing chickens and goats.

The non-Rais, however, are numerically insignificant; the village is clearly a Rai village with almost 80 per cent Rais. But far from being a coherent group, the basic distinction between 'genuine' Chintang and immigrant Rai, who usually also speak Chintang, cuts across the whole Rai population of Chintang village. This becomes most evident when considering their major festival, Wadhangmi, which is, like the *sakewa* rituals described above, basically a ritual to guarantee the territory's prosperity.

Wadhangmi (also Wadangmi) is an important ritual event which expresses the unique status of the 'genuine' Chintang clans in a celebration lasting six consecutive days. A similar festival is common throughout the Khalsa area (especially in Chintang, Khoku, Akhisalla) and is also celebrated by the Athpahariya Rais. In Chintang it is closely associated with Budhahang, a legendary tribal king in the region who heroically fought the invading Shah army by using his magic. According to one story, people suffered from severe hardship, such as disease, famine, draught, etc. when Budhahang once disappeared from his residence. To protect them against such misfortune, people started to worship Budhahang as a protective deity in order to secure his presence.

The festival begins on the Tuesday following the full moon of Kartik (October/ November) and it lasts until the following Sunday.

On the first day, called *thâlni* (N.), literally 'starting', the offering called *sakewa pūjâ* is the major ritual event. This takes place at the *sakewa thân*, a shrine at a rock formation on the margin of the core settlement. A special ritual expert, called *picami*, who belongs to the Khukkang clan, acts as priest—and this is the only function in this category. On this occasion, the *tupla phema*, 'opening the banana leaf', is performed. It is a very important

Photograph 1: Picami priest during Sakewa on the first day of Wadhangmi (Photo courtesy CPDP)

ritual because the spreading of offerings on a banana leaf marks the beginning of the festival.

On the second day, Wednesday, two other ritual functionaries, the *nangsuba* (tribal priest) and the *chambak* (shaman) perform offerings to the domestic hearth and in particular to Budhahang, as well as a small divination in their own houses. Relatives join in and dance in the courtyard to the sound of drumming.

On Thursday, the same ritual sequence as described above at the house of the *nangsuba* is performed at the house of the *nakchong*, the temple priest of the goddess Chintang Devi. The *chambak* and the *nangsuba* come to the house of the temple priest where they become possessed by their tutelary divinities and perform the main divinations on behalf of the whole village, diagnosing any possible tension and announcing the prospects for the coming year. Then, the *nangsuba* and *chambak* dance outside the house. On this day, all villagers join in on this occasion and dance all night long.

On Friday and Saturday, the same basic sequence of rituals is performed in the house of all genuine Chintang clan households. The senior-most households of Ananda and Tuprihang are a special focus on these two days. This appears to be linked to the fact that they belong to the 'oldest' stratum of settlers who descend from Chendahang (see above).

On Sunday, the villagers, led by the *nangsuba* and *chambak*, gather at Pancakanya, a gateway into Chintang. Pancakanya is named after the goddess Pancakanya ('five virgins'). Sunday is called *turni* (N.), 'finishing', as it is the last day of the Wadhangmi festival. The main priest of the *turni* is the temple priest of Chintang Devi. There are three altars: offerings are first performed at the permanent altar of Pancakanya Devi. Offerings are then made at the permanent altar of Budhahang. Lastly, offerings are given at the permanent shrine called *deorali*, at Sahid park ('martyr's park').

The festival is clearly a communal one; all villagers take part in the public offerings and the dances, and the celebrations often last all night. Yet on examining the social roles and privileges, it is apparent that the leading parts in the celebrations are reserved for the original, "genuine" Chintangs, and here more than in any other social event the implicit hierarchy of clans becomes visible. As the key rites during the ancestral offerings and any divination regarding the village's good fortune are performed by the members of these clans, it would appear that the prosperity of the village rests with them. Again, the village, in a distinctly hierarchical sense, is a social entity which is reconstructed every year in a grand ritual event.

The Sakewa festival as an ethnic symbol

Since the foundation of the ethnic organisation Yayokkha in 1990, the Rais have asserted their cultural identity in a more and more standardized manner. They have encouraged the documentation and study of the various Rai cultures and languages, and have set up a vast network of district offices which represent the interests of this minority group. Yet, in spite of this acceptance of cultural diversity within the ethnic group, the Rai leaders of Yayokkha have promoted the re-fashioned Sakewa festival as a ritual common to all Rais. Sakewa/Sakela is now celebrated as the major Rai festival, drawing large numbers of participants, usually with several guests of honour from public life. This festival is especially popular among the urban Rais and those living outside Nepal.

In the process of promoting this ritual, Yayokkha claims that all the different Rai festivals are basically one and the same:

> "'SAKKHEWA' is one of the main cultural festivals of the Kirant people of Nepal, which has, indeed, a long history that goes back as the Kirant civilization [...] SAKKHEWA is known with various names viz. 'SAKLE', 'SAKELA' and 'TOSHI'." (Yayokkha leaflet, undated)

In fact, similar standardization and symbolization have emerged among the Rais on the Indian side, in Sikkim. There a pamphlet clearly states:

> "The main Kirat Khambu Rai festivals are Sakewa and Sakela. [...] Sakewa [...] is celebrated with great enthusiasm by all Kirat Khambu Rai" (Akhil Kirat Rai Sangh Sikkim 2005)

In cities like Kathmandu the festival is organised on a grand scale, partly in Jawalakhel but also at the Khula Manch, i.e. right in the centre of the capital. A political figure, like a Rai minister (e.g. Bal Bahadur Rai in the 1990s), is usually selected to act as the dance group leader, *silimakpa*. Interestingly, all the different Rai groups dance in separate circles, thus expressing the 'unity in diversity' principle. The style and speed of the dancing differ between the different groups.

Photograph 2: Kirat Rai Yayokkha members on the way to a Sakhewa celebration in Beltar, Udaypur district. (Photograph: courtesy CPDP).

So is this the same ritual as the one performed in villages? If one looks at the dance, there is no doubt that this is largely the same performance as the one in the local context. Today, ethnic organisations take great care to teach the proper dances to younger people: this is one of the most elaborate forms of transmitting a special habitus to the next generation.

However, there are a number of obvious differences in the festival as celebrated in the modern context:

- The multi-ethnic context is rather unique: although there is sometimes more than one Rai group in the hill villages, one group usually dominates the scene. Here all Rai groups are seen as equal, though the Bantawa have managed to generalize their model of the festival.

- In the modern context participants are generally middle-class people, well educated Rais who have accumulated a certain wealth in the modern economy (e.g. as Gorkha soldiers, foreign workers, development institutions, etc.). For them the festival is a symbol of their ethnic heritage, and a substitute for the lost village context.
- The religious significance of the ritual is simplified. The complex interconnection between the earth cult, territorial claims and legitimacy, the link to the local political system, its connection to household rituals: all this is lost in the new context. What remains is a watered-down imagery of Rai mythology. Today the couple Sumnima-Paruhang are emblematic of Rai identity, but few know the detailed mythic background still prevalent in the villages.
- Whereas in the village context the ritual is entrenched in local kinship and the political system, in the urban context it takes on new political meaning: the Rais assert their ethnic identity as a minority with particular claims in the national context. Thus it is logical that a modern leader, a minister or party chief, acts as the leading dancer (originally a semi-religious functionary).

The question remains: is the new ritual merely a variation on the old one or is it something qualitatively different? For a villager, the urban celebration is no real substitute for the one at home, though he does take part when he happens to be there. For the fields and animals to be fertile, the traditional practice has to be continued, but for the urban Rais (who no longer have any fields) this is not the issue.

Thus there is a shift in the intended efficacy: the new festival is not about fertility but about strengthening the ethnic identity within a nation state. The dance in honour of their origins therefore has a different meaning in the two contexts:

- In the village, it is a renewal of the link with ancestral forces (as emphasized in the household clan rituals) as well as with the space and fertility of the territory.
- In the urban context, this link is replaced by a generalized image of heritage as "culture", encoded in such things as dance, music, dress, food habits, and a particular mythology.[3]This also implies a more abstract notion of a pan-Rai, or rather pan-Kiranti territory. As the Constituent Assembly, elected in April 2008, has been dissolved and the task to draft a new constitution is yet to be fulfilled, the question of whether the eastern hills will be defined as a federal state remains a burning issue. Many Kiranti ethnic activists are even demanding a Kirant Autonomous Region! Indeed, such rituals of a newly defined ethnic unity are a significant statement with a political message. They express claims related both to ethnic kinship and territorial belonging.

In the final analysis, the ultimate aims of rituals, both old and modern, are not so different: they both concern prosperity. The idea of prosperity, of course, has changed. It no longer relies on the fertility of any particular village land which is imbued with ancestral memories. But it is still based on an image of the past, the founding ancestors at the beginning of an unbroken line of transmitted knowledge, and the notion of a homeland.

REFERENCES

Allen, N. J. 1976. 'Approaches to illness in the Nepalese hills', in Loudon, J.B. (ed.), *Social anthropology and medicine*, London and New York, Academic Press, pp. 500-552.

Burghart, R. 1987. 'Conclusion: The perpetuation of Hinduism in an alien cultural milieu', in Burghart, R. (ed.), *Hinduism in Great Britain: The perpetuation of religion in an alien cultural milieu*, London, Tavistock Publications, pp. 224-251.

Gaborieau, M. 1982. 'Les fêtes, le temps et l'espace : structure du calendrier hindou dans la version indo-népalaise',*L'Homme*, 22: 11-29.

Gaenszle, M. 1997. 'Changing concepts of ethnic identity among the Mewahang Rai', in Gellner, D., J. Pfaff-Czarnecka, J. Whelpton (eds.), *Nationalism and ethnicity in a Hindu kingdom: the politics of culture in contemporary Nepal*, Amsterdam, Harwood Academic Publishers, pp. 351-373.

– 2000. *Origins and migrations: kinship, mythology and ethnic identity among the Mewahang Rai of East Nepal*, Kathmandu, Mandala Book Point and The Mountain Institute.

– 2002. *Ancestral voices: oral ritual texts and their social contexts among the Mewahang Rai in east Nepal*. Münster, Hamburg, London: LIT Verlag.

– 2008. 'Shamans at the lake', in Marschall, W., P. von Wyss-Giacosa and A. Isler (eds.), *Genauigkeit: Schöne Wissenschaft*, Wabern-Bern, Benteli.

– 2010. 'Political aspects of the territorial cult among the Mewahang Rai', in Lecomte-Tilouine, M. (ed.), *Nature, culture and religion at the crossroads of Asia*, New Delhi, Social Science Press, pp. 190-221.

Gaenszle, M., Balthasar B., Goma B., Elena L., Netra P., Ichchha R., Manoj R., Novel K. Rai and S. Stoll. 2005. 'Worshiping the king god: a preliminary analysis of Chintang ritual language in the invocation of Rajdeo', in Yadav, Y., G. Bhattarai, R. R. Lohani, B. Prasai and K. Parajuli (eds.), *Contemporary issues in Nepalese linguistics*, Kathmandu, Linguistic Society of Nepal, pp. 33-47.

Lecomte-Tilouine, M. 1993. *Les dieux du pouvoir. Les Magar et l'hindouisme au Népal central*,Paris, CNRS Editions.

McDougal, C. 1979. *The Kulunge Rai: A Study in Kinship and Marriage Exchange*, Kathmandu, Ratna Pustak Bhandar.

Needham, R. 1975. 'Polythetic Classification: Convergence and Consequences', *Man* 10: 349-369.

Schlemmer, G. 2004a. *Vues d'esprits. La conception des esprits et ses implications chez les Kulung Rai du Népal*, Université de Paris 10.

– 2004b. 'Inventing a past for inheriting a future: New visions of history among the Kirant intellectuals of Nepal', *European Bulletin of Himalayan Research* 25/26 ('Special Double Issue: Representing Local Histories in the Himalayas'): 119-144.

Sundar, N. 2002. 'Village histories: coalescing the past and present', in Chatterjee, P. and A. Ghosh (eds.), *History and the present*, New Delhi, Permanent Black, pp. 144-182.

Notes

1. On Rai ethnicity in the modern context, see Gaenszle 1997, 2000, and Schlemmer 2004b.
2. The project was funded by the Volkswagen Foundation under the DOBES program (Grant No. BI 799/1-2 and II/81 961, 2004-2009, PI B. Bickel). I am grateful to this agency for its support and to my colleagues in Zurich and Kathmandu for their collaboration. In particular I wish to thank Balthasar Bickel, Judith Pettigrew, Arjun Rai, Ichchha P. Rai, Novel Kishor Rai, Shree Kumar Rai and Vishnu Singh Rai. The data on which this paper is based are accessible at the DOBES archive (www.mpi.nl/DOBES). For further information on the 'Chintang and Puma Documentation Project' see www.uni-leipzig.de/~ff/cpdp
3. On the ideological value of essentialist notions of 'culture' see Burghart 1987: 246ff.

1.3.3.

Defining Community: A Historical Study of Territory and Transformation in the Western Himalaya

Chetan Singh

> "In the majority of communities, the bulk of their populations remained in the area of their original or adopted territory [...] Their modes of production, patterns of settlement and folk cultures spring from their diurnal round of work and leisure, itself formed out of their ceaseless encounter with a particular environment" (Smith 1987: 183).

> "Places in the past were peopled. Their changing landscapes contained within them social conflict and social cooperation among individuals and groups; and these relations should be seen as a vital part of the making of these landscapes" (Baker 1984: 27).

Conceptualizing territory

We need to acknowledge at the outset that a set of physical and geographical conditions does not, by itself, constitute a territory; just as the mere grouping together of people at any particular time or place is insufficient to create a community. A geographical area is transformed into a 'territory' as a result of social thought and activity. Conversely, the material realities of geography also influence in great measure the nature of activities and social organization that a particular society adopts (Kobayashi 1989: 164, 177).[1] Consequently, a more explicitly recognized community is brought into existence. It appears, therefore, that the emergence of community and the configuration of territory often occur simultaneously. By discussing one in the absence of the other we risk ignoring their essentially interdependent and complementary nature. The processes through which ideas of territory and community come into existence, and the manner in which they are perpetuated or altered, appear therefore, to be inextricably connected.

How a community constructs its territory depends on how it structures itself. Individuals and families, Godelier suggests, have necessarily to be part of a community in order to "reproduce the conditions of existence of this community while producing their own" (Godelier 1986: 54). For this purpose, he argues, "Space is distributed among these communities, each of them exploiting a territory whose bounds are known to the neighbours" (Godelier 1986: 55). Godelier, nevertheless, recognizes that some societies do not claim a territory of their own while some even share territorial resources with others (Godelier 1986: 86). It seems, therefore, that the nature of a geographical area, and the necessity of evolving methods of procuring a livelihood from it, caused societies to organize themselves differently. Yet, from another viewpoint, it can also be argued that it was the diverse forms of social organization that compelled societies to search for and claim the resources of different kinds of territories.

Mountain topography, by its very nature, provides observers with an apparent—clearly visual—depiction of interaction between human activity and physiographical space. One is tempted to believe the eye as it traces the likely contours within which community and territory might have interacted over the course of their historical development. But obvious appearances are often deceptive. Ideas and beliefs are no less significant influences upon the community than compulsions of the geography. In organizing its physical surroundings, a community's response is often implicitly ideological–encompassing polity and social structure. What a particular landscape or territory means to it is as much a matter of physical realities, as it is a question of mentalities and interpretations.

Small communities, especially in isolated areas, probably needed to develop a high degree of economic self-sufficiency. In pre-modern times this self-sufficiency was, perhaps, combined with a great amount of political autonomy. The same might not be true for individuals. Compelled by medieval ways of thinking, the individual found it difficult to separate himself from the society in which he lived. Robert Nisbet reiterates the point that allegiance to the "family, the village or the gild" preceded the notion of individualism (Nisbet 1962: 81). Medieval law and polity recognized and interacted, not with the individual, but with the group (family, community, etc.) as the primary and most significant entity. A village, therefore, would be represented by a community whose members acted collectively. It was only through the community, or the social group, that life in the village was possible. An individual could hardly have survived otherwise. Quite possibly, the ability of the community to retain its sphere of influence was inversely related to the central authority's ability to assert its own power and restrict that of the community (Nisbet 1962: 83, 84-5; Smith 1987: 31).[2]

Nevertheless, the authority of a community could hardly have been constructed entirely as an ideology of opposition to a hypothetical or real central authority. There were other means of creating authority. In settled agricultural or agro-pastoral villages, landholding families usually formed the core around which the community was consolidated (Habib 1999, Kasturi 2002: 45).[3] Local histories were constructed to legitimize the dominance of landholding groups. A large number of pre-modern societies adopted this method of rationalizing social control, and many communities across the Himalayas were no exception. Rooted, as they were, in a mixture of fact, myth and legend, historicized 'memories' of the community formed the basis of its world-view. Customs, norms, social hierarchies and a host of other notions central to the integrity and survival of the community were derived from this world-view.

These created 'memories', or histories, also became the channel through which the local community linked and described itself to a world that lay beyond the limited confines of the village (Levine 1988: 21). They enabled the community to locate itself in a larger—invariably unequal—grid of social and political relationships.

For the present, it may not be relevant to dwell upon the internal functioning of the village community. Of greater interest is the relationship that existed between the community and its constituting lineages and clans on the one hand and the larger polity within which they operated on the other. It may be useful to understand how these lineages or clans effectively functioned as political organizations and thus successfully asserted explicit territorial claims—however small.

Majumdar's detailed study of polyandrous Jaunsari society (in present-day Uttarakhand) has attempted to reveal how the "lineage, locally known as *aal* or *thok*" served as the criterion for dividing society into groups. It was, he says, "on the basis of these groups that the villages have been established and maintained" (Majumdar 1962: 70).[4] Within the village, families belonging to a common sub-lineage resided in a common locality called *bhera*. Several villages were grouped together in a politico-administrative unit called *khat*. Even as the affairs of the *khat* were managed primarily by the *khat sayana*, he usually did so in consultation with the council (*khumri*) of village elders. Furthermore, and most importantly, the temple of the village god was the central institution through which the collective identity of the community was constructed and reinforced (Majumdar 1962: 25, 35).[5]

In the adjoining mountains of Himachal, peasant society was strikingly similar. Far-flung hamlets, with adjoining patches of cultivated land, were too small to form viable administrative or political units by themselves (Fraser 1982: 217, Kashyap 2000: 120-1).[6] They were, therefore, grouped together to form a larger unit called the *ghori* (Sharma 1990: 136, *SHSG 1910 Bashahr*: 42-3).[7] Apart from the fact that in British times each *ghori* had a *lambardar* to look after its affairs (especially the task of revenue collection) a common grazing ground, shared by several hamlets of the *ghori*, was often an important integrating factor (*SHSG 1910 Bashahr*: 65). For an agro-pastoral society, the control and efficient management of pastures was a matter of survival. Not surprisingly, therefore, the *ghori* appears to have been a territorial unit dominated by a community of peasant-pastoralists. It was a community that could, and did, mobilize from within itself a group of warriors or *khund* to protect its interests (Sax 2006b: 120-34; Sutherland 2006).[8]

Legitimizing territorial control

The aggressive assertion of territorial claims by competing clans or communities through the use of force was only part of the story. Permanent control over an area—even at local level—required legitimisation through a long-drawn but more widely acceptable social process. Gradually, the cosmic view of the community began to reflect the physiographical features and natural peculiarities of the area in which it resided. Occupation of territory was ideologically reinforced and legitimized through the mediating role of deities. Stories about the landscape and supernatural actions of local gods were woven together into mythical sagas that passed for history.

Through the power of myth and legend, communities legitimized their claim to territories. This simultaneously gave deities 'an identity and a place of their own'. Western Himalayan myths about how deities acquired a domain for themselves often begin with their accidental discovery in a forest or a field. This was followed by a period of search, migration and successful struggle for a territory and a people to rule over. Then came the delineation of territories and subject village communities, and the creation of hierarchies. The new (i.e. now prevailing) socio-political order is thus seen as the outcome of a contest between gods: something that ordinary mortals could question only at great peril (Rose 1970, Singh 2006: 328-35, Sutherland 2006: 91).[9] Moreover, the naming of many community gods after local physiological features converted the surrounding space into a meaningful place: one that *belonged* to a particular community and the resources of which could be justifiably used by it (Lovell 1998: 55). Man, god and nature were thus inextricably linked in a relationship that was—and remains—inherently political.

The institution of village gods thrived across the entire Indian subcontinent. Each local deity resided amidst its followers, often in a tiny village domain. There were, thus created, thousands of territorial spheres that consisted of either a single hamlet or a few closely associated villages (Singh 1901: 49, Sec. 29).[10] The ubiquitous village god (*gram-devta*) was concerned primarily with the everyday 'facts of village life' such as local rivalries, personal misfortune and disease. It exercised no control over the more distant 'great world forces' that impersonally shaped the destiny of multitudes (Whitehead 1921: 16-17). The small gods exhibited entirely human characteristics—quick to take offence and to anger, but also easy to please through the sacrifice of animals (Whitehead 1921: 17, 30). Most importantly, the priests of the village gods were only occasionally drawn from Brahmin castes (Klass 1995: 83; Srinivas 1965: 181; Whitehead 1921: 30, 43-4).[11] Yet, if exigencies required, Sanskritic symbols and rituals were used to portray these clearly local deities as manifestations of powerful Brahmanical gods (Srinivas 1965: 184) There existed, as it were, two easily discernable socio-religious territories. One was the locality of the village god, the non-Brahmin priests and the constituency of small peasant communities. The other was the larger world of the Sanskritic gods, of Brahmin priests and ideologues and of the prosperous and powerful dominant classes (Srinivas 1965: 60, 181, 185).[12] Because of the unequal scale and significance of the concerns they addressed (and the kind of people they involved) the two 'territories' seemed to emphasize their difference.

We know, of course, that these were not exclusive zones. They interacted, merged or separated from time to time and in different situations. So difficult, indeed, were they to demarcate, and so remarkable was the flexibility and dynamism they possessed, that it seems only appropriate to regard them as part of a hierarchical continuum (Singh 2006). These socio-religious spheres were redefined as they followed the changing concerns of the 'community'. Conversely, the notion and description of community could also be altered to correspond to a delineated territory. In effect, community and territory were both redefined as one moved up or down the social and spatial continuum. A common linking factor, however, appears to have been the use of a religious idiom to articulate socio-political objectives.

There were, at village level, countervailing forces contesting for domination. Srinivas has argued that while "caste has a tendency to stress horizontal ties", the economic compulsions of agrarian life pushed small communities in the direction of 'vertical', village-based solidarity. While caste loyalties encouraged links between similar caste groups in different villages, the need to organize agricultural labour required that the collective interests of the village community be accorded priority. Long-lasting feuds between communities may possibly have been rooted in the primacy accorded to village loyalty. Srinivas suggests that only with the political integration brought about under the rajas did caste become an important political factor (Srinivas 1965: 43, 69). Community and territory, therefore, responded to historical and political developments occurring at a higher level, and were both gradually transformed in the process.

This shifting relationship between community, territory and state can be illustrated by examples from the oral history of the Shimla hill region. Local accounts are embedded with collective memories of how village communities established new village gods, replaced old ones or carried them along as they settled in distant places. The most common method of creating, or legitimizing, a socio-religious territory was for a community to establish in its area the image of the principal state deity. But the new image would often be known by a name other than that of the principal deity—preferably after the person responsible for its establishment. In Keonthal state, Deo Chand, one of the ancestors of the Khanoga clan of Kanet peasants installed a newly made image of the state god, Junga, in his village. This new local *deota* came to be named 'Deo Chand' after the Khanoga ancestor, and was ranked as one of the many gods subordinate to Junga (Rose 1970: vol. I: 447).[13] Similarly, the Chhibar clan of Kanets in Jatil *pargana* installed a Junga image in a separate temple, as did the Brahmins of Bhakar at a place called Koti and the Rawal clan at Gaum (Rose 1970: vol. I: 447, 448).[14]

Equally fascinating was the 'displacement' of one god by another. Rose recorded the legend of how *deota* Junga became the state deity of the small principality of Keonthal, after successfully displacing the now forgotten older god (Jipur) and taking over his temples and territory (Rose 1970: vol. I: 443).[15] Even at village level, folk legends record instances of such replacement. By implication, this was also a usurpation of the displaced god's domain. The cult of Kalaur in *pargana* Ratesh (Keonthal) that was associated with Brahmins was displaced by a *devi* (goddess) that was brought by two Kanets from the state of Sirmur. A god from Suket replaced *Deota* Dharta in Jamrot *pargana* (hill area of Patiala state) and this marked the beginning of the popular cult of Manuni. The appearance of new gods and their growing popularity are, perhaps, indicative of important underlying social change (Rose 1970: vol. I: 445). The growing popularity of a new cult at the expense of an older deity was not simply a cosmic reordering. It involved the shifting of social and political allegiances, and the redistribution of territories and real communities (Rose 1970: vol. I: 446).[16]

The story of people and deities migrating from one place to another is also commonly told throughout the hills. For most village communities, their initial arrival at the place where the village stands seems to mark a new beginning. When Kaneti *deota* travelled from distant Dodra to the territory of Keonthal, an entire community of followers—Kanets, Kolis and Turis—accompanied him. These castes represented virtually all social sections that were to be found in a Western Himalayan village (Rose 1970: vol. I: 446). The state of Koti, adjoining Shimla, was quite characteristically, home to gods brought there by their followers from different places. Klainu Deo of Kiar belonged originally to Kulu, while Sip Deo (Sipur) came with the ruling family of Koti from the Kangra region. Sharali Deo (of Sharal village) and Dhanu Deo (of Chhabrog) travelled along with their followers from adjacent Keonthal state where they had formerly resided. Korgan Deo (of Chhabalri village) was brought from Sirmur by one of the princes of Koti and Nual Deo accompanied the people of Kogi village when they migrated to Kogi *pargana* from Suket state (*SHSG*, Koti: 8). This is not merely a recounting of gods who migrated. More importantly, it signifies the movement of groups of people from one place to another. Such movements often required the reiteration of old, or the reconstruction of new, social identities. Local and community gods were used not only for this purpose, but also for legitimizing the occupation of new territory. In the process, new communities emerged and new territories were delineated.

In large parts of the Western Himalayas, it was through the local *deota* that village communities asserted a great degree of political autonomy. The local *deota* was often seen as a 'ruler' in his own right, reigning over his followers, as would a king over his subjects. Collective—albeit unequal—participation in the periodic ceremonies of the village temple forged the community into a political unit (Sax 2006a: 7-13, Sutherland 2006: 84).[17] For the Kulu villages we have the detailed description that Lyall made in 1872. Many of the important temples owned extensive and fertile agricultural land. They also possessed a *bhandar* (granary) where the *deota's* share of the grain from his land and other contributions of his followers were deposited. Several temples had large establishments. Lyall notes:

> "Some of the large shrines have large fixed establishments, a *kardar* or manager, an accountant, one or more *pujaris* or priests, several musicians, several *gur* or *chelas* i.e. interpreters of the oracle, standard-bearers, torch-bearers, blacksmith, carpenter, florist, watchman, messenger, carriers of loads &c., to all of whom *barto* or land rent-free in lieu of pay, are assigned out of the temple endowment. Most have a *kardar*, a *gur* and musicians" (Lyall 1889: 83-4).

Lyall observed further that the village temple in Kulu was "owned, served and managed" by the peasantry of the area over which its *deota* wielded influence. This could range from a small hamlet to an entire *kothi* (a revenue division consisting of several villages). Notably, however, the influence of Brahmins—so common in the lower hills and the plains—was almost entirely missing. Most of the temples had, for centuries, been recipients of revenue-free land (*muafi*) from the *raja* or state. What this meant in actual terms was clearly understood by Lyall when he prepared his revenue settlement report for the Kulu territory of Kangra district. He commented perceptively that, "the *zamindars* of a hamlet or hamlets, who are themselves the only worshippers of the shrine and who distribute the office of *kardar*, *pujari*, *chela* &c., among themselves, eat up the proceeds in periodic feasts. The *zamindars* themselves are in fact, in some degree, the real *muafidars*" (Lyall 1889: 84). Through their close association with the *deota* and the temple, the dominant sections of the peasantry not only controlled the *muafi* land, they effectively held sway over the entire territory of the *deota*. In small states where the position of the ruler was somewhat weaker, the proportion of *muafi* land controlled by the temples may have been considerable. For example, the land revenue settlement report of 1901 of Keonthal records that of the total of 75 revenue-free (*muafi*) holdings in the state, 42 were held by temples. In terms of revenue, these temple lands accounted for 71 per cent of the total revenue assigned to *muafi* (Singh 1901: 34).

Control over temples, associated offices and institutions, as well as over the large temple land holdings, gave to the Kanet peasantry considerable political influence. The recognition thus accorded to leading Kanet clans bound them ever more closely to the monarchical state. In Keonthal, the heads of six large Kanet clans acted as chief functionaries of the state by rotation (Keonthal SR 1901, Singh 1901: 19 para. 29).[18] The system in Bashahr state was quite similar, where the office of the *wazirs* was hereditary and held by Kanet families (Bhuj 1928: 23). Their control over land and their ability to act collectively with the approval and support of their deity made the Kanet peasantry rather difficult opponents to deal with (Singh 1901: Para 29; Sutherland 2006: 88).[19] Peasant protest movements in the western Himalayas during pre-colonial times were, therefore, likely to have been well-organized and highly effective. It was only subsequently, in colonial times, that the *rajas* could—with British support—successfully resist the pressure that their influential Kanet subjects had traditionally exerted on them (Singh 2002). Yet, more often than not, the hill rulers would have found it wiser to obtain the silent approval of their assertive, yet obedient, peasantry. It was the support of their subjects in normal times that gave to many petty *rajas* a prestige and primacy that they could hardly have otherwise acquired. Their military power and wealth—or rather the lack of it—rested upon a polity that ultimately drew its strength from territorially defined communities.

Political context of village communities

While such communities enjoyed considerable political autonomy, it would be simplistic to see them merely as a collection of autochthonous village kingdoms ruled by their respective gods (Sutherland 2006: 88).[20] The 'theistic sovereignty' of the local deity emphasized by recent work on the subject is somewhat exaggerated (Sutherland 2006: 91). The image that this seems to create is one of segregated local communities engaged in ruthless and unending feuds. Robbery and bloodshed are projected as a way of life and the absence of a dominant political authority is, by implication, seen to be the cause of this (Sax 2006b: 123). Fascinating though it is, this picture is only partly correct. We need to recognise that there existed a larger system within which gods, *khunds* and communities in general negotiated a complex network of relationships and hierarchies. Deities between whom there was believed to be a familial relationship (usually as brother or sister) often visited each other along with their followers. Those that were not 'related' in this manner, gathered at larger congregations of *deotas* that were periodically held. In order to obtain a broad semblance of order, it was not always a precarious *equality* that was "ritually made and unmade" as Sutherland has argued (Sutherland 2006: 110). A similarly tenuous *inequality*—in the form of a hierarchy—also needed to be intermittently negotiated. Sutherland's suggested 'equality among peers' may have arisen—in the first instance—not so much because *khunds* admitted an inherent equality *per se* amongst themselves, but more because they were commonly subjected to a larger system or a higher authority. Within this broad equality of a shared subjection were conducted the more nuanced ritual and political transactions of *inter se* status between *khunds*.

The relationship between monarchy on the one hand, and local communities and their deities on the other was an extremely complex one. It was also one that influenced how territory and kingship came to be perceived by different groups. At local level, the *deota* took on the veneer as well the authority of a ruler (*raja*). Not only did the deity possess the material regalia of the *raja*—umbrella, mace, flags, drums, trumpets, palanquin, etc.—it actually functioned as one. The deity's will and command were expressed through its *gur* (oracle or medium). In important temples this royal status of the *deota* was further strengthened by a host of functionaries who acted as its ministers, courtiers and officials for enforcing the divine writ. Interestingly, while the *deota* asserted his rule by adopting the appearance of a *raja*, the *raja* legitimized his authority by claiming to be divinely ordained—or being some form of divinity himself.[21] There was, thus, a close conceptual interdependence between *deota* and *raja*. Contrary to what is sometimes suggested, it appears unlikely that local communities—little kingdoms—ruled by *deotas* represented a form of political organization that preceded the emergence of the larger kingdom. In fact, the forceful social enactment of the *deota* as a ruler assumes a familiarity with (and the prior existence of) the monarchy. It seems to be a local imitation of the power of the *raja* whose political influence prevailed over the domains of many village *deotas*.

Subtle divisions of territory and varying levels of allegiance could exist within a monarchical system without creating serious contradictions. Folklore in the region occasionally indicates how loyalties shifted in different circumstances. The legend of 'Ḍhar Deshu' recalls the invasion of the state of Keonthal by *raja* 'Mahi' Prakash of Sirmur. As Mahi Prakash prepared to attack the village of Balag, he was approached by a Brahmini of Balag who offered him a goat and a necklace to ward off the attack. Mahi Prakash then inquired, "Who are your enemies (*boiree*) here?"

The Brahmini replied: "In Keonthal, enemies walk around everywhere. The people of Palwi–they graze their sheep and goats in (my) hayfields. When (these animals) go to (drink) water they tear up the pathway (with their hooves)" (Ramdayal 1973: 230-1).[22]

She then lamented that one Homiyan Mian (a local village chief) had even built a platform (*thada*) in the village of Sainj. This act of constructing a *thada* was apparently symbolic of creating a separate centre of power–an assertion of authority and autonomy. For the Brahmini, therefore, quarrelsome and inconsiderate neighbours or an overbearing village chief occupied the place of *boiree* (enemies; *vairi* as derived from *vair* or feud) (Ramdayal 1973: 250-1; Sax 2006b: 124, 126).[23]

But this dispute with her neighbours and her unhappiness with the village head were matters to be resolved by herself or by the ruler of Keonthal. When Raja Mahi (of Sirmur) offered to teach these "enemies" of the Brahmini a lesson, the latter's feudal loyalties came to the fore. She advised the *raja* to return to his home (Nahan), because the *rana* of Keonthal would not submit and was all too eager for battle. She described the kingdom of Keonthal (where she had moments earlier said, 'enemies walk around everywhere') as a place where complete peace prevailed:

> "(Where) the lion herds the sheep and goats,
> (Where) the cat churns the butter-milk.
> The small kingdom of Junga (Keonthal),
> is like the Delhi of the Turks." (Ramdayal 1973: 232-3)[24]

It is evident then, that for the Brahmini of Balag, feuding neighbours were lesser enemies than an invading *raja*. Mahi Prakash and his army were clearly seen as outsiders—'*dakkhaniye*' (Southerners)—against whom the *rana* of Keonthal successfully mobilized not only the different clans within his state, but also warriors from neighbouring states of the north and towards the Satlej river (Ramdayal 1973: 240-1, 244-5). The battles between the two contending forces—with the active participation of numerous *devis* and *deotas* who fought (and fled!) alongside the opposing armies—are graphically recounted.

Territorial hierarchies and contemporary issues

Despite its apparent economic self-sufficiency and considerable political autonomy, the mountain peasantry was never insular. Its peculiar construction of village polity—as part reaction and part appendage to a larger Sanskritized world-view and overarching monarchical system—gave to it an ambivalent position. This ambivalence became explicit when the relationship between village polities (that was premised on a broad notion of equality) was altered into a politico-religious hierarchy whenever they interacted with the state. The emphasis on equality between communities prioritized local social and economic issues, but the recognition of a hierarchy underlined the political and cultural dominance of a more extensive, more complex world.

Hierarchical systems—where peasant communities are usually situated quite low down the ladder—emphasize the fact that peasant culture is not complete in itself. It is, as Redfield argues, "an aspect or dimension of the civilization of which it is a part. As the peasant society is a half-society, so the peasant-culture is a half-culture" He further elaborates:

> "the peasant village invites us to attend to the long course of interaction between that community and centres of civilization. The peasant culture has an evident history; we are called upon to study that history; and the history is not local; it is a history of the civilization of which the village culture is one local expression." (Redfield 1989: 40, 41)

In India, particularly, apart from serving as a "safety valve mechanism for regulating relationships between villages", the regular convening of congregations, *jatras* and other forums of collective activity symbolized "the weaving together of local, regional and pan-Indian traditions" (Beals 1964: 105, 110). By their very nature, therefore, large gatherings of a socio-religious nature reiterated, as well as transcended, territorial boundaries. Small village communities asserted distinct local identities, yet perceived themselves as integral and active constituents of a vast social network (Berti 2006, Conzelmann 2006, Goswamy 2006, Goswamy 2007, Luchesi 2006, Sutherland 2006: 115-18).[25] This consciousness of a hierarchical (sometimes theoretical) system of social relations did not entail a negation of a 'sense of community' that was rooted in its immediate surroundings. Unlike the experience of some other societies, a 'breakdown' of traditional territories and identities was not, here, a prerequisite for the transformation of linkages "beyond the immediate locality and into the regional, the national and even international arena" (Baker 1984: 25, 26).[26] Linkages of a fairly extensive nature were, in fact, embedded into the structure of village communities. Smith had argued that for '*ethnie* to move towards nationhood' required 'a triple movement: from isolation to activism, from quietism to mobilization and from culture to politics' (Smith 1987: 154, 156).[27] But what we need to consider is that societies might also disguise activism as isolation, present mobilization as quietism and practice politics as culture!

By collectively withdrawing from all dealings with the state and its functionaries, the hill peasantry's customary method of protest (despite its outward appearance of withdrawal) represented a high degree of organised activism. The *dhoom* or *dum* (as such protests were called) gave to the peasantry a decisive influence in the affairs of the pre-colonial state (Singh 2002). It also temporarily transformed, for purposes of mobilization, numerous territorially defined village communities into a larger integrated political entity. More importantly, such methods of mobilization resurrected the underlying idea that despite their scattered nature, local communities—especially Kanet clans—communicated with each other in a common socio-religious idiom. They conceptualized the larger epochal world in the language of the great Brahmanical tradition, but told the humbler, intimate tales in the dialect of local beliefs and identities.

There were other, more peaceful, occasions during which perceptions of territory underwent a transformation. Perhaps the most conspicuous were the customary celebrations of Dashehra and Sivaratri, at which the centrality of the monarchy and the state deity were asserted.[28] These were also occasions when representatives (and *deotas*) of distant communities reaffirmed their political and ritual allegiance to the monarchy. Small territorial domains were, for the moment, subsumed by the Brahmanic state. The tiny divine kingdoms of the *deotas* became, as it were, a part of the empire of Visnu. Even as they acknowledged their subservience to the raja and the principal deity, however, the large number of village communities assembled tried to acquire a higher political standing through this annually enacted cultural drama. It was their proximity to, or distance from, the ritual centre that determined their status. Despite the disappearance of princely states in independent India, the social importance of this ritual enactment has persisted. An interesting example of such a contest for a higher ritual status is the ongoing dispute between two Kulu *deotas*—Balu Nag and Shringa Rishi—for occupying a position to the right of Lord Raghunath during the Kulu Dashehra procession. The assertion of local identities and the political significance of this long-lasting confrontation are still all too evident to local observers (Chauhan 2004; *Tribune* 3.10.2006; *HT* 3.10.2006, 22.10.2007).[29] Most interesting, however, is the fact that even in entirely changed political conditions the area that once constituted the former kingdom of Kulu remains the territorial context of this struggle.

Conclusion

Post-independence India has witnessed the emergence of new political regions and provinces. The creation of Himachal Pradesh as a state of the Indian Union has been one such development. It would appear logical, therefore, that the territorial dimensions of politics should have been radically transformed in this western Himalayan region. Socio-economic processes of subcontinental dimensions have prompted rapid and unprecedented change in this area that was once regarded as peripheral in almost every respect. The possibilities arising from accompanying these processes are larger and more complex than those that western Himalayan society has ever confronted before. Have these developments prompted a reorganization of local political and social territories? Has the language of politics changed or have the methods of political mobilization been altered?

The prolonged Shringa Rishi-Balu Nag dispute in Kulu points towards the persistence of the traditional socio-political concept of territories defined through a hierarchy of *deotas*. Rivalry over contemporary political issues is the underlying reality of what has taken the appearance of a disagreement between *deotas* over status. Cultural celebrations have, perhaps, always carried political content. But this certainly does not mean that society and polity in the region has defied change. While outwardly, the boundaries of territorial units and methods of local mobilization display the continuity of tradition, often the nature and content of the debate engages with political polemics that are almost entirely contemporary.

A case in point is the proposal by an American company to establish the Himalayan Ski Village (a winter sports resort) in Kulu. Numerous village communities and their *deotas* opposed this proposal at a large congregation (*jagati pat*) convened by the former raja of Kulu. As in the case of the Shringa Rishi-Balu Nag disagreement, once again religion does not appear to be the most important issue at stake in the ski resort project. Neighbouring communities have instead raised questions about the likely economic and environmental impact of a tourism project of such huge proportions in a fragile mountainous area (*HT* 10.1,

20.1.2006).[30] Certainly, some long established interest groups in the region, too, feel threatened by a more powerful, rival. The promoters of the project and their supporters have argued, on the other hand, that it will contribute substantially to the state exchequer and also create (rather than take away) a large number of jobs for the local people (*HT* 9.2.2006).[31] For our present purpose it is a significant fact that local communities continue to be territorially defined by their traditional allegiance to a *deota*. They are, at the same time, engaged in a contemporary debate involving crucial issues of development and government policy. More importantly, the collective decision of community members is taken through the *deota* and his *gur*, and then communicated not to a government or elected representative, but to the former raja of Kulu within whose kingdom the area of the concerned villages once fell. It appears, therefore, that while the territorial units and methods of political mobilization retain pre-modern characteristics, the essential nature of political debate is clearly modern. The contemporary language of developmental and environmental discourse is used with considerable sophistication by the *deota*'s functionaries.

More recently, local opposition to a large hydro-electric project in Kinnaur has also been articulated through the *deota*. Local communities have resisted the construction of the 1,000 megawatt Karcham-Wangtoo Hydel Project on the Satlej river on the grounds that the tunnelling of the mountain would have an adverse environmental impact and thus affect their villages. At the forefront of this agitation has been Maheshwar Devta of Chagaon an important deity of Kinnaur. In fact, in October 2006, the *deota* 'summoned' the Chief Minister of the state through the District Magistrate of Kinnaur to 'discuss' the matter. It was reported in *The Tribune* that the District Magistrate "went to the deity temple at Chagaon [...] to hold talks with the deity on behalf of the state government seeking grant of permission to the company to start the construction work on this mega project" (*Tribune* 14.10.2006). Not much progress seems to have been made in these negotiations, and the agitation of the Karcham-Wangtoo Sangharsh Samiti led by the *deota* continued for several days. It finally turned violent and the police resorted to firing at the agitators to prevent matters from getting worse.[32] Escalating tension drew political parties into the agitation. Subsequently, the Bharatiya Janata Party (BJP) attempted to make it a political issue in the state assembly elections of December 2007 (*HT* 19.12, 11.11.2006; Sharma 2007). However, it is not the environmental implications of the mega power project that were sought to be raised by the BJP in this confrontation—for it seems unlikely that even the BJP government (that was formed in January 2008) would reverse the decision to build the project. An attempt was made to arouse local sentiment on the question of the "humiliation" and "insult inflicted a year ago on Maheshwar Devta of Chagaon" (Sharma 2007). This recognizes the fact that traditional loyalties to *deota* and locality remain strong and relevant.

How has a remarkably diverse range of complex questions come to be effectively debated and decided upon by traditional village communities: communities that remain territorially defined along distinctly pre-modern criteria? Logically, the socio-economic transformation experienced by modern India should have compelled radical social and territorial restructuring at every level. In Himachal Pradesh, however, other than the political merger of the princely states, such a transformation does not appear to have occurred on any great scale. One possible reason for this may be that shifting notions of community and territory have always been historically articulated for varying purposes. Culturally established norms of interaction between different levels of the political hierarchy may have something to do with the continued existence of traditional territorial divisions in Himachal.

It goes without saying that South Asian society has constantly changed through dynamic exchanges between–what scholars have called–the Great and Little Traditions. But there are few areas where the legitimising process of state formation itself has been so powerfully influenced by the Little Tradition as is the case in Himachal. Long-term historical processes combined with environmental circumstances to give small mountain communities the political autonomy to engage with the state from a position of considerable strength. Even after the emergence of the monarchical state, village communities remained important centres of power and exerted considerable collective influence over state functioning. The pre-colonial polities of the region carried this socio-political structure along with them when they merged to constitute Himachal Pradesh. As was the case earlier, the collective political influence of local communities has ensured that they continue to remain at the centre of government policies. This strength of village communities is indicated by the fact that despite the remarkable socio-economic development that Himachal has witnessed over the last decades, it remains even today the least urbanized state of India.

REFERENCES

Baker, A. 1984. 'Reflections on the relations of historical geography and the *Annales* school of history', in Baker, Alan R. H. and Derek Gregory (eds.), *Explorations in Historical Geography*, Cambridge, Cambridge University Press, pp. 1-27.

Beals, A. 1964 'Conflict and Interlocal Festivals in a South Indian Region', in Harper, Edward B. (ed.), *Religion in South Asia*, Seattle, University of Washington Press, pp. 99-113.

Berti, D. 2006. 'Ritual Kingship, Divine Bureaucracy, and Electoral Politics in Kullu', *European Bulletin of Himalayan Research*, 29-30 (Summer): 39-61.

Bhuj, Pt. C. 1928. *Report Tarmim Bandobast Tahsil Chini* (*Tahsil Chini Settlement Report*), copied from original by Daulat Ram Tegta in 1970, Shimla, Himachal Pradesh Secretariat Library, Mimeo.

Chauhan, K. 2004. 'Feud between devotees intensifies', *The Tribune*, Chandigarh, 25.10.2004.

Conzelmann, E. 2006. 'A Royal Ritual of Mandi State', *European Bulletin of Himalayan Research*, 29-30 (Summer): 14-38.

Fraser, J. B. 1982. *The Himala Mountains* (First published 1820), Delhi, Neeraj Publishing House.

Godelier, M. 1986. *The Mental and the Material. Thought, Economy and Society*, Martin Thom (tr.), Norfolk, Verso.

Goswamy, K. 2006. 'Narsingh/Narsingh Bir and Kulu: A Preliminary Inquiry into Cultural

Processes', unpublished paper presented at the Fellows' Seminar, Shimla, Indian Institute of Advanced Study.

– 2007. 'Some Thoughts on Kulu: The Nature and Formation of a State', unpublished paper presented at the Fellows' Seminar. Shimla: Indian Institute of Advanced Study.

Habib, I. 1999. *Agrarian System of Mughal India, 1556-1707*, New Delhi, Oxford University Press.

HT, Hindustan Times. 2006. 'Annul MoU in Ski Village: NGO', Chandigarh, 10.1.2006.

– 2006. 'Another NGO raises voice against Ski Village', Chandigarh, 20.1.2006.

– 2006. 'Ski village project: Ex-MLA opposes 'jagati conclave'.', Chandigarh, 9.2.2006.

– 2006. Chandigarh, 3.10.2006.

– 2006. 'No vested interests behind agitation: CPM', Chandigarh, 11.11.2006.

– 2006. 'Tribal Commission's intervention sought', Chandigarh, 19.12.2006.

– 2007. Chandigarh, 22.10.2007.

Kashyap, P. C. 2000. *Living Pre-Rigvedic and Early Traditions of Himalayas*, Delhi, Pratibha Prakashan.

Kasturi, M. 2002. *Embattled Identities: Rajput Lineages and the Colonial State in Nineteenth Century North India*, New Delhi, Oxford University Press.

Keonthal Settlement Report. 1901. (Mimeograph) H.P. State Archives, Shimla: File no. PHSA/1/15.

Klass, M. 1995. *Ordered Universes: Approaches to the Anthropology of Religion*, Boulder and Oxford, Westview Press.

Kobayashi, A. 1989. 'A critique of dialectical landscape,' in Kobayashi, Audrey and Suzanne Mackenzie (eds.), *Remaking Human Geography*, Boston, Unwin Hyman, pp. 164-84.

Levine, N. 1988. *Dynamics of Polyandry: Kinship, Domesticity, and Population on the Tibetan Border*, Chicago, University of Chicago Press.

Lovell, N. 1998. 'Wild gods, containing wombs and moving pots: emplacement and transience in Watchi belonging', in Lovell, Nadia (ed.), *Locality and Belonging*, London, Routledge.

Luchesi, B. 2006. 'Fighting enemies and protecting territory: deities as local rulers in Kullu, Himachal Pradesh', *European Bulletin of Himalayan Research*, 29-30 (Summer): 62-81.

Lyall, J. B. 1889. *Kangra Settlement Report, 1875* (Published together with G. C. Barnes, *Settlement Report of Kangra District*), Lahore, Civil and Military Gazette Press.

Majumdar, D. N. 1962. *Himalayan Polyandry. Structure, Functioning and Culture Change. A Field Study of Jaunsar-Bawar*, Bombay, Asia Publishing House.

Nisbet, R. A. 1962. *Community and Power*, New York, Galaxy Book.

Ramdayal (tr.). 1973. 'Dhar Deshu', in Ramdayal (ed. and tr.), *Himachali Lok Gathaein*, Shimla, Public Relations Department, Himachal Pradesh.

Redfield, R. 1989. 'Peasant Society and Culture', in Redfield, Robert, *Little Community and Peasant Society and Culture* (First published 1960), Chicago and London, Midway Reprint, University of Chicago Press.

Rose, H. A. 1970. *A Glossary of Tribes and Castes of Punjab and the North-West Frontier Province* (First published 1881), 3 volumes, Patiala, Punjab Languages Department, rpt.

Sax, W. 2006a. 'Divine Kingship in the Himalaya', *European Bulletin of Himalayan Research*, 29-30 (Summer): 7-13.

– 2006b. 'Rituals of the Warrior *Khund*', *European Bulletin of Himalayan Research*, 29-30 (Summer): 120-43.

Sharma, A. 2007. 'Deities not to take sides, put BJP on backfoot', *Indian Express*, Chandigarh, 11.11.2007.

Sharma, B. R. 1990. 'The Institution of Village Gods in the Western Himalaya', in Rustomji, N. K. and Charles Ramble (eds.), *Himalayan Environment and Culture*, Shimla: Indian Institute of Advanced Study.

SHSG. Simla Hill States Gazetteer. 1910. Lahore, Civil and Military Gazette Press. (The gazetteer of each Hill State constitutes a separate chapter in this volume. The pages of each gazetteer/chapter are numbered separately.)

Singh, C. 2002. 'The *dhoom* in Himachal Pradesh: Community consciousness, peasant resistance or political intrigue?', in Thakur, Laxman S. (ed.), *Where Mortals and Mountain Gods Meet*, Shimla, Indian Institute of Advanced Study.

– 2006. 'Long-Term Dynamics of Geography, Religion and Politics: A Case Study of Kumharsain in the Himachal Himalaya', *Mountain Research and Development*, 26 (4) November: 328-35.

Singh, D. 1901. *Keonthal Settlement Report* (Urdu version), Shimla, Army Press.

Smith, A. D. 1987. *The Ethnic Origin of Nations*, New York, Basil Blackwell.

Srinivas, M. N. 1965. *Religion and Society Among the Coorgs of South India*, Bombay, Asia Publishing House.

Sutherland, P. 2006. 'T(r)opologies of Rule (*Raj*): Ritual Sovereignty and Theistic Subjection', *European Bulletin of Himalayan Studies*, 29-30 (Summer): 82-119.

Tribune. 2006. Chandigarh, 3.10.2006.

– 2006. 'Power project: devta summons Vibhadra for talks', Chandigarh, 14.10.2006.

Whitehead, H. 1921. *The Village Gods of India*. Madras and Calcutta, Oxford University Press.

Notes

1. This argument has also been made in the context of 'landscape' by Audrey Kobayashi. She argues that "a historical understanding of landscape formation is necessary to link the qualities of a particular landscape to both their social meanings and the technological means by which they are controlled".
2. Robert Nisbet stresses that "the solidarity of each functional group was possible only in an environment of authority where central power was weak and fluctuating". Anthony Smith sees pre-modern *ethnie* as being "usually made up of numerous clans or villages who practice nomadic pastoralism or a local, self-subsistent agriculture".
3. A 'landholding brotherhood' (*biradari*) frequently referred to related families within the village, but it could also include those that were not related but possessed land in the village. The studies on village '*biradari*' are too numerous to be quoted here. Irfan Habib provides a detailed description of the classic village community in medieval India. Rajput lineages in the villages of north India have been studied by Malavika Kasturi.
4. Majumdar makes a distinction between clan (for which he uses the term *gotra*) and lineage (what he calls *aal*) and argues that in Jaunsar the clan was completely absent. It is by mistranslating '*gotra*' as 'clan' that Majumdar comes to this conclusion. On the contrary, it appears that the organization of society into clans was important in the Himalayan region, especially considering the fact that large parts of it were politically dominated by a clan-based Khas population.
5. A more appropriate spelling of the word *bhera* –as used by Majumdar–would be *berha*. In the Himachal region, the village council is called *khumbli*.

6. This does not, however, rule out the existence of some form of individual leadership in the village. Fraser, who travelled into the Shimla hill region in the early nineteenth century, refers to the fact that, "In every village there is a man to whom they pay great deference, and to whom they refer all disputes; who, in short, is their chief, and who goes by the name of the *Seana*". Kashyap has argued that the smaller unit, in terms of social organization, below the *khund* was the *barind*. This *barind*, he suggests, was headed by a *sayana*.
7. In Bashahr, several hamlets grouped together were called a *ghori*. But in smaller states, such a unit was called a *pargana*. B. R. Sharma suggested that a *ghori* consisted of ten or twelve villages.
8. Sax studies in detail the institution in the Garhwal region. In Kulu a comparable unit (though more explicitly for revenue purposes) was the *phati*. Sutherland discusses the role of *khunds* in eastern Himachal.
9. Numerous such instances were recorded by Rose in the detailed survey of the Western Himalaya carried out in the late nineteenth century.
10. It is mentioned in the section on the 'Customs of *Deotas*', that "every village has a temple of the *deota*. One *deota* cannot go into the area of another *deota*". This is, perhaps, an exaggeration though it would certainly be true for *deotas* belonging to feuding village communities.
11. In the context of South India, Whitehead has, in fact, argued that "in the worship of the village deities, the *pujaris* are drawn from all the lower castes indiscriminately". Furthermore, that "In the Telegu country the potters and the washermen, who are Sudras of low caste, often officiate as priests".
12. There were also lower castes in the village who had their own deities. This is true of large parts of rural India.
13. It is significantly recorded that the new *deota* that was established after the loan of the original Junga image for conducting a religious ceremony was refused to this ancestor of the Khanoga clan. This might be a typical instance of the peasantry conflicting and conforming at the same time.
14. We also have the instance of a smaller village god (Shaneti) being similarly established by the Shainti clan of Kanets as their own separate god. There would, undoubtedly, be many more such cases in the entire region.
15. It is likely that the displaced god was Sipur, not Jipur. The temple of Sipur is located in the neighbouring principality of Koti with which the state of Keonthal has close ties. Legend has it that *deota* Junga was once a prince of Koti who disappeared in the forest, and was discovered later as a *deota*. There would, therefore, be reason enough for Junga to make a complete break with Sipur (who was closely associated with Koti) in order to establish his independence in the separate principality of Keonthal.
16. An image of Kaneti *deota* was stolen from Kawar and brought to Dodra (bordering present-day Uttarakhand). Thereafter it was carried to the village of Dagon in Keonthal state. Here the new god, Kaneti, took over the temple of Jipur who was the older *deota* and also won over the *wazir* of Keonthal and the Bhaler clan that had previously been followers of Jipur.
17. Sutherland terms the process "the ritual construction of outer landscapes of collective subjection by figure of space." By this, he explains, it implied "not only the production of political subjects, but also the formation of gods as rulers".
18. Though the mimeographed English version uses the word '*wazir*' for the office they occupied by rotation, the published Urdu version of the settlement report uses the

phrase '*virasat ka kam*' (traditional work). It also mentions, however, that the influence of these Kanet clans was considerable and that they are greatly respected in the state.

19. Singh notes in the report that, 'They sometimes take oath in the temple of their god to accomplish a certain matter and having done so, they can with difficulty be induced to give up their oath till they have succeeded in the object. The other castes also have to join them, otherwise the Kanets would go against them and the other castes are always at their mercy'. This point is also implied by Sutherland when defining the *khund* as a 'political community' that included Khas-Rajputs, Brahmins and Kolis.
20. The simplification goes thus '…rural caste assemblies called *khund*, resident in multi-village domains called *ghori* are understood to be ruled by local gods called *deota*'.
21. This question has been recently explored in the context of the states of Mandu, Kulu and Bashahr by contributors to the Summer 2006 issue of *EBHR*. What has not been seriously considered here is that if the 'divinity' of the raja was constructed with the help of the local *deotas*, the 'kingship' of the local *deota* was created in the image of the *raja*.
22. The legend has numerous chronological contradictions and seems to refer to a much later period than that of Raja Mahi Prakash. It is not, however, the dates but the ideas expressed in the legend that we are primarily concerned with here.
23. In his discussion on *khunds*, William Sax has pointed out two very relevant parts. The first is that the '*khund* is closely associated with pastoralism', and that 'Much of the feuding between the various territories centred on pastoral disputes'. The other is the importance of the *that* which was built in large villages and regarded as being associated with a *that* goddess, that was 'the source of the warriors' martial power and energy'. The protective power of the *thod* is also referred to in the legend of Dhar Deshu.
24. The reference to the "Delhi of the Turks" emphasizes the position of the town of Junga as a power centre.
25. The case studies of the Kulu and Bashahr *Dashehra* and the Mandi *Sivaratri* are the few instances that have been studied by scholars.
26. An entirely different development seems to have occurred in nineteenth century France. The aggravation of a 'class struggle', and the growth of a 'heightened *sense of class*', is believed to have brought about the disintegration of rural communities and the breakdown of *pays*.
27. Smith argued that "Any *ethnie*, then, that aspires to nationhood, must become politicized …"
28. This practice was common in various parts of the country and the studies on this subject are too numerous to be all enumerated here. For Himachal some of the recent works have already been mentioned above.
29. The importance of Shringa Rishi of Chehni Kothi (Banjar) began growing after Maheshwar Singh, the former raja of Kulu, was elected from Banjar legislative constituency as a BJP member of the State Legislative Assembly in 1982. The neighbouring Ani legislative constituency—the territory of *deota* Balu Nag—was apparently a Congress Party stronghold. It appears that the '*haryans*' of Balu Nag were at a subsequent date prevented by the administration from participating in the Dashehra *rath yatra*. In 2006, even Shringa Rishi did not participate in the procession. In 2007, the Kulu Dashehra Committee did not invite either of the deities for the *rath yatra*, so as to avoid a confrontation.

30. NGOs opposing the project have raised questions pertaining to the destruction of forests, soil erosion, pollution, loss of traditional rights of local communities, the danger to folk culture and a host of other related issues.
31. Dile Ram Shabab, a former member of the State Legislative Assembly representing the Banjar constituency criticized Maheshwar Singh for convening a '*jagati* conclave' of *deotas* for this purpose and demanded that the issue be discussed by 'political leaders' and 'elected members' of village bodies.
32. 'People with vested interests instigating locals: District Magistrate', *HT*, Chandigarh, 19.12.2006. From the observations of the District Magistrate it appears that villagers who failed to participate in the agitation were fined Rs. 500 to Rs. 1,000 by community leaders who apparently were acting on behalf of the *deota*.

Part II: Territories of Modernity, Territories of Exclusion

2.1. DEVELOPMENT PROGRAMMES AND EMERGENCE OF NEW TERRITORIES

2.1.1.

How Environmental Policies Reshape the Himalayan Area. New Environmental Territories, New Environmental Borders; the Example of Kaziranga National Park (Assam)

Joëlle Smadja

In the Himalayan area[1] changes in the management of natural resources over the last decades have led to significant territorial restructurings. A differentiation and specialization of space and milieus that bring about a clear separation between 'nature' and 'society' have been at work since the 1970s on various scales i.e. the scale of the village, of the state and of the Himalayan region as a whole. On a village territory scale, there is a growing divide between, on the one hand, areas considered 'natural' such as forests and pastures and, on the other hand, inhabited and cultivated areas. On a state scale, the split between protected 'natural' milieus and society has become even more marked since, in order to preserve biodiversity and unique landscapes, nature conservation policies that have introduced dozens of protected areas, that is national parks, biosphere reserves, conservation areas and the like have been implemented in each Himalayan state. Parallel to this, numerous dams have been built, also contributing to changes in the territorial organisation. As a result, the Himalayan area as a whole is more and more limited to the role of a nature reserve for the Indian sub-continent and of a water tower to supply the rest of India. Thus, like fractal figures, whatever the scale of observation, we are witnessing the emergence of new specific spaces due to the further split between nature and society. This is a current trend throughout the world. It is based on an environmental paradigm built at international and national levels and shared by various actors who consider, by and large, that 'nature' has to be protected from societies that can only destroy it and who claim therefore the need for a clear separation between these two ontologically redefined entities. The effects of the objectivization of the environment and consequently of the split between 'nature' and society[2] take on a particular aspect in the Himalayan region, for at least two reasons. First, the interrelation in time and space between agriculture and so-called 'natural' milieus—such as forests, pastures, ephemeral islands in the case of rivers, etc.—has been an essential component of the dynamics of farming systems. More generally, for Himalayan rural communities (80% of the population), far from being differentiated and specialized, space is a territory where farming, grazing, gathering and hunting, fishing, social and religious activities, etc. are closely interlinked (Campbell 2010, Nightingale 2010, Smadja 2009 [2003]b). The agricultural and religious calendars, for example, echo each other, and in India (see Robbins 2001), as in Nepal (see Meyer and Koppert 1983), studies have shown that over 40 per cent of the fruit and vegetables eaten by populations relying on a subsistence economy come from forests or other so-called 'natural' spaces. Secondly, the split between nature and society which was already clearly brought to the fore as early as the nineteenth century, as in Western countries, has been reinforced since the period 1970-1980 in keeping with the Theory of Himalayan Environment Degradation.[3] Though highly criticized since its inception and played down over the years, this theory still determines a large part of the environmental policies in the Range and, consequently, significant changes. Combined with many other changes, the process has led to territorial restructurings that confer a specific identity on the Himalayan area.

In this paper, after a short overview of the territorial restructurings on a village scale, the focus will be on protected areas in order to show how, as enclaves within states, and with new environmental borders, they constitute new environmental territories at the heart of debates that are not only limited to nature protection issues. The social consequences of the environmental limits separating 'nature' and 'society', and the way they are instrumentalized by various actors are illustrated here using the meaningful example of Kaziranga National Park in Assam. This paper supports the hypothesis that internal boundaries related to the environment are of relevance and importance in the structuring of space and societies at different levels, and can generate issues that fuel the numerous conflicts already raging in the Himalayan area.

1. Territorial restructurings on a village scale

With regard to a village community scale, many territorial restructurings are at work in the Himalayas and here I will briefly point out those taking place mainly in Nepal, which are representative of the situation in the Central Himalayas (i.e. Nepal and part of India).

In the Central Himalayas, the great majority of today's population settled at least one or two centuries ago.[4] Up until the 1980s, the utilisation of the complementarity of milieus was an essential component of peasants' way of life. Though sedentary, they travelled extensively over a given slope. They went from their village to different kinds of fields, to water mills by torrents where they also fetched drinking water and where animals watered, to the forest to collect firewood and to cut timber as well as fodder for their animals, to pastures on the slope or to pastures located several days' walk away where they led animals in transhumance, either higher up in altitude to the foot of the Upper Range or into the plains, etc.

Quite a number of them had both a winter dwelling (often a house in the village) and a summer dwelling (often mobile shelter-cum-stable). Every milieu was put to some kind of use. This did not include trips to various places of worship on the slope or in neighbouring villages, to the district headquarters or to the nearby bazaar for trading or shopping. These places were linked up by footpaths and it took some time to travel along them. Thus, the territory of Himalayan villagers, that is the frequented, used, appropriated space made up of these places and links, was for most of them both limited to the region and scattered over the mountainside. Moreover, trade has always thrived between the mountains and the plains and, beyond the borders, between the Tibetan plateau and the Indo-Gangetic plain. Some people have migrated to work in the plains or in India. The Central Himalayan economic system—though rural—has never been entirely agricultural, and land has never been used in a set manner.

As far as the example of Nepal is concerned, the changes that have occurred since the period 1970-1980 have led to major territorial restructurings and redefinitions associated with a fundamental transformation of society. Peasants have found themselves faced with two opposing processes. While the population has increased[5] the already limited space available for farming (cultivated space represents 20 per cent of the country's total surface area) has dwindled: part of it has been turned into protected areas and dams, and moreover the surface area reserved for farmland has been defined by registering land in a cadastre and by protecting forests, thereby preventing access to a number of resources. All spaces have now been clearly accounted for. In addition to this comes the low income from agriculture, the damage to crops by well protected wild animals that can no longer be hunted, a drop in manpower due to the schooling of young people or to political conflicts such as the Maoïst insurrection during the period 1996-2006, the development of modern means of communication, and an opening onto the world. One of the consequences of all this is out-migration: migratory flows from the mountains to the plains are on the rise, as well as those from villages to towns and cities in Nepal itself, but also in India, the Middle-East and the Asia Pacific (see Bruslé 2006). Today, with thousands of Nepalese migrating, territorial restructurings on different scales that lead to a new geography of the country are often both the cause and the effect. This rural exodus might be regarded as a syndrome of modernity such as industrialised countries underwent in the nineteenth century: it would have arrived here a century later. However, contrary to nineteenth-century Europe, there is hardly any industry here to absorb the demographic growth, or any mechanisation to compensate for the drop in agricultural labour in the mountains. In a country where land is only worth the hands that work it, the future of farming and of Nepalese peasants is very much on the line here. In the Himalayas, the restructuring of village territories is being accompanied by a feminisation of the world of farming: in many places, women are the only ones to still work in the fields and maintain this system.

All this has led to an important trend in abandoning farmland as shown by Narendra Khanal and Teiji Watanabee (2006), as well as to a decline in animal husbandry, but also in many places to a restructuring of village territories and in any case of villagers' territoriality. In order to reduce the time needed to access resources, farming activities are now grouped around the farmstead. Fields have been grouped together (see Buslé *et al.* 2009 [2003], Ripert 2000, Smadja 1995); common grazing is on the decline and animals are kept in stalls; in some places water tanks have been installed to water animals and drinking water is supplied by pipes; trees have been planted close to farms to provide fodder, fire wood and timber; diesel-run or electric mills have been built in almost every village. People travel much less than before for their farming activities—short journeys are further reduced—yet, at the same time, they are now linked to the world by roads, the radio and television. Long-distance migration, more often at international level, is on the rise. Part of the family moves far away, to the Emirates, Malaysia … Some links between places have been reduced, while some have disappeared. New places have appeared and most often they are not linked by footpaths but by road or plane, and for those who stay there, by phone, cellphone, internet and the like. Thus, the very nature of the links has changed.

The territories used by populations have been redefined. This reorganisation leads, on the one hand, to a contraction of space used in daily life and, on the other hand, to an opening onto the world. And, as is the case for protected areas, spaces are more and more specialised and separated from each other. The forest is used less and less for domestic use and a more distinct separation can be observed between forests—which are protected—and inhabited or cultivated areas. Thus the split between 'nature' and 'society' is accentuated.

This is also the case in the eastern part of the Range, where among more mobile populations there is a trend towards a sedentary lifestyle. In Arunachal Pradesh for example, where there is no land register and where shifting slash and burn cultivation (called *jhuming*) is predominant, fields and populations are mobile, and territories have so far always changed. In many places, houses and cultivated areas have moved every twenty years or so, and cultivated plots of land have changed every two years. Over the last decades, projects aimed at abandoning this practice, which is thought to destroy the environment, and at settling populations by terracing slopes have also led to a total redefinition of populations' practices and consequently of their territories. Here again, these changes introduce a new kind of division between society and nature with a specialisation of spaces. As a matter of fact, the status of forest and the status of *jhum* (that is to say, a plot where the farmer practises shifting cultivation) is supposed to alternate: the forest becomes a *jhum*, it is then cultivated and within a few years the *jhum* goes back to a forest state, to a wilderness. Yet, when the *jhum* is turned into permanent paddy fields, there is a complete separation between what is supposed to be natural and what is supposed to be cultural. Even the *mithun* (*bos frontalis*), the semi-domesticated animal which, in this area, is a symbol of this link between man and nature, is in the process of being completely domesticated for new farming activities in the newly defined spaces, and the rituals where it is traditionally sacrificed abandoned.

In a way, the trend is the same in the far western Himalayas where, for example, the last nomadic herders from Ladakh sell their herds and set themselves up on the outskirts of Leh (see Dollfus, in this book).

All these restructurings have many consequences that will not be developed here. The effects of separating societies from the so-called 'natural' environment will be studied here by taking as an example protected areas such national parks where these are particularly evident.

2. Protected areas and territorial restructurings on a Himalayan area scale

In the area encompassing the Himalayan range and its bordering plains, which is home to a wealth of endemic fauna and flora and to a number of ethnic groups whose economy depends largely on the use of the natural resources in their environment, over a hundred protected areas now cover more than 13 per cent of the total surface area[6] (18% in Nepal, 29% in Sikkim, 31% in Bhutan) without taking into account the corridors linking protected areas or the numerous projects yet to be implemented (Smadja 2005a). These protected areas both constitute new territories in themselves and reshape national territories.

Setting up these protected areas has meant withdrawing many so-called 'natural' spaces from common use. They are thereby the direct product of the theory of Himalayan Environmental Degradation. The main issues motivating the creation of protected areas are: to avoid rare species from becoming extinct and the degradation of unique landscapes, but also the fight against rural poverty and the implementation of sustainable development. They may also prove to be political and geopolitical tools for controlling spaces and may be used as a means of pacification in the case of transborders parks or 'peace parks', for example.[7]

By causing major changes to the geography of the region, to the division of space, to the definition of milieus and territories, the nature conservation policy related to protected areas has impacted patterns of land use and resource management. It has led to milieus, once used in a versatile way by populations and which used to be social spaces, being reclassified as 'natural areas' from which villagers have been partially or totally excluded. In some places, this process has been achieved by establishing (real or fictive) barriers, 'new environmental borders' (see Guyot and Mniki 2008), delimiting new environmental territories which, though closed worlds like islands of nature (Smadja 2005a), have the peculiarity of combining a multitude of actors, both human and nonhuman, at different scales:

> "managers and politicians at all decision-making and active levels, from the most local to the International Union for Conservation of Nature, UNESCO, or the European Union; conventions, laws, treaties, directives and decrees which govern all aspects of life of parks including the ontological qualification of entities that occupy them; modes of production and consumption whereby the practices of animal breeders, farmers, naturalists, guards, poachers, activists, tourists and educators are superimposed in the same space. All these actors, whether human or nonhuman are caught up in the web of conflicting interdependence the threads of which extend as far as ministries, research laboratories, parliaments and the confines of international organizations" (Descola 2007).[8]

In this paper, territories and borders are addressed in the same manner, borders being a fundamental reference for relationships to territories, for the expression of

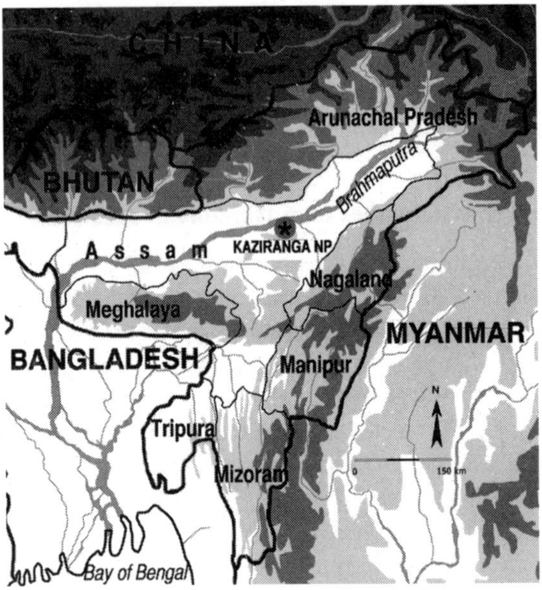

Figure 1: Kaziranga National Park in Assam, North-East India

identities and the perception of Otherness. It will be demonstrated how these protected areas possess features of administrative and political territories, and to what extent their limits possess characteristics of international borders working on various registers: the real, symbolic and imaginary (see Foucher 1991 [1988]). Yet, when not appropriated by populations who do not feel that they belong there (Smadja 2011), protected areas can hardly be considered to be real territories that people can identify with and build projects in. They end up being a source of conflict.

This is the case of Kaziranga National Park, located in North-East India in the heart of Assam along the Brahmaputra (Fig. 1), and it is taken here as an example. It is representative of the policy implemented in some of these protected areas and of the ensuing problems; problems that are exacerbated due to it being set up in a very densely populated and physically and politically unstable milieu. Here one finds a whole range of problems stemming from protected areas, sometimes even verging on caricature.

3. New environmental borders and territories, the example of Kaziranga National Park

3.1. A park in a milieu subjected to heavy constraints, where mobility of land and men is the rule

Mobility is one of the specificities of this region. In addition to mobility for political and economic reasons (Partition between India and Bangladesh in 1947, migrations for work from Orissa, Bihar, Rajasthan, Nepal, etc),[9] to mobility in relation to agricultural practices (such as slash and burn shifting cultivation in the hills), the region's dynamics and economy have been profoundly marked by the changing course and the flooding of the Brahmaputra River. Each year, during the monsoon between June and September, floods wash away land belonging to both farmers and the National Park, and regularly force entire villages to move. Furthermore, and this is an interesting point as far as this paper is concerned, in the bed of the Brahmaputra, there are shifting, ephemeral sandy islands, characteristic of braided rivers. These are called *sapori* in Assam, and are occupied by riverine populations in the dry season, mostly by the Misings, a Scheduled Tribe originally from Arunachal Pradesh, and by the Nepalese who are mainly herders here. Before the next floods, they cultivate cereals or vegetables on these islands. They also collect for domestic or commercial purposes branches that are used as firewood or for roofing. They graze buffaloes that are reared for dairy production, they fish in the ponds and the arms of the river ... Far from being unproductive, these islands are used by populations highly adapted to this shifting environment and who were once very mobile themselves. They are far from being barren, uninhabited deserts.

Access to land is a crucial issue in this region and is the source of most conflicts. Therefore, the mass immigration of Muslims from Bangladesh—to understand it one look at the population density is enough: more than 1,000 inhabitants per square kilometre in Bangladesh, 340 in Assam, about 500 near Kaziranga—has created a real psychosis here. Parallel to this, or in connection with these migrations, claims for autonomy have been expressed, sometimes violently. They are based on Appendix 6 to the Indian Constitution which, after Independence of India, was drafted to give administrative autonomy to certain tribal populations (registered as Scheduled Tribes). Since the 1970s, several ethnic groups have been granted territorial autonomy (the Khasi and Garo, Dimasa, Karbi, Bodo). Others who already have their own autonomous councils such as the Misings, Tiwas, Thengal-Kacharis and Deuris are now demanding autonomous districts. As for the United Liberation Front of Assam (ULFA), it is fighting an armed struggle for the total independence of Assam and advocating the exclusion of migrants.

Kaziranga National Park was set up in this particular context.

3.2. An expansionist nature conservation policy

The Park was established in 1974 in an area covering 430 sq km, between the Brahmaputra in the north, and the hills of Karbi Anglong to the south, from where populations have been completely excluded. The Park authorities may well boast of its success in protecting wildlife since its number has doubled in the space of 30 years and, in addition to the hundreds of one-horned rhinoceroses, wild buffaloes and elephants, it now has the highest density of tigers in the world at 32 per 100 sq km. However, problems have been mounting since its inception, which risk jeopardizing this success story. Indeed, the number of wild animals is not regulated; with their number on the rise, resources in the protected area no longer suffice, which means that they venture outside, causing more and more damage to fields and villages. In order to ensure that the size of the Park is in accordance with the needs of the animals, the Assamese Government has not stopped extending it since 1977. The Park now encompasses the *sapori* from where populations are also evicted.

The growing divide between so-called 'natural' milieus and society, accompanied by the exclusion and marginalization of populations stems from a process that began with British colonial rule when the protected area was created. In 1908, it was a reserved forest to protect the rhinoceros, then in 1916 it was turned into a Game Reserve, and in 1950 into a Wildlife Sanctuary when, as Arupjyoti

Saikia (2005) puts it, "guns were exchanged for cameras". It became a national park with double the surface in 1974 and was inscribed on the UNESCO World Heritage List in 1985, then it was recognized as an important area for bird conservation by Birdlife International. Since 2006 it has been included in a Project Tiger. From 1977 to 1999, the park underwent six extensions and its original surface was doubled yet again (Fig. 2). Within the Project Tiger, the surface of the park will further increase, while a Karbi Anglong-Kaziranga landscape project, encompassing the hills south of the Park, is in the pipeline. This boundary, which is forever being pushed back is the result of an expansionist nature protection policy which has led to a requalification of spaces.

This obviously raises the question concerning the limits to this policy. What surface area will yet be needed for the forever increasing wildlife population? Or, to put it in other words, "how much is enough?"[10]

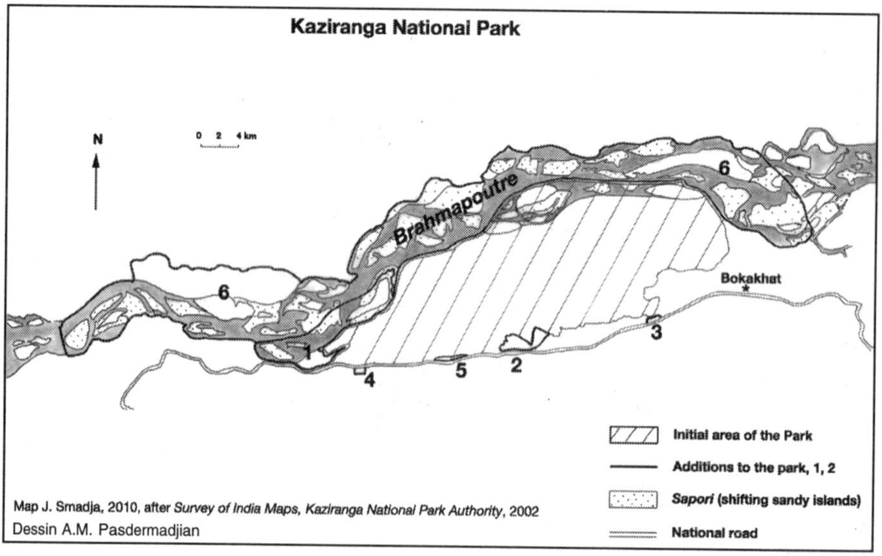

Figure 2: Kaziranga National Park and its additions

As for other borders in the world, this expansionist policy and the aim of separating 'nature' from 'society' have led to the displacement of thousands of people. Only those holding property titles have been given new land, though often in milieus where they cannot pursue their activities. The others, of whom there are many, remain in some sectors illegally. They are now considered outlaws, poachers or encroachers. However, they are often only "passive, not active encroachers" (see F. Landy),[11] in as much as they merely have the new borders forced upon them. They now live in an area forbidden to them, but which previously made up their territory and was a vector of their identity (Fig. 3). As Willem Van Schendel puts it regarding the Indian/Bangladesh border, "the border took people by surprise"…"A deep fissure suddenly separates one half of the landscape from the other. And then it is all over. An eerie silence hangs over a land that is forever scarred, broken, double." (Van Schendel 2005: 1).

Photograph 1: Boundaries between Kaziranga National Park, on the left and a Mising village settlement on an embankment on the right (Bokakhat, Assam). Photo B. Ripert, March 2006 (in Smadja 2011).

An expansionist policy and the displacement of populations are both characteristic of border situations. Many other features of borders are associated with the delimitation of the Park, since it demarcates different territories, the Park itself being like a state within a state.

3.3. The new territories and their borders

A territorial border marks the separation between two territories falling within different jurisdictions and, indeed, the boundaries surrounding National Parks demarcate new territories that have their own characteristics recognizable throughout the world. Their limits are fixed in agreement with international organizations according to ecological criteria; they have their own legislation within their own territorial subdivisions: 'core zones', 'buffer zones', 'transition zones', 'corridors' and the like. Places are reinterpreted, and as S. Depraz puts it, these are no longer spaces for local life, but elements of humanity. They fit into a new semantic network formed by all national parks or reserves within the worldwide network for the programme Man and Biosphere. In short, they are one visible manifestation of a collective representation of nature and echo other remote locations that take part in a globalization process (see Depraz 2008: 277).[12] They are new spatial productions, a product of globalization.

These territories come with their own identity symbols, such as boundary pillars—symbols of territorial sovereignty—, and with their sites of memory, their own ceremonies and festivals, such as the Elephant Festivals in Kaziranga for example.

Like other borders, the border of the new environmental territory both demarcates and excludes and includes, designating what is alien. In Kaziranga, the park includes wildlife but excludes populations, except tourists. These kinds of parks are mainly animal territories.

Tourists visiting the Park travel from one park to the next with the aim of contemplating an 'untouched nature' and, possibly, of discovering the folklore of the surrounding indigenous populations. For them, protected areas constitute territories where they can still find their bearings: places of identification that are very similar throughout the world and which we could call (after the French geographer Bernard Debarbieux, 1995) 'generic places', such as paths and road infrastructures, observatories, shops, restaurants, hotels (sometimes with a swimming pool and helipad), of which there are many around the park.

By contrast, the protected area is surrounded, at the 'borderland', by villages lacking any basic facilities, barely benefiting from tourism and where peasants ('borderlanders') are deprived of access to the natural resources they need. For them, the border is the symbol of exclusion, of the loss of part of their identity, as well as of a threat to their property and even to their lives. In Kaziranga National Park, the boundary is all the more marked as no form of participatory management has been introduced. The only proposed forms of participation are public awareness campaigns and training courses in nature conservation, which come in the form of guilt-inducing speeches, during which farmers are merely requested to take part in the arrest of poachers. Villagers are asked to look after spaces that are qualified as a 'Gift to the Earth'. As for development, the only form proposed is the setting up of eco-villages in which their folklore is put to the fore, while, if need be, they reinvent their traditions (see Hobsbawm and Ranger 1992). This is in agreement with the goals of environmental agencies and development programmes that reify culture and consider that culture has become identified as a potential tool for effectively promoting conservation initiatives (Campbell 2010). However, they are only supposed to retain certain traits of their own culture, 'the proper customs' (housing, costumes, dances, songs ...), while ridding themselves of what would be prejudicial to 'nature' (animal sacrifices, for example, or slash-and-burn farming practices, or simply agriculture in places that are considered to belong to wildlife). 'Tradition' is redefined depending on the public's expectations, and for villagers ethnic belonging has become a resource, a resource linked to these new territories. If we consider these new territories according to the notion of 'agency',[13] we might also imagine that the populations use, in a very cunning way and to their advantage, all the new features at their disposal. However, we will see in the following paragraphs that this is far from being generalized and that this can also have unexpected consequences.

This new territory, once demarcated, is protected at its limits against smuggling, poaching, encroaching, which are frequent activities around borders. The protected area is kept under heavy surveillance. This is the case for many national parks. More army personnel can be found around certain parks than on the country's own borders.[14] In some places barbed wire fencing encloses them, and the wooden fence around Kaziranga is gradually being replaced by an electrified one. Groups of forest guards, and sometimes the Border Security Force and the army are posted around and inside Kaziranga. Some of their operations are spectacular. For example, after the announcement was made to build a 6th extension to the park in 1986, the Forest Department started to evict herders who had settled in the area since the beginning of the twentieth century by setting fire to their stables.

However, and this has been verified everywhere in the world, 'environmental barriers' cannot be clear-cut separations between men and nature. This is confirmed in Kaziranga. Confining nature by trying to freeze land, landscapes, populations and even animals within milieus that are highly mobile proves problematical whatever the case, since wild animals themselves do not respect park boundaries. This nature protection policy creates paradoxical situations, to say the least. On the one hand, as evoked by Bruno Latour (1991) or Paul Robbins (2001), by accentuating the divide between nature and society, proponents of a 'pure' and 'protected' nature then have to fight against hybridisation and undesirable species. In Kaziranga, park authorities are spending a lot of time and money on an endless fight against hybridisation between wild buffalo and domestic buffalo, for example, or the invasion of plants such as Mimosa (*Mimosa rubicaulis* and *invisa*) from nearby tea plantations where it is planted to regenerate the land.

On the other hand, villagers suffer damage caused by wildlife venturing outside the park's perimeter. Indeed, near the park, hordes of elephants (which may include more than a hundred animals) and rhinoceroses regularly destroy crops, homes and sometimes kill people. Tigers slaughter dozens of domestic animals each week. Peasants receive no or very little compensation for this damage and cannot replace the animals they have lost. Their only means of control is to light up their fields at night and to make noises to try to ward off wild animals which they cannot kill because hunting is prohibited. However, any human being entering the park can be shot on sight by a ranger.[15] A few deaths are reported every year.

Finally, this divide creates a sharp contrast between, on the one hand, the resources made available for nature protection and for tourism—since a large part of the Park's budget is earmarked for putting an end once and for all to poaching, for recovering lost wildlife roaming about outside parks, for the annual Elephant Festival, for tourist facilities—, and on the other hand, the utter destitution of the populations.

As in South Africa (see Rodary 2008: 215), the takeover of protected areas by hunters and foresters, then the spread of parks, the development of international tourism aimed to a very large extent at a wealthy clientele, have systematically marginalized the poorest populations. Some researchers, such as Sylvain Guyot who works in South Africa, talk of "green apartheid" when referring to these parks (Guyot 2006).

Conflicts on borders

Ultimately, this conservation policy has adverse effects since protected areas become territories in defiance of the state, lands of 'resistance' (see Laslaz 2008: 25) where conflicts on both sides of the border are not uncommon.[16] In the Himalayas, the forestry administration has always been regarded as hostile by populations, it has never had the means to implement its policy, it represents the government, and is therefore often contested. Moreover, in the Assamese political context, which is marked by demands for autonomy, the Forestry Corps plays a role of controlling populations in conjunction with the army. For example, in Manas National Park in Assam at the border with Bhutan, during the Bodo insurrection, "the destruction of the forest was part of the fight against the state" (Vandekerckhove 2010:54); the forest service personnel became primary targets and thirteen of them were killed between 1988 and 1991 (Vandekerckhove 2010: 52). Similarly, in Nepal, the first actions carried out by the Maoists during the insurgency that started in 1996, were aimed at destroying all the forestry offices because, more than any other institution, they were the symbol and representative of the government in office (Smadja 2005b). This whole situation leads to the ineffectiveness of nature protection.

Indeed, in Kaziranga, encroachments on the park and poaching persist and are even on the rise. A large amount of wildlife is killed by the population, while damage due to wildlife is also on the increase, including the killing of people and of cattle. And, on both sides of the border, as on a battle field, the victims can be counted in both camps. Between 1995 and 2005, 62 persons and 56 elephants were killed, 857 houses were destroyed along with hundreds of hectares of crops in Golaghat district alone (Di Fonzo 2007). In the whole of Assam, 567 rhinoceroses were slaughtered between 1980 and 2005, while between 2005 and 2009, 209 persons were killed by elephants (70 in 2009) and 60 elephants were slain (42 in 2009). Between April 2008 and March 2009, 171 domestic animals (for the most part cows and bullocks) were killed by tigers on farms around Kaziranga (Bora *et al.* 2009).

The situation has significantly worsened since 2005. Two units of an Anti-Depredation Task Force have been created by the Assamese government, yet with no success.

The conflict has been qualified as a 'man-animal' conflict and up to now it has been mostly a question of a man-elephant conflict. Since 2008, there has also been talk of a man-tiger conflict and most villagers around the Park now vehemently oppose the Project Tiger, at the same time questioning the relevance of the Park itself and organising their struggle via new farmers committees such as the Krishak Mukti Sangram Samitee (KMSS).[17]

Finally, ceremonies during which poachers surrender—as was the case in 2009 for poachers from villages surrounding Kaziranga National Park—are yet another illustration of border situations.

To sum up the situation, we can borrow Willem Van Schendel's words about the border between Bangladesh and India:

> For borderlanders, there were three main aspects to the upheaval. First, the world as they had always imagined it was gone forever, their universe torn in two; and half of it lost. They were subjected to the violent rescaling that results from state formation, and that rescaling took place in their backyard. Their geographical imagination abruptly violated, they needed to reinvent themselves as people with new identities as borderlanders, as citizens of a new state and as inhabitants of a divided landscape. The strategies they employed were complex and variable, ranging from outright defiance to acceptance, accommodation and innovation. (Van Schendel, 2005: 12)

From nature conservation to communalism: from one border to another

The setting up of national parks does not simply create a separation between nature and society and a 'conflict between humans and animals'. Policies implemented on the borderland outside parks may generate or strengthen borders delimitating other territories, and one may wonder to what extent these policies are related to the increasing claim for autonomous territories: claims that are under way throughout the Himalayan region, where identity constructions are emerging based on ethnic particularities which are more and more entrenched in the territory. In this context, it appears that when combined with conflicts over access to land and natural resource management surrounding these protected areas, the promotion of 'ancient cultures' and of ethnicity for tourists' sake that goes hand in hand with the nature protection process—as seen in eco-villages and with endogenous tourism, cultural tourism, ethnic tourism etc.—may fuel identity claims, communalism and demands for autonomous ethnic territories.

This is the case for the Misings near Kaziranga, the group most affected by the Park's policy. Elsewhere in Assam, this has been the case of the Bodos who obtained their autonomy in 2003 after Manas National Park (on the border with Bhutan) was inscribed on the World Heritage List in 1985 and following the violent conflict over resource management that it triggered. The Forestry Department finally entrusted the Park management to the Bodos.

This is also the case in other parts of the Himalayas and particularly in Nepal, where ethnic promotion linked partly to tourism—promoting for example in Mustang 'the land of ancient culture and unspoiled wilderness' or in any eco-tourism village 'traditional houses, songs and dances'—seems to fuel demands for a federalism based more and more on ethnic criteria.

Now, in their quest for autonomy, communities can also appropriate the environmental discourse held on a global scale and use it to obtain funding to carry through their projects and to legitimize their territory control strategy. The texts written by the Misings or by the Bodos, for example, include notions of sustainable development, conservation, environmental degradation, biodiversity, etc. In their Memorandum of Settlement on Bodoland Territorial Council, the Bodos plan "to establish a cultural complex to promote and develop Bodo tradition and cultural heritage" and "to develop adequate infrastructures to promote Manas sanctuary as an international tourist spot". Of course, these communities do not accept modernity passively, but they may take advantage of it as has been shown in many places of the world within the notion of 'agency'. However, this is only true for societies that are still structured and still have strong social links. Destructured societies where people are too uprooted and where individualism is put to the fore do not have these capacities (see Beck *et al.* 2006). Moreover, in North-East India at least, autonomous territories have been set up to curb injustices and inequalities (within them, people claim land rights, the right to a bilingual education, to self-management of local resources and of the environment), though it often transpires that they have not improved the situation of the populations themselves. Women's representation in administrative bodies is not compulsory (unlike in the Panchayat), the population is not involved at the grass roots level, and there may be substantial corruption, enriching local elites. In addition, they also serve to resist migrations: they all have their own legislation, in particular denying persons other than those from their community access to land. This process has also been observed in South Africa (Guyot 2006) or in Latin America: for example, and in a very similar manner, among the Kunas from Panama as studied by Martinez Mauri (2007).

National park and nationalism: one territory/frontier may represent another

Finally, on another but somewhat similar register, nature conservation can also be instrumentalised by activists in order to fight against settlement by foreign populations. At Kaziranga, for some people, 'national park' still rhythms with nationalism. Hence, the student movement, All Assam Student Union (AASU),[18] which is a nationalist movement, plays on the emotional charge the park carries to organize nature defence demonstrations. Its clearly displayed goal is to protect the rhinoceros and at the same time promote Assamese culture, yet its main purpose is to exclude Bangladeshis, and through them Muslims, who are accused of poaching (though the great majority of poachers are not Bangladeshis) and encroaching, and who are considered a threat to Assam.

The lyrics of the following song, sung and distributed at one of these events, illustrate this point; particular attention should be paid to the last verse.

"On a carpet of greenery, silvery star
Is golden Kaziranga
The one and only, unique
Beautiful, lovely Kaziranga
Where the one-horned rhino lives
Exceptional, exceptional park
Kaziranga National Park

You are pure and kind
What a vast greenery you are
From your womb are borne
Elephants, tigers, rhinos, primates
You are a mother, dearest mother
Thou have borne and bear
Deer, fish, buffalo, boar and reptalia
Plants and creepers and fruits and flowers

Thou have borne and bear
Thou gives shelter to migrated birds.

You are the daughter of mother Axom [Assam]
And the pride of India and Asia

Today Kaziranga is in danger
That is why we are on the alert
We take resolution in life and death
To defy the destructive activist
And to eliminate them all."

(Text recorded at an AASU meeting in Kaziranga National Park in February 2008. Translation from Assamese)

AASU members place the purity of the national park dedicated solely to nature on a par with that of Assam. In actual fact, they defend the boundary of the National Park just as they defend the border between India and Bangladesh and in a way, for them, the fence surrounding the National Park is a replica, on another scale, of the wall between Bangladesh and India. The process is in keeping with the 'Geography of Anger' from Appadurai (2006) who shows how tensions between India and Pakistan appear in mutant forms at various levels and scales (i.e. from global to local). It is worthwhile noting that the All Assam Student Union calls for DNA tests to be carried out on 'suspected Bangladeshi infiltrators' in order to update the National Register of Citizens (NRC) and then to complete the process of sealing the international border with Bangladesh. This mixture of styles echoes the study conducted by Emma Mawdsley (2010) related to the case of Tehri Dam and of the World Hindu Council, a militant nationalist organization, whose rhetoric which is used to protest against the dam's construction is linked to anti-Muslim action. This association between 'the Green and the Saffron' (see Sharma 2001, 2002) is not uncommon with regard to environmental issues in India. As shown by Rademacher (2010) and Mawdsley (2010), ecological degradation has become a metaphor for cultural degradation (Guneratne, 2010).

In Assam, whether by groups claiming territorial autonomy or by nationalist activists, resistance to migrations and more specifically to Bangladeshi migration shows how, as explained by Evans-Pritchard (1940) in his 'segmentary model', collective identities that are opposed at a territorial level may be complementary or even similar at a highest territorial level when faced with an Otherness that is considered stronger, hereupon represented by Bangladeshis.

Thus, the situation on the Kaziranga National Park borderland appears somewhat chaotic and any foreseen solution to remedy it leaves one perplexed. The government is encouraging the creation of "village protection forces", some sort of militia armed with torches to drive away wildlife from the villages and fields at night, while student nationalist movements advocate the creation of "rhinoceros protection forces" which would be armed with rifles, trained militarily and would be in charge of protecting the rhinoceros and ... "of eliminating" poachers believed to be

Bangladeshi. The park authorities are putting up electric wire fences in the hope of stopping wildlife in one direction and domestic animals and poachers in the other, like the separation walls along certain state borders to stop the flow of migrants; the suggestion of surrounding farms with electric fencing has also been put to villagers. Hence, the time has not yet come for doing away with barriers, but rather for increasing their number.

As for farmers, who are powerless, they attend meetings totally dumbfounded where park officials, NGO environmentalists, forest wardens, etc. explain to them that man must cohabit with wildlife. Some farmers, still rare—though their number is on the rise—, coordinate their efforts to make a case for taking their expulsion before the courts. The High Court of Guwahati and sometimes the Supreme Court in Delhi have been hearing complaints made by associations of breeders and cultivators who are to be expelled from the park extensions. They claim their right to this land that had been allocated as grazing land, then converted into reserved forests or which they have always cultivated and which now falls within the Park perimeter. The 'Scheduled Tribes and Other Forest Dwellers (Recognition of Forest Right) Act' passed by the Indian government in 2006 should help local people regain access to the resources they are deprived of. Yet this has been disputed by advocates of strict nature conservation and has so far hardly been implemented, notably in Assam where foresters are totally hostile, no doubt for fear of losing their authority.[19]

In 2011, most of the conflicts around the park were on the rise, and farmers organized themselves under the impulse of the KMSS. The social aspect of the fight was starting to prevail over the ethnic one: a social fight in which most of the people involved are now totally against the National Park itself and against the Tiger Project.

Conclusion

While one aspect of modernity is the freer movement of goods, persons and ideas in an increasingly open space, another is the differentiation and specialisation of space, leading to fragmentation, enclosure and segregation. In the Himalayas, as far as the environment is concerned, this process is expressed by a growing separation between nature and society on different scales, from the village to the mountain range. And it is clearly not yet the end of territories or the end of borders. Globalisation leads just as much to the disappearance of certain limits as it does to the building of new ones. On a state scale, the setting up of protected areas illustrates this phenomenon particularly well. Indeed, as a result of this globalisation process, protected areas experience both of these trends. Here, I have chosen to focus on only one of them, showing how these protected areas divide, introduce discontinuities, exclude, and how they can enforce inequalities. That being said, these spaces may also lend themselves to experimenting with new forms of governance. They bring together different protagonists at global and local levels and may be beneficial to populations that can then free themselves of the hold of local elites, providing that real participatory management is implemented (Smadja

2011). This does seem to happen in some places, such as in the Kangchenjunga Conservation Area, for example. On a village scale, Forest User Groups (FUG) in Nepal in the 1980s and, to a lesser extent, Joint Forest Management (JFM) in India may also improve villagers' livelihood, though it is not always the case as demonstrated by Nightingale (2010) and Metz (2010).

This study shows that the new boundaries separating 'nature' and 'society' are not merely limits, neutral ones, intended only to protect nature. By asserting a new form of power, that of nature conservation, governed partly by international policies, these boundaries represent political choices and choices of society. One of the inherent problems of this process is the wide gulf that exists between official nature conservation policies that bestow a purely ecological dimension on milieus considered as natural and the point of view of populations not belonging to an ecosystem, but to a territory in which certain milieus, such as the forest or pastures, are both a special biotope and a material or immaterial resource (see Smadja 2009, 2011).[20] Nature conservation measures therefore raise questions regarding the definition of the resource that is conjunctural and that only exists when perceived as such, when endowed with a usage value and socialized. Spaces such as forests, pastures or riverbeds, which for some, only have an ecological or industrial value and where uncontrolled human activities are forbidden, may be vital to others. It is therefore essential to know who can legitimately define a resource, what the purpose of its protection is, for whom it is protected, from whom it is protected and with what aim in mind: whether for the survival of populations, the economy of a region, world heritage, or the tourist industry? As for national parks, the struggles expressed on both sides of the border reveal clashes of a cultural, political and economic nature. They cannot be summarized as a conflict between humans and animals, as some conveniently qualify the situation, but they illustrate a conflict between different points of view on what the protection of nature should be, a conflict between men about what resource management should be, as acknowledged by Jacques Weber (1995) or by Emmanuel Bon (2005) in studies on India.

In the end, it seems that doggedly seeking to establish a radical break between nature and society without involving people in resource management leads to multiple aberrations and that such a nature protection policy, which is at work in many places in the Himalayan region, fuels—if not actually causes—part of the conflicts that are shaking this region of the world.

If we place the new territories formed by national parks and their borders in a regional and national context, it must be said that in Assam, and more generally in North-East India, the process of fencing and of rescaling is in keeping with a history where boundary/territory-making has been permanent, radical and has had major consequences on the structuring of space and on the displacement of populations (see also Vandekerckhove 2010). To cite just a few examples, one can go back to colonization and to the creation of game reserves, then of reserved forests, followed in the late nineteenth century by the setting up of tea estates that are also enclaves with their own laws, economy, population and even language, and which are a particular feature of Assam. Then came Partition in 1947 and later the formation of autonomous territories. These National Parks seem to be another step in the process of boundary-making and of enclosure.

The process of closing and fragmenting space, of building fences and enclaves also comes across as being a strong trend in the recent production of space in South Asia and particularly in India. It is a recurring figure if we consider the mushrooming of 'Gated communities' and 'Urban colonies' in Indian cities, or of Special Economic Zones and industrial zones. It is a process that actually began during British colonial rule, a process of exclusion of the other, exclusion of what is alien, which has been addressed in urban studies, but so far not in environmental studies. We are therefore witnessing the genesis of new territories which, for many, are marked by closure, privatisation, and by functional and socio-spatial segregation. Thus, globalisation can also be viewed as a juxtaposition of closed worlds. While urban enclaves protect a more or less wealthy population against foreigners, 'savage', natural enclaves protect the 'savage' from populations whether local or alien. Whatever the case, the process consists in excluding undesirables. The territorial policy at work throughout the world is often de facto a policy of exclusion.

Willem Van Schendel (2005) and James Scott (2009) have defined an area, ranging from the highlands of the western part of Northern India to those of the lower end of peninsular Southeast Asia, they call Zomia, which is not characterized by the states and the Centre of Power but by their periphery. The Himalayas are part of this entity. As for Sarah Schneiderman, while commenting on Zomia, she has emphasized that "scholars will be well served by continuing to develop the concept of the Himalayan Massif as a regional unit of analysis, particularly when empirical studies from the region are placed in comparative conversation with those emerging from the Southeast Asian Massif and other such broadly defined world areas." (Schneiderman 2010: 312). To supplement the statements made by these authors, I would add that internal borders in the Himalayan area,[21] such as those delimiting ecological and ethnic territories, also confer a particular identity on it and strengthen its status as a distinct cultural area that deserves to be studied via its internal borders. As suggested in a French book entitled *Jeux d'échelles* (which can be translated as 'Playing with scales') edited by Jacques Revel (1996) and related to microhistory, borders should not only be studied at the limit between two political states, but also in their multiple expressions within the states themselves, at different scales. And, if "much can be learned about centres of power by looking at their peripheries" as Willem Van Schendel wrote in *Geographies of knowing, geographies of ignorance* (2002: 62), we might well add: by looking at their internal borders.

REFERENCES

Appadurai, A. 2006. *Fear of Small Numbers: an Essay on the Geography of Anger*, Duke University Press.

Bahuchet, S., P. de Maret, F. Grenand and P. Grenand. 2001. *Des forêts et des hommes : Un regard sur les peuples des forêts tropicales*, Brussels, APFT-ULB, Editions de l'Université de Bruxelles.

Beck, C., Y. Luginbühl and T. Muxart (eds.). 2006. *Temps et espaces des crises de l'environnement*, Paris, Éditions Quae, coll. 'INdiSciplineS'.

Berque, A. 1986. *Le sauvage et l'artifice. Les Japonais devant la nature*, Paris, Gallimard.

Bon, E. 2005. 'Le drame des communaux. Appropriation publique, privée, commune ou accès libre aux espaces-ressources naturels indiens ?', Ph.D. thesis, Université Paris X-Nanterre.

Bora, P. J., A. Sharma, R. Bhatta, G. M. Das, S. N. Buragohain and D. D. Gogoi. 2009. *Human and Tiger Interface in Kaziranga Tiger Reserve, A Technical Report, May 2009,* WWF and Department of Environment and Forests, Government of Assam.

Bruslé, T., 2006, 'Aller et venir pour survivre ou s'enrichir. Circulations de travail, logiques migratoires et construction du monde des Népalais en Inde'. Ph.D. thesis, Université de Poitiers, Migrinter.

Bruslé, T., M. Fort and J. Smadja. 2009. 'A Bocage Landscape, Masyam and the Hamlet of Kolang', in Smadja, J. (ed), *Reading Himalayan Landscapes over Time. Environmental Perception, Knowledge and Practice in Nepal and Ladakh*, Collection Sciences Sociales, no. 14, Pondicherry, Institut Français de Pondichéry, pp. 509-551 and 602-605.

Campbell, B. 2010. 'Beyond cultural models of the environment : linking subjectivities of dwelling and power', in Guneratne, A. (ed.), *Culture and the Environment in the Himalaya*, London and New York, Routledge, pp. 186-203.

Debarbieux, B. 1995. 'Le lieu, le territoire et trois figures de rhétorique', *L'Espace géographique*, 2: 97-112.

Descola, P. 2005. *Par-delà nature et culture*, Paris, Gallimard.

– 2007. 'Les coulisses de la nature', *Cahiers d'anthropologie sociale, Gouverner la Nature*, Paris, Editions de l'Herne: 123-127.

Depraz, S. 2008. *Géographie des espaces naturels protégés. Genèse, principes et enjeux territoriaux*, Paris, Armand Colin.

Di Fonzo, M. I. 2007. 'Determining correlates of Human-elephant conflict reports within fringe villages of Kaziranga National Park, Assam', Masters of Research of the University of London and Diploma of Imperial College.

Evans-Pritchard, E. E. 1940. *The Nuer. A description of the modes of livelihood and political institutions of a Nilotic people*, Oxford at the Clarendon Press.

Foucher, M. 1991 (1988). *Fronts et frontières. Un tour du monde géopolitique*, Fayard, Paris.

Guille-Escuret, G. 1989. *Les sociétés et leurs natures*, Paris, Armand Colin.

Guneratne, A. (ed.). 2010. *Culture and the Environment in the Himalaya*, London and New York, Routledge.

Guyot, S. 2006. *Rivages Zoulous*, Paris, Karthala, coll. 'Tropiques'.

Guyot, S. and L. Mniki. 2008. 'Les parcs nationaux sud-africains, entre frontières raciales et 'frontières environnementale'. Les nouveaux enjeux de la conservation de la nature sur la Wild Coast', in Héritier, S. and L. Laslaz (ed.), *Les parcs nationaux dans le monde. Protection, gestion et développement durable*, Paris, Ellipses, pp. 227-242.

Héritier, S. and L. Laslaz. 2008. 'Introduction. Les parcs nationaux dans le monde. Protection, gestion et exigences d'acceptation sociale', in Héritier, S. and L. Laslaz (ed.), *Les parcs nationaux dans le monde. Protection, gestion et développement durable*, Paris, Ellipses, pp. 11-31.

Hobsbawm, E. and T. Ranger. 1992. 'Introduction: Inventing Traditions', in Hobsbawm, E. and T. Ranger (eds.), *The Invention of Tradition*, Cambridge, Cambridge University Press, pp. 1-14.

Ives, J.D. and B. Messerli. 1989. *The Himalayan Dilemna. Reconciling Development and Conservation*, London, Routledge.

Khanal, N. R. and T. Watanabee. 2006. 'Abandonment of Agricultural Land and its Consequences. A case study in the Sikles Area, Gandaki Basin, Nepal Himalaya', *Mountain Research and Development*, vol. 26, no.1, Feb. 2006: 32-40.

Kollmair, M., U. Müller-Bökker and R. Soliva. 2003. 'The Social Context of Nature Conservation in Nepal', *European Bulletin of Himalayan Research*, no. 24: 25-62.

Laslaz, L. 2008. 'Les parcs nationaux alpins français ou les rattrapages de l'acception perdue', in Héritier, S. and L. Laslaz (ed.), *Les parcs nationaux dans le monde. Protection, gestion et développement durable*, Paris, Ellipses, pp. 243-265.

Latour, B. 1991. *Nous n'avons jamais été modernes*, Paris, La Découverte.

Martinez Mauri, M. 2007. 'Une réserve de la biosphère non achevée. Les Kunas et la conservation de la nature', *Cahiers d'anthropologie sociale*, no. 3: 97-107.

Mawdsley, E. 2010. 'The abuse of religion and ecology: the Vishva Hindu Parishad and Tehri Dam', in Guneratne, A. (ed.), *Culture and the Environment in the Himalaya*, London and New York, Routledge, pp. 151-165.

Metz, J. 2010. 'Downward spiral? Interrogating narratives of environmental change in the Himalaya, in Guneratne, A. (ed.), *Culture and the Environment in the Himalaya*, London and New York, Routledge, pp. 17-39.

Meyer, F. and G. Koppert. 1983. 'Anthropology of Health and Nutrition', in *Ecologie et développement, Journées franco-népalaises, 27-30 novembre 1983*, typescript, Villejuif, Centre d'études himalayennes.

Milian, J. and E. Rodary. 2008. 'Les parcs nationaux dans le monde, un aperçu géographique', in Héritier, S. and L. Laslaz (ed.), *Les parcs nationaux dans le monde. Protection, gestion et développement durable*, Paris, Ellipses, pp. 33-44.

Nightingale, A. 2010. 'A forest community or community forestry? Beliefs, meanings and nature in North-western Nepal', in Guneratne, A. (ed.), *Culture and the Environment in the Himalaya*, London and New York, Routledge, pp. 79-99.

Rademacher, A. M. 2010. 'Restoration and revival: remembering the Bagmati Civilization', in Guneratne, A. (ed.), *Culture and the Environment in the Himalaya*, London and New York, Routledge, pp. 166-185.

Ramirez, Ph. 2009 (2003). 'The Nepalese State and the Transformation of Landscapes according to Administrative Documents Dating from the XVIIIth and XIXth Centuries', in Smadja, J. (ed.), *Reading Himalayan Landscapes over Time. Environmental Perception, Knowledge and Practice in Nepal and Ladakh*, Collection Sciences Sociales, no. 14, Pondicherry, Institut Français de Pondichéry, pp. 325-350.

Revel, J. 1996. *Jeux d'échelles. La micro-analyse à l'expérience*, Paris, Seuil and Gallimard.

Ripert, B. 2000. 'Dynamiques spatiales et transformations de la société, l'exemple des Tamang au Népal Central', Ph.D. thesis, Université Paris 7.

Robbins, P. 2001. 'Tracking Invasive Land Covers in India, or Why Our Landscapes Have Never Been Modern', *Annals of the Association of American Geographers*, 91, no. 4: 637-659.

Rodary, E. 2008. 'Les parcs nationaux africains, une crise durable', in Héritier, S. and L. Laslaz (ed.), *Les parcs nationaux dans le monde. Protection, gestion et développement durable*, Paris, Ellipses, pp. 207-226.

Schneiderman, S. 2010. 'Are the Central Himalayas in Zomia? Some scholarly and political considerations across time and space', *Journal of Global History*, 5, London, London School of Economics and Political Science: 289-312.

Saikia, A. 2005. *Jungles, Reserves, Wildlife. A History of Forests in Assam*, Guwahati, Assam, Wildlife Areas Development and Welfare Trust.

Scott, J. C. 2009. *The Art of Not Being Governed. An Anarchist History of Upland Southeast Asia*, New Haven and London, Yale University Press.

Sharma, M. 2001. 'Nature and Nationalism', *Frontline*, 18 (3), February 3-16, available at: www.hinduonnet.com/fline/fl1803/18030940.htm

– 2002. 'Saffronising Green', *Seminar (New Delhi): Shades of Green, a symposium on the changing contours of Indian environmentalism*, August 2002, available at: www.india-seminar.com/2002/516/516%20mukul%20sharma.htm

Smadja, J. 1995. 'Sur une dégradation annoncée des milieux népalais : initiatives villageoises pour remplacer les ressources forestières', *Nature Sciences Sociétés*, vol. 3: 190-204.

– 2005a. 'L'Himalaya des grands espaces' in Sorgues, P. and J. Smadja (eds.), *Géo, dossier 'Himalaya'*, no. 311, Jan. 2005: 42-89.

– 2005b. 'De la forêt utilitaire, à la forêt loisir, à la forêt symbole de 'pays qui se meurent'', in Taabni, M. (coord.), *La forêt : enjeux comparés des formes d'appropriation, de gestion et d'exploitation dans les politiques environnementales et le contexte d'urbanisation généralisée, Colloque international, 16-17 octobre 2003*, Poitiers, MSHS Université de Poitiers, Laboratoire ICoTEM, EA 2252, pp. 295-297.

– 2009. 'Entre Nature et Société : de nouveaux territoires en Himalaya', in *Actes de la 'Journée Himalaya', 8 janvier 2009*, INSU (CNRS) and Réseau Asie-Imasie (UPS 2999 CNRS/FMSH), http://www.reseau-asie.com/article/journees-d-etudes/entre-nature-et-societe-de-nouveaux-territoires-en-himalaya/

– (ed.). 2009 (2003)a. *Reading Himalayan Landscapes over Time. Environmental Perception, Knowledge and Practice in Nepal and Ladakh*, Collection Sciences Sociales, no. 14, Pondicherry, Institut Français de Pondichéry.

– 2009 (2003)b, 'A Reading of the Salme Tamangs' Territory and Landscape', in Smadja, J. (ed), *Reading Himalayan Landscapes over Time. Environmental Perception, Knowledge and Practice in Nepal and Ladakh*, Collection Sciences Sociales, no. 14, Pondicherry, Institut Français de Pondichéry, pp. 199-239 and 563-589 + map.

– 2011. 'Belonging, Protected Areas and Participatory Management. The Case of Kaziranga National Park (Assam) and of the Misings' Shifting Territory', in Pfaff-Czarnecka, J. and G. Toffin (eds.), *The Politics of Belonging in the Himalayas. Local Attachments and Boundary Dynamics*, Delhi, Sage Publications, pp. 246-271.

Van Schendel, W. 2002. 'Geographies of knowing, geographies of ignorance: jumping scale in Southeast Asia', *Environment and Planning D: Society and Space*, vol. 20: 647-668.

– 2005. *The Bengal Borderland. Beyond State and Nation in South Asia*, London, Anthem Press.

Vandekerckove, N. 2010. 'No land, no peace : dynamics of violent conflict and land use in Assam, India', Ph.D. thesis, Ghent University.

Viveiros de Castro, E. 2010. *Métaphysiques cannibales*, Paris, PUF.

Weber, J. 1995. *Gestion des ressources renouvelables : fondements théoriques d'un programme de recherche (rapport)*, Paris, CIRAD Green.

Notes

1. Here an area that is 2,500 km long between the summits of Nanga Parbat (Pakistan) and Namche Barwa (China) and 300 km wide between the Tibetan plateau and the northern fringes of the Indus, Sutlej, Ganges and Brahmaputra plains, covering 750,000 sq km.

2. Which have been studied by several scholars such as Berque (1986), Guille-Escuret (1989), Latour (1991), Bahuchet *et al.* (2001), Descola (2005), Viveiros de Castro (2010), etc.
3. We will not develop this theory here again. For more information on this topic, see Ives and Messerli (1989), Smadja (2009 [2003]a), Metz (2010) among others. Suffice to say that a catastrophic scenario predicted that given the surge in demographic growth after the 1950s, growing demand for tree resources (that is to say firewood, timber, fodder) as well as for farmland, would lead to massive deforestation, thus to a catastrophic increase in soil erosion, to the loss of productive land through accelerated landslide incidence, to a massive drop in soil fertility, all processes said to cause devastating floods in the Ganges and Brahmaputra plain. Himalayan peasants have been held responsible for this environmental degradation, thereby leading to the foregone conclusion that nature has to be protected from society in the name of sustainable development.
4. See *Reading Himalayan Landscapes over time* (ed. J. Smadja 2009 [2003]a) and in particular the text by Ph. Ramirez (325-350).
5. The Nepalese population amounted to 5,639,000 in 1911, 9,413,000 in 1961 and 23,151,423 in 2001. It will exceed 26 million in 2011 (Central Bureau of Statistics, Nepal).
6. In 2005, the world average was 11.6 per cent when only taking into account land areas (see Héritier and Laslaz 2008, Milian and Rodary 2008).
7. Protected areas may be regarded as tools for managing international conflicts with the setting up of 'peace parks' such as the Limpopo Park between Mozambique, Zimbabwe and South Africa, or the 'peace park' project on Sachien glacier between India and Pakistan. However, these are only very marginal attempts and would be the subject of another study.
8. Translated from the French.
9. Indians from Orissa arrived in this region in successive waves, to be employed by the British on tea plantations, while others arrived from Bihar to take up activities involving fishing, river navigation, etc., others from Rajasthan for trade, and Nepalese migrants practising cattle breeding, etc.
10. As Roe and Hollands (2004), quoted in Héritier and Laslaz (2008: 291), would say.
11. F. Landy, paper presented at the International Workshop 'Espaces protégés, acceptation sociale et conflits environnementaux', 16-18 September 2009.
12. Summarized and translated from the French.
13. 'Agency' is defined as the capacity of an agent to act on its destiny.
14. This is the case, for example, in Chitwan Royal National Park in Nepal where the army is posted all around and inside the Park, while a large part of the country's borders is left opened.
15. This is the case of Kaziranga according to the Wildlife (Protection) Act 1972, as was the case in South Africa with the 'shoot on sight' guidelines (Neuman 2004, cited by Rodary 2008).
16. The same statement was made by Kollmair *et al.* (2003) in Nepal: "All over Nepal, the protection of crop-damaging animals and predators that kill domestic animals meets with limited acceptance among the farming population. As long as farmers are not compensated sufficiently for such damage, 'conservation education' is not likely to succeed in convincing them of the necessity of wildlife protection. In other words, a smallholder in the Teraï region is not likely to understand the 'immense national

and international importance of the rhinoceros', so long as these animals keep destroying his crops and he does not receive any compensation".

17. Translation: Freedom for Farmers Committee.
18. It allegedly played a role in 1980 in the massacre of Muslims in Assam, and members of this movement are suspected of having links with nationalist fundamentalist movements such as the RSS (*Rashtriya Swayamsevak Sangh*).
19. Criticising the position of Assamese foresters with regard to this law, an official at a ministry in Guwahati once said: "Like at the time of the British, they like to rule, not to serve" (personal paper).
20. Ben Campbell recalls that, for example, Tamangs "perceive their mountain environment as a field of social agency (in which villagers, the state, the supernatural, and wildlife are participant subjectivities of power), and not as a separate domain of nature in the exclusive power of others." (2010: 200)
21. Including bordering plains, contrary to these authors who only take into account mountainous areas.

2.1.2.

Nature Reserves, National Politics and Local Resistance: Why the Lepchas from the Dzongu Reserve in Sikkim starve themselves to death for their motherland

Brigitte Steinmann

One remarkable feature of the mythology[1] of the Lepchas inhabiting the northern Dzongu[2] reserve in Sikkim lies in the close links they have established between human artefacts and nature, especially the origin of musical instruments based on perceptions of hunters on the lookout in the forest, stalking their prey. Through these links there emerges a broader synthesis between their perception of natural phenomena and their explanations about the origin of cultural artefacts, a synthesis that goes towards understanding, among other things, the underlying reasons behind their commitment to the Dzongu reserve—a territory that was for a long time isolated and placed under royal guardianship, but is now threatened with destruction due to the building of hydroelectric dams for the production and sale of electricity to India. Let us consider some of the links between the Lepchas' ideas of territory and cultural heritage through some examples taken from hunting tales.

Hunting and the origin of musical instruments

One of these stories retraces the origin of the *satsang*, a guitar similar to the Nepalese *sarangi*. It is described as follows:[3]

> "Following the reign of the Lepcha King Ritat Panjyum, King Rongbung Punu ascended the throne around 800 BC. During his reign, there lived a hunter by the name of Sato. Once, when Sato went hunting, he saw the extra large footprints made by a wild boar. Instinctively, he tracked down the animal and shot it dead with a poisoned arrow. The boar staggered along for sometime and died shortly afterwards in a bamboo grove. Soon, it grew dark and Sato was not able to return home with his trophy. So, he cut off one of the boar's hind legs and put it in his hunting basket. He climbed up a bamboo for the night. The hunter sat silently trying to make out something in the dead of the night. He could hear lovely music made by the bamboos rubbing against each other. After a

> while, he saw a man of large build moving towards the dead boar. The man chided the dead boar for disobeying him and for meeting such an unfortunate end. Then he cut off a piece of bamboo and fitted it to where the foot was missing. He then chanted six prayers and tapped the bamboo foot three times; thereupon the dead-boar got up from its deathly sleep and followed him out of the bamboo grove. As they both walked away, there was a strong gust of wind and sweet music could be heard again. Sato spent the whole night totally engrossed in this music. At dawn, he slipped down from the bamboo and the first thing he did was to inscribe the six prayers he had heard in the dry bamboo sheath. As he was about to set off home, he once again heard the sweet music coming from somewhere. On closer examination, he saw where the music was coming from and how it was made; two bamboos rubbing against each other were making the sweet sound. This gave Sato the idea of making an instrument that worked on the same principle. On reaching home, Sato split a bamboo and with another thinner bamboo, he made a bow-like accessory and using it, he played the instrument. The instrument, like a violin, played as sweetly as the music he had heard in the bamboo grove [...]. Since Sato had developed this instrument, it was given the name Satsang. This instrument was played in the court of King Rongbong."

Many other musical instruments came into being thanks to a close bond between a hunter or king and a natural element, for instance: the *tungbak*, a three-stringed instrument invented by a man while hollowing out a tree who heard a musical sound from the dug out cavity; the *bangho*, a percussion instrument that a King imagined and made while leaning on a hollow tree trunk when the tree began to resonate; the *pupatik*, an instrument made of split bamboo and a rope that was originally used to tie up pigs under a house, which was discovered when the animal pulled on its lanyard to escape a tiger. This natural element is distinguished by its particular proximity with Man: a bamboo grove sheltering the hunter from his night terror, a tree providing support for the tired king, a lanyard enabling the pig to save its skin.

"We are animists"

These links are at the heart of a very strong sense of nature that the Lepchas have forged and that they readily maintain by describing themselves as 'animist'. They are defined by their specificity and their stance against all other religions to which they have historically been converted, including Christianity. They use the term 'animism' to include a set of beliefs claimed to be their true religion. However, a political argument has become the leitmotif of resistance and rebellion against the degradation of their territories and their landscape caused by the building of dams in the heart of the reserve. It is clear in the myth how special links between Man and nature can animate Lepchas with particular strength, such as deployed in their exemplary struggle[4] against hydroelectric projects and policies to enhance the 'environment'. For former hunter-gatherers, such as the Lepchas, the very idea of 'environment' is nonsensical. Nature is their home. An environmental policy represents the very dissociation of their bearings, a blow to the profound harmony of their existence with the conditions of this existence. The relationship described here between artefacts and natural elements perceived by man, speaks of the true sensitivity of the Lepchas and how they relate to a territory in which humans and nonhumans coexist.

The wild man who appears before the hunter at night and repairs the leg of the wild boar endorses this general disposition of populations who claim that animism "imputes to non-humans the same understanding as their own."[5] For the hunter to find prey, so the myth says, counter powers must also be able to constantly repair the effects of human predation. In its restorative role, mythology compensates for Man's lack of attention and explains which actions to take to prevent further disasters. The warnings given by the wild man to the boar which disobeys him and by the wind in the trees to alert the hunter, results in an invention for human use: the guitar comes from the tree trunk, which resonates, and is used to accompany songs, to describe the boar's lament as well as the wind's melody heard at night by the hunter. Yet the musical instrument is not only designed to mimic the sounds of nature: it must also sublimate the feelings of the hunter who has been taken by surprise by the nocturnal powers. The musical instrument becomes an offering by the forest spirits to the predator.

The similarities between hunting stories therefore explain the care the Lepchas have always taken in creating their artefacts without depleting their natural resources. The shape of trees available in the forest has to be used to invent instruments. According to the myth, it is the tree that forms the shape of the guitar. This idea is eloquently illustrated in Lepcha cinema productions, and in particular in a film[6] directed by Dawa Lepcha, a Dzongu resident, main organiser and driving force behind the Lepcha hunger strike in 2007-08 to protest about the building of dams. The film, which describes the building of a traditional bamboo bridge emphasises the strong sense of familiarity that exists between men, creepers, bamboos, trees and animals that live in the vegetation (Photo 1). The Lepchas boast about being capable of climbing trees as fast, if not faster, than monkeys. The monkey is regarded as an emblematic animal, a keeper protecting the doors of houses which are often decorated with a monkey skull. Lepchas readily say: "We are aware of all the forest's resources. If one day we were forced to live alone in the forest, we could survive on our knowledge of plants. And in the end, we could also live with the monkeys."

Photo 1: Woman in her garden (Dzongu area, the last inhabited village in the upper sector)

In the nineteenth century, Joseph Dalton Hooker[7] widely praised the Lepchas' naturalist qualities and in particular their extraordinary ability to take advantage of the vegetation, bamboo and wood, and to design a shelter or instruments in record time: "A more interesting and attractive companion than the Lepcha I never lived with [...] with the simple resource of a plain knife he makes his house and furnishes yours, with a speed, alacrity, and ingenuity that wile away that well-known long hour when the weary pilgrim frets for his couch".

In 1910 Francis Younghusband wrote: "They are great lovers of Nature, and unequalled as collectors", while also strongly despising the Lepchas' natural qualities, their character that he described as "frank and open, good-natured and smiling", but to whom he preferred the "naughty little Gurkhas" who would no doubt have willingly "swallowed up the poor Lepchas had the English not guaranteed their protection"![8]

And H. H. Risley in the *Gazetteer of Sikkim*: "The Lepchas find innumerable things to eat in the jungle in the shape of fruit, leaves and shoots, pith of stems, roots and flower-buds. They also eat an enormous variety of fungi and seldom make a mistake in collecting them but occasionally a whole family gets poisoned [...]. In case of famine, the Lepchas will survive longer".[9]

Exploiting forest resources: ancient and modern collecting methods

Today, the Lepchas' natural qualities are being undermined or manipulated through projects to create 'botanical gardens' or 'herb gardens' born of a desire to engineer resources for the sake of national and international development. On the opening of the Dzongu Reserve, NGOs were set up to develop, among other things, 'ecotourism' projects. This was about protecting a reserve that was in fact already widely exploited by commercial projects, and in particular cardamom and ginger farming (Photo 2).[10] Lepchas were hired by some NGOs in an attempt to implement external development guidelines that clashed with the actual policy of forbidding access to Dzongu because of hydro-electric projects.

In 2004, in the middle of a thick jungle, Urgyen Lepcha from Hee Gyathang,[11] a primary school teacher and employee of an NGO specialising in developing tourism in Lower Dzongu, imagined landscaped parks incorporating both food crop farming, aesthetic and recreational aspects, medicinal plants and plants of use to Man with his own inherited knowledge of traditional skills, which have long been part of the young Lepchas' education. The Lepcha language, which includes a series of suffixes for the names of plants, provides a classification that is a form of spontaneous reference in ordinary usage. The suffix *kung* denotes, for instance, plants in the 'tree' category; *muuk*, the one for herbs;

Photo 2. Trees and plants in the garden-like forests of Upper Dzongu (including cultivated cardamom)

rik, climbers; *pot*, fruit; *buuk*, tubers; *be* any edible plant. In the categories for use, it is very common to be able to distinguish between poisonous plants for hunting weapons; plants for making pigments and colours; poisonous plants for fishing; plants to be used for textile fibres; plants for building materials, those for paper; plants for forage; plant substitutes for tea; medicinal plants. In addition to this general know-how, there is a mine of specific knowledge on how to move around the forest and on a daily basis find a wealth of edible plants used in the preparation of meals. Here, for example, in the village of Tingvong, is a typical day for a mother of seven children, who every day travels the rugged paths covered in mossy rocks that twist through the forest among the cardamom plants.

The mist bathes the entire horizon and turns epiphytes, tree ferns and bamboo groves into ghosts, whilst a woman climbs the slopes faster than the group of young monks who come to the *gompa*. On the way she is able to describe all the edible plants and fruits of all the trees she passes; she ferrets around in the undergrowth for bamboo shoots which, once stripped of their bark using a sickle make the most exquisite dish. She is careful not to leave visible white bark on the road, so as not to attract the attention of the *mungs*, those harmful spirits who would interpret this as a desire to harm their offspring. She therefore hides the peelings in the undergrowth and continues on her way, just as she sees two mushrooms that she hastens to pick. Higher up, she fills her bag with betel nuts, as she points out other edible plants around her. Then there are the nettles under the tall trees to complete the collection. Her bag is now full of bunches of foliage for the animals, side by side with the edible plants. Peeping out from between these spoils, young fern shoots collected with a small swift movement add the finishing touch to the meal: "Our fathers never died of hunger," she said, "they knew all these plants." Thashe Thing,[12] an old Lama, once said that "the Pachen plant, a kind of fern tree whose leaves are edible, was Shiva's stick planted in the ground: a coded language had to be used to discuss it. Similarly, for banana: when you wanted to use their leaves, you had to say: bring me the eyes of a frog, otherwise the plant would be destroyed by envious spirits. It was forbidden to laugh at some plants, especially fungi, because thunder and lightning might strike."

Then the woman spotted her cousins in the forest on their way back from gathering wood, she greeted them joyfully and picked up a tree trunk she was taking to the village together with to her crop, then ended her jaunt with a quick splash in the waterfall, demonstrating—as if really necessary—that Lepcha women are indeed born of water.

Dual opposition between meteors and cultures, artefacts and nature

Among other artefacts borne of this encounter between natural elements and Man or animals, there are what the Lepchas call 'thunder stones' (*soderlong)* fallen from the sky, which they consider to have come from meteors, thunder and lightning. These stones have become the subject of considerable ancestral cults and are specially kept in houses to guarantee their solidity:[13]

"We sincerely believe that these 'Sadaer Longs' are the creation of the Thunder God sent to us humans as blessings, and sometimes as chastisement. Among qualities attributed to this supernatural stone, the first and the foremost is that the wrath of the Thunder God will not fall upon the house possessing a bit of his own self. In other words, it acts as a lightning conductor. Also, the different evil spirits will not have easy access to the house; they will be afraid of the weapon of the Thunder God [...] This 'Sadaer Long' is used both as medicine and as a charm, because to us who are yet to cross over the primitive stage, it is a sure cure [...] The base of our Lepcha culture has emanated from the original 'Big Stone', the Kongchen-Kongchlo or Mount Khangchendzongu. To us, it acts as the very embodiment of the almighty-omnipotent, omniscient creator Herself."

The author then provides other more informed explanations about these thunder stones which allegedly date from the Neolithic period. This overall bond they establish between the thunder stones which ensure that their houses and mountains remain standing, attribute another meaning to the relationship between nature and artefacts by basing it on an inversion of ordinary terms. It is not society that is produced from projections onto nature, but it is the whole of nature that emerges from Man's ideas about how to keep material bonds together: stones produced by thunder, which keep houses standing, are compared to these other 'stones' erected under the sky; the mountains; creepers taken from the forest and wound together to make bridges are an extension of the forest onto the river; bark and tree trunks dug out to make musical instruments spontaneously bring to the hunter's mind the idea of producing a sound with them. In the same way, social links are also defined as being closely related to the types of mountain, sources of life and the Lepcha culture. Following on after Foning, another Sikkimese author has emphasised this relationship between 'thunder stones' and 'standing stones' (*long chok)* or sorts of obelisks that can be found in particular on the site of Kabilongchok in Sikkim. This is a historic location where a sworn vow of cooperation and friendship between the first Shaman Lepcha and a Tibetan chief is said to have been sealed. According to Pema Wangchuk,[14] the stones erected on the site symbolise the kinship links between the different summits which make up the five peaks of the Khangchenjunga Mountain, conceived by the Lepchas as the elder brother born of the mother of creation (*Itbu Mu*) that stands before the Land of Rums (*Rum Lyang*), Land of the gods. Traditionally, the origin of each male clan (*putso)* is indeed explained through a bond linking it to a peak, whereas each female clan is associated with a lake. Therefore, the Lepchas would never have had to move on since the mountain has provided them with everything they need. The very first couple allegedly emerged from the snow on Kangchenjunga and the Lepchas have remained bound to it by an umbilical cord. The Lepchas still readily refer to themselves as Mutanchi Rong Kup Rum Kup, 'well-loved children of their mother Mu, of the Rong peaks and of the Rum spirits', whilst the shapes of the mountains symbolise both the lineage relationship and an army of soldiers (*vik*) surrounding their leader. These human dimensions of the territory, which divide the population's opinion, are commonly evoked in the form of noted similarities between, for example, the line of a mountain and a woman carrying her baby on her back. This is how a mountain peak on Mount Longmyong or Longmyong that stood out to the north in the Dzongu Reserve was described by the young Ongit Lepcha, while she made her way with her load. Like the mountains, torrents, snakes and birds suggest as many analogies with women and their dress: torrents are described as the hair of women and a number of accounts can be found where a hunter discovers a golden thread in the river, which points him to the forest where he comes across a woman bathing in a grotto. He wishes to marry her and discovers the young woman's father, a being covered in snakes who challenges him with a series of tests before agreeing to tie the knot between the hunter and his daughter.[15]

The relationship between artefacts and nature therefore inversely refers to the links between meteors and culture: genuine artefacts in fact refer to meteors, to a gift from heaven, while Man must contrive to make nature culturally acceptable, to domesticate it, to appease and to pacify it. These dual bonds are indissociable. If you touch just one of the interlinked elements, you destroy the integrity of Man as it specifically relates to a particular land, which is both procreator and vessel. Real 'nature' would in fact be made up of a set of artefacts, whose existence is conditioned by an invisible world populated by auspicious or hostile spirits and in which specialists have a permanent role to play: bamboo bridges over rivers only hold thanks to the shaman's intervention as he celebrates the spirits of the forest; houses only remain standing in the storm because within themselves they possess some elements from this storm that are worshipped; beer only ferments because it contains yeast, a crucial issue between the human and animal world, a permanent object of discrimination that signs the alliance between the divine, human and animal worlds; this is demonstrated by the myth told by Chöden Lepcha to Hee Gyathang:

> The first marriage was in fact an incestuous relationship between Narip and Torbong. Near the Khangchendzonga Mountain, there was an enormous lake called Narip Torbong Partham. Torbong had gone hunting and had caught a bird, but someone stole it from him. He waited to see what would happen and in the morning he saw a beautiful woman whom he captured so that he might marry her. The gods then gathered for the celebration, but no alcohol was to be had: where could they find any yeast? An old woman was hiding some yeast in the bun on her head. A cockroach saw it and wanted it, but the woman, Nyikung, prevented it from doing so. She hid the yeast under the nape of her neck, then caught the cockroach and stuffed it into her basket. The cockroach shouted: 'I've seen your medicine, I've seen your medicine!' Furious, Nyikung turned the doko upside down. The cockroach shouted: 'I can't see anything, I can't see anything!' Yet in fact, he could see through the meshwork of the bamboo doko. When she took the basket, the cockroach hypnotised the nits on the old woman's head and she fell asleep. It then took the yeast and fled to Partham. The old woman woke up and realised that her yeast had been taken. Nyikung was very angry. She cast a spell: 'If you know which drink you can make with yeast, then it will be good; otherwise, it will be poison!' Since then, if you become dependent on the drink, you die, otherwise it proves to be good medicine. So the gods gathered at Partham and prepared millet beer. What sort of medicine is that, they asked. They decided to get the thunder to test the drink. The thunder fled. So they called the king of the cobras; Pomol drank, then started writhing on the ground. He

> became a poisonous snake. The second test was carried out on the cobra Porwikbu. It started vomiting. It was called 'the snake that vomits'. Then came the red snake, Polongbu, 'the one that goes very fast'; it slithered away very quickly. Next the drink was tested on Potsongbu; it went into hiding in a bamboo pot. Then the drink was tested on Bulyokbu. It became blind. The last to try it was Man: for him, the drink was good provided it was drunk in moderation; otherwise, it became poison. If you laugh at goats, lizards, dogs,or mushrooms, thunder strikes. When you laugh by mistake, you must throw alcohol everywhere since thunder was the first to taste alcohol and was poisoned.

Men and insects share the same ambitions, the same point of view and furthermore the same substances. These substances then become different through various artefacts (food grain, sorts of offering, beverages, meat or non-meat products, sexed flowers and fruits) with which these substances are closely associated and which continue to govern their very nature. This is particularly true for everything that touches women's bodies, especially their hair, and which is closely linked to nutritive substances, as underlined in this other myth about 'Thing Gwokmu and the woman from the lake':

> Thing Gwokmu was a holy Lepcha man. One day he saw a woman emerge from the lake and start sunbathing. He was unable to capture her, so he went to ask his mother how to get the woman. His mother gave him her trousers to help catch her. Prior to that, Thing Gwokmu had tried divination: he had thrown an enormous eight-sided rock towards Hee Gyathang. With that he had found a way, but that had not worked either. So he hid with his mother's trousers and waited. When the woman came out of the lake, he caught her with the trousers. She agreed to go away with him on one condition: she had to leave the nits from her head in the lake. They became fish, and since then, you cannot eat that sort of fish.

Analogies may be made between human and non-human substances based on rules whereby one should respect the bodyliness of invisible beings. As a result, the rites primarily consist in obeying these substance-based analogies by simple deduction. This can be observed, for example, when offerings are prepared for the different *rum* and *mung* spirits: sharp instruments will not be used to make bamboo and flower offerings for *rum* spirits; care will be taken to not drop a single drop of blood from the sacrificed yak offered to the Kangchenjunga Mountain; pains will be taken not to leave scraps of plants on the ground, since the latter primarily belong to the forest spirits. It goes without saying that such rules are not taught in books.

The relief as a founding social event

From an 'animist' perspective, the existence of man, as well as the use of the nutritive substances he takes from nature, is closely conditioned by other series of existences and *discontinuous* usages that he has no control over. P. Descola presents the theory behind this form of animist consciousness which admits any metamorphosis between the different worlds: "Metamorphosis is not a revelation or a disguise, but the culminating stage in a relationship where each one, by altering the position of observation their original physicality imposes on them, works towards matching the perspective in which they think the other sees themselves."[16] (Photo: 3)

Yeast is central to all sorts of prohibitions which shape the alliance and serve as a mirror which reflect how humans and non-humans find themselves at crossed purposes. Many variations on this myth of origin about yeast can be found throughout the Himalayas, which fixes the proper rules governing the alliance, which always refers to an incestuous relationship between brother and sister.[17] Generally speaking, the vital links between peaks and men, lakes, torrents and women, lineages and mountains, reveal a conception of the human and social body as being literally doubled with aerial, terrestrial, liquid and solid presences (Photo 4). The *rum*, forest and mountain spirits, are omnipresent like the *mung*, evil presences that are blood-thirsty for human blood and flesh.[18]

Photo 3: *Bonhting*, Lepcha shaman, who specialises in relations with *rum* spirits.

For the Lepchas, just as the death of the forest means the death of men—the difference being that the timed death of a tree to make a musical instrument according to specific natural codes is seen as being the very origin of culture, in mythology—,[19] likewise, degrading mountains by digging mines and galleries to extract building materials for building dams represents the very negation of all their beliefs and their social system. This is based on the idea that if the outer shapes of peaks can be associated with human clans, they also contain inner shapes made of hollows and of underground canals through which the spirits of the dead return to their place of origin, the Kangchenjunga Mountain (Photo 5). Hence the entirely natural idea of an apocalypse that would strike with the collapse and arbitrary digging of underground galleries.

Photo 4. Harvesting in Dzongu. Lepchas farmers' liking for mountains and peaks.

Photo 5. Mount Kangchenjunga

Chöden Lepcha, who recounted the myth of origin about yeast, conveyed his concern about the building of dams by evoking in fabulous accounts how foreign predators: "sought to grab the country's resources, gold and diamonds, by digging galleries in the mountainside." He assimilated the placing of mines in rock excavations to the mining company's feverish quest for gold and diamonds. Chöden alluded to the account about the origins of Mayel Lyang, ancestral land given to the Lepchas by the *rum* mountain divinities, by describing miners as thieves, the total opposite of the first hunting spirits that had arrived from behind the mountains to provide men spontaneously with elements of civilisation: cereals, plants and seed. Mining excavations turned this narrative of origin around, they created a second artificial death by destroying the underground canals of the dead, by permanently closing the Mayel Lyang pathway. Here is one of the versions about the closure of this pathway, recounted by Chöden in the context of dam construction work under way in the lower part of the reserve:

> There was a mother and her son. The mother died and the son had to go out and work in the fields. One day, he noticed that the rest of his work had been done without him. It was the same the following day. He wondered what had produced this miracle. One night, he stayed up to keep an eye on the field. His mother appeared at midnight. The son held her and started to weep. His mother said to him: leave me, I must go now. The son wanted to go with her. So his mother took him to Mayel Lyang. There, he saw all the mothers who had become tiny. There was also a group of tiny hunters and the mothers shouted after the hunters because the son was sitting on the animal tracks. A cicada arrived and the son caught it; he showed it to the hunters saying that it was a deer. The hunters shared the prey between them and appointed the son as a great hunter, by giving him the leg of the cicada. The son took it to his mother. Where is the meat, she asked him. So he threw the leg onto his mother's leg, which broke. One day, they went to pick some ferns with axes; some of them were injured. The son thought: if I stay with these people, I'm done for. The little men had been wondering amongst themselves how they were to take the boy back to his home. But nobody knew the way. Then an animal arrived and said: I will deal with this! It was the porcupine. But an agreement had to be found. If humans exchange their teeth with mine, then I'll agree to take you back. Since then, human teeth belong to the porcupine and the porcupine's quills belong to humans. Yet the nyangbu insect took the opportunity to enter a tree and cut away at it on the inside, thus closing the pathway. The boy returned home and told the story. Men started to worry because the mountain pathway was now closed.

The physical relief as a founding social event is central to Lepcha mythology. It is characterised by the possibility that men have of travelling along it both outside and inside. Mount Kangchenjunga, Kongchenbu for the Lepchas, is both massive and hollow. Its underground is more important than the relief itself in explaining their origins and ultimate ends. In reality, everything happens as if the origins of society, obscure and complex, were the exact opposite of the origins of the physical and natural world, which would be clear and straightforward: men are only capable of producing absurd rules, such as incest between brother and sister, whilst only proper reference to the physical world may confer the right prioritization to the laws of culture. The conceptions of the world's origin as presented by the Lepchas in their mythology perfectly illustrate the generation's necessary inversions.

Among the many accounts of the origins of Mayel Lyang, Chöden Lepcha evoked a non-Buddhisized version of the underground creation of the first woman, *Itmu*, without mentioning Guru Rinpoché's intervention, as reported by H. Siiger.[20] This myth tells how *Itmu*, the mother of creation, had created *tak bo thing* and *na zong nyu*, the first incestuous couple, made up of a brother and a sister, who were to give birth to *mung* demons. In order to separate this couple, the gods divided all domestic animals into two groups which they shared between the brother and sister. *Tak bo thing* put all his animals into a large basket and left for Tibet to breed them. *Na zong nyu* put hers in a broken basket and all the animals escaped into the jungle, and became wild. She thus lived alone on a high mountain and gave birth to two children, *rel bu* and *sen bu.* The first was a *rum* god and the second a *mung* demon. She cherished the two equally, but the *mung* demons, jealous, poisoned the children, who died. The gods gathered to celebrate the first funeral rites. They ordered two birds to pour water into the tombs, taking care that the *rum* water went into *rel bu* tomb. But the birds switched over the water and *mung* water was poured into the *rums.* Two lakes formed from these "inverted" waters and since then, no new *mung* demon has been born of the waters of the *rum* spirits. Henceforth, birth and death are the lot that has fallen to Man. However, another intervention by Itmu, a sort of superior feminine principle very different from carefree *Na zong nyu*, is needed for humanity to be able to develop. Indeed, because of *Na zong nyu*'s fault, all animals with edible flesh become wild; the goat becomes a doe, the sheep a mountain goat, the bull a wild boar, the pig a bear.

'Inversion' which causes disaster at the world's creation is also a recurring theme in the myth about the birth of the rivers Teesta and Rangit, the country's two main rivers flowing from lakes Naho and Nahor. A major earthquake damaged the lakes. According to the myth, the king of snakes, *Payel bu,* emerging from underground accompanied by Tufo, the national bird, guides the two rivers to their confluence at Pozok. Unfortunately, the river led by the bird arrived there first and the snake, furious, wanted to return to its source. The river he was leading totally flooded the land of Mayel Lyang. Several versions of this story exist. The Lepchas who lived near Tengdong Mountain escaped by climbing onto the mountainside and praying to *Itbu rum* to save them.[21] *Kohomfo*, the bird, appeared and made an offering of millet. There was another earthquake which produced several mountains and valleys, allowing the water to run off to the south, thus saving the Lepchas from the floods and from drowning. Since then, celebrating Mount Tendong Lo (Tendong Hlo Rum Faat) in the south of Sikkim has become the Lepchas' national feast day.[22]

Animism and museography: resisting conversions

Sonam Tsering Tamsang is the founder of the first Lepcha Museum in Kalimpong (West Bengal), an amazing collection of cultural and artistic Lepcha artefacts that was compiled in the 1990s. Thanks to his unique knowledge of the Lepcha culture,

Tamsang himself has collected and catalogued a large amount of objects that give an extensive yet precise idea of life in a traditional Lepcha village, as well as of their history and cosmology. All the objects concerning their material life and culture have been described in their context and catalogued ethnographically according to people's daily usage of agricultural and domestic instruments, houses, medicinal plants, hunting and fishing techniques and above all, musical instruments and a number of these strange meteorological phenomena, the stones fallen from heaven (*soderlong*). All kinds of useful plants have been described, including the famous versatile nettle plant or *sisnu*. A large collection of domestic and agricultural instruments include Lepcha knives (*banphok*), old bamboo arrows and lances with handles made of deer antlers.

Ritual life, such as funerary rituals, was also described in detail, with the names of the most important places for the Lepchas: Nampa Longchok, near Phodong; the Tumlong Fort in North Sikkim; Daling Longchok near Renok at the southern border; mythical places such as the cave on the way to Mayel Lyang, and thirteen kinds of most precious resources, including a tree with huge roots on the way to the hidden god's place Dechen phuk. Guru Rinpoché's footprints can be seen on the roots of the tree.

Symbiosis was achieved between mythology and history, daily life and the supra-natural sphere in a way that most accurately conveyed the idea of how people think and perceive their universe. Thus the origins of the Lepchas' kingdom or Mayel Lyang were retraced.

Several events, described below, have taken place there.

The anniversary of King Gebu Achok (20-21 December)

The Ashuley Association, chaired by Lyansong Tamsang, organised large-scale celebrations in the village of Takna. These were staged to allow Indian television to shoot a film and to portray a Lepcha village. A representative from the Bhopal museum came as sponsor to the Ashuley association. Celebrations to commemorate the 266th anniversary of the birth of the last Lepcha king, Gebu Achok, were launched on 20 December 1994. At Pedang lie the ruins of the fort considered by the Lepchas to be their own territory, their stronghold. At dawn, the association gathered its members who numbered about a hundred Lepchas. The procession in jeeps and buses rattled along as far as Pedang.

The path here climbs through the forest up to the fort. The men were dressed in their *dampa*, the women in their colourful saris. Everyone gathered next to the fort where a banner was hung and a portrait of Gebu Achok. The chairman of the association made a speech, which was essentially a reiteration of the vow made by all those present to belong to a Lepcha nation. Mayel Lyang's name is *Mutanshi Rongku*, 'beloved son of Mother Earth'. The chairman's every sentence was punctuated by loud cries from the crowd: "Ashuley, ashuley". Everyone then made their way down the slope, near the ruins of a vast well called 'the queen's bath'. At the foot of this well made of close-fitting stones covered in moss, lied an altar, on which are images of Hindu divinities (Vishnu). A *mun* priestess arrived and began reciting in a semi-trance state. This woman was originally from Mirik, near Darjeeling; she was the association's official priestess even though she also celebrated ordinary rituals. Yet her discourse corresponded in many respects to the ambitions of the association: to reassert the Lepchas' right to occupy and manage their territory themselves. Whilst she officiated, a *bongthing* shaman remained at her side to provide political explanations about her discourse. Nearby, another shaman, in a cataleptic trance, dozed off. The priestess went on with her speech punctuating it with calls to "Mutanshi Rongkup". They all made their way to a third place, a chorten situated level with the fort, just a little further away. The *mun* recited again, accompanied by the *bongthing*. Everyone returned to the fort. The priestess went down to the altar set up in the ruins and yet again addressed the spirits of the place and the crowd.

Chairman's speech

This was a nationalist event aimed at reaffirming the Lepchas' inalienable ownership of this place, the residence of the Lepcha kings who resisted an attack by Bhutan invaders. It was also understood that there was strong antagonism towards the Buddhist Bhotia, who had in fact dominated the Lepchas and who had converted them. This anti-Buddhist theme is taken up on several occasions by members of the association. At the end of the speech by the chairman of the association, everybody returned to Kalimpong for the festivities.

Street party

Celebrations took place next to the museum, which has constantly been reorganised and extended by Sonam Tsering Tamsang,[23] the chairman of the cultural association who has proved to be the most genuine and fervent museographer. S. T. Tamsang has edited and translated many texts, has made opuscules presenting the Lepchas' customs and songs in the form, for instance, of small booklets presenting their musical instruments. On the occasion of the party a platform to stage the various acts which alternated between speeches, girls' dance performances, songs, sparring matches and archery demonstrations was set up.

Sparring matches

Villages presented groups made up of three persons, both men and women, who debated about a subject. The chosen subject that year was: 'Does religion have a good or bad influence on Lepchas?' In these debates it was not a question of openly defining religions as 'Christianity' or 'Buddhism,' but by proclaiming oneself an 'animist,' yet fighting the dominant influence of Christians and Buddhists. The village of Rion won the competition. It turned out that this was in fact a very Christianised village, where the young girls from the Christian schools proved to be far better than the boys at debating. For the heads of the association, it was a question of challenging the stereotypes the Hindus had of the Lepchas: 'animist, sorcerer, tribal.'

Although the Lepchas themselves very commonly used the term 'tribal' to refer to themselves, they insisted on the fact that they formed a politically, linguistically and ethnically autonomous group. What the competitions wanted to demonstrate was that in the end, as in Tamsang's theory, Christianity was a bad influence on the Lepchas, the only real religion being what they called Bongthinism and Munism. These attitudes came 'from inside', an inside that contrasted with 'outside' practices, and those of Christian sects in particular. Christian sects developed positions similar to those of caste Hindus who recommended a separation and a hierarchy. As opposed to the classes of Christian and Brahman priests, the Lepchas had to define their specialists based on two simple classes:

- *Guolikmun*, a Lepcha priest who has supernatural powers, and therefore who knows everything in this world, in heaven and hell, and can lead the souls of the dead to heaven.
- *Yuk mun*, a Lepcha lama, a Lepcha priestess who guides the spirit or soul of a deceased person to heaven by chanting hymns and prayers and who performs religious ceremonies and rituals.

Gathering at Takna

This was organised by the Achulay association.[24] Lepchas came from Sikkim and Darjeeling (including a few prominent persons from Dzongu) in order to reinstate all the Lepcha rituals, as well as their main domestic, craft and medicinal activities. Tamsang commented on the different short scenes in front of an Indian film producer, while describing the inside of the houses. This type of museographic and folklore activity that took place in the 1990s among more urbanised and developed populations in the south of Sikkim partly resulted from the Lepchas' recently acquired dependence via the education systems in Christian and Hindu schools. This dependence started to be felt through conflicts and rejections that Lepchas from the reserve in the north, suffering from political and economic isolation, begrudged their more fortunate neighbours in the south. Hydroelectric projects developed by the Indian National Hydro Project Cooperation (NHPC), situated in New-Delhi, that planned the building of seven hydroelectric dams, from the south of Sikkim to the very sources of the rivers Rangit and Tista in the Tibetan region of northern Sikkim, were to crystallise the conflicts and raise the Lepchas in Dzongu to an international position well above these local ethnic and nationalist leitmotivs. They were to lend new meaning to 'combatant animism' and to the romantic image of lovers of a declining nature that they had seen themselves attributed by bestselling authors, who had long before got themselves out of the underdevelopment rut.

Weeping Sikkim: the ACT, an exemplary struggle demonstrating how the Lepchas are ready to die for their motherland

Organized resistance and protests emerged in 1995 against a state hydro-electric power station project at the heart of the most sacred place of the ancient Buddhist Kingdom of Sikkim (Demojong) at Yuksum (West Sikkim) (Photos 6 and 7). Prior

to this date, many Lepchas had already begun to express alarm at the consequences these projects were having on their own territory, Dzongu. "A group of Lhopo activists called the Concerned Citizens of Sikkim (CCS) decided to take the matter into their own hands" in 1995[25] and oppose the government and the threat this project represented to the sacred Khanchendzonga landscape.

A few years later, after much internal struggle between Lhopos and Lepchas within the movement, and due to the development of the initial phase of the project despite government announcements and the relative success of CCS actions, a small number of Lepchas from Dzongu decided to mount a huge show of resistance and a hunger strike on 20 June 2007. It was a question of attracting the world's attention to the mega-hydel project on the Tista River that had already been planned in the course of previous projects in Yuksum. The Lepchas, with their own conceptions of the place, created the 'Affected Citizens of Tista' (ACT), a peaceful yet radical organization on a Gandhian model. Friday 27 July 2007 was the 38th day of the hunger strike led by Dawa Lepcha, a film maker from the Institute of Tibetology in Gangtok, and Tsering Gyatso, a young 20-year old Lepcha, the younger brother of Urgyen in Hee Gyathang, and who was immediately joined by a number of concerned Dzongy citizens. That day, the website[26] that had been launched at the beginning of the hunger strike announced:

> While the beautiful river Teesta swollen and angry washed away and damaged ongoing construction of dam at Kalijhora […] today incidently is the 38th day of the indefinite hunger strike by Satyagrahis of Act led by Dawa Lepcha and Tenzing Lepcha who are admitted in STNM Hospital at the moment. The rest of the members of ACT continued their relay hunger strike at BL House.

A large number of monks from the Sangha in Dzongu supported the movement, fasting in turn and performing exorcism rituals by making dough effigies (*gtor gyap*) in the temples. Actions were publicly announced from one day to the next on the web site and in the local press (NOW). Geological information about the dangers of dams built near seismic places was provided, together with the political points of view, now referred to as 'ecological' and 'environmental' considerations borrowed from the world of media, although psychological and biological aspects were still predominant, encompassing the whole territory and the Lepcha national soul in a form of transcendental nature. Nature was now able to oppose the destruction of its sites. It reflected the kind of religious and supernatural links that people had with the mountains and the rivers. The spirits of the Tista reacted angrily to the decision of the government, which had earlier fixed an ultimatum to bring the hunger strike to an end within 24 days; failing this, action would be taken by the police. Here are some extracts from the local press in Gangtok (*The Daily NOW*, ed. Pema Wangchuk):

- SIBLAC [Sikkim Bhutia Lepcha Apex Committee], convinced that the divinities guarding the mountain are angry, suggest taking precautions at a spiritual and material level.
- He expressed serious concern about recent flooding in the north of Sikkim. A press release from SIBLAC announces that a meeting of the executive committee will be held today to discuss the actions to be taken against the state.
- The meeting took the sudden eruptive activity in the glacier waters of Sacred Zemu very seriously. One of the deities guarding Sikkim, on the eve of the feast of Lhabsol Pang, therefore gives a warning. This is a SIBLAC press release under the authority of Tseten Tashi Bhutia.
- This means that the guardian deities are angry following the different activities that have polluted several places in Sikkim, such as the building of the Dzongu dam. The SIBLAC therefore draws the government's attention to these strange phenomena. It calls for urgent precautionary measures. SIBLAC also addressed its compliments to everyone in Sikkim on the occasion of the auspicious festival of Pang Lhabsol; it voiced the hope that the guardian deities of Sikkim be encouraged to protect and guide the population, to bring it peace and prosperity, while ridding it of the evil spirits that harm living creatures and have always tried to cause disasters in Sikkim's natural environment in opposing the spread of Dharma.
- ACT evokes Guru Rimpoche's prediction about warning against plans to build a hydro-electric plant.

Citing 'national security and territorial integrity' reasons threatened by 'terrible guardian deities of Mother Nature,' the ACT (Affected Citizens of Teesta) movement has ordered the state and central government to cease forthwith work on Dzongu dam. It referred to the prediction made by the patron saint of Sikkim, Guru Padmasambhava, who said that after the fall of the Tibetan monarchy, the Indians would govern Sikkim, but when they would end up disturbing the guardian deities and nature, a new government would be set up in Sikkim.

According to ACT's secretary general, Sherap Lepcha, in yesterday's press release concerning the delicate question of borders, the hydroelectric projects would be a threat to territorial integrity and international trade in Sikkim.

The statement called on the Union Government to administer the state based on a thorough study of Padmasambhava's predictions to avoid an irreparable political catastrophe. Bearing in mind the recent floods in northern Sikkim, ACT stresses that the ecosystem of Tolung and Tista rivers will be completely destroyed under the eyes of the government.

It recommends that employees in these projects take precautions to protect their homes and their families against the wrath of the divinities of fire, water and wind as well as their own lives since we are all Sikkimese. It calls for a complete stop to Teesta Urja Ltd. Himagiri Hydro Energy Pvt. and NHPC, and for such unfeasible projects to be immediately abandoned.

Conclusion: why the Lepchas are ready to die for their motherland

This Lepcha territory 'struck' by its future destruction does not meet the ordinary anthropological definition of territory as 'a space inhabited by men' or 'a political space'. It is thought to be closer to primary notions of territory taken from ethology and sociobiology, since it is the vital dimension that prevails in describing its inhabitants: rivers cry about being diverted, mountains tremble, rocks fall, and the whole of nature is consumed by a kind of frenzy that is communicated to men. The proposed naturalization of the human being, which generally refers to the notion of territory,[27] is allegedly achieved through the Lepchas' bias of considering their living space as consubstantial to their group. However, this radical ethnocentrism common to all societies that make their place of residence the best of possible worlds, or their group the only real human group, is different in the Lepchas' conceptions by the very diversity of people who share their conceptions of the inhabited earth. A possible opening onto other eventual worlds is clearly indicated in the distinction the Lepchas make between objects of culture from the sky and meteors, and natural phenomena understood through the effects they produce on men. The real 'natural territory' becomes a virtual world made of invisible presences, populating a sort of up-side-down world travelled daily by its inhabitants. It may be deduced not by measurable borders or acres of land, but by a residual substrate, formed from what men have designed and made with their hands by borrowing directly from the 'non-human' world. The 'non-human' world, indeed this world of the *rums* and *mungs*, spirits of the forest and mountains, precisely makes this 'difference', this otherness according to the nature of various places where *padim* and *mun* shamans encounter them. The construction of concrete dams, the roar of caves collapsing and the explosions accompanying them have, in the eyes of the inhabitants, resulted in the flattening of the worlds, the disappearance of the infinite diversity of animals and plants which governs the lives of people and the depletion of reservoirs of wealth that was stored behind the mountains.

If the non-human world, the world of spirits, is a fundamental idea in the Lepcha mythology and of animist populations in general, there is nevertheless a 'natural' physical domain dissociable from the human body: no-one confuses the materiality of a tree or a plant with that of his own physiological body. Then yet again, the heavy materiality of trees, plants and gardens has to be put aside to describe this kind of existential lining of the social body: certain ideas must be taken into account about how phenomena such as lightning, thunder or certain living creatures, such as insects, crawling animals, fish, lice and parasites, appear to men and metamorphose, in order to qualify this natural world, a world made up of a 'worrying strangeness' for which we are forever having to tame the different ways of being and appearing. In this respect, Lepcha ideas of living space are totally opposed to those of Western political philosophy, such as Locke's,[28] for example, who made the issue of territory an intrinsic element of property, the juncture between political and economic liberalism. The Lepchas do not assert that they 'own a territory' when they say they are the 'beloved children of Mayel Lyanga' or when they starve themselves to death in order to protect the forests and mountains where their ancestors hunted. It is precisely this repetition of non-possession that makes them suspect in the eyes of the state and which relegates them to a non-place, a place for 'border minorities', who can be bought and indefinitely displaced.

Sacrificing indigenous populations for the sake of development and profit goes back a long way in history. In 1784, the British paved the way for this in India by executing the first hero of the *adivasi* resistance, Tika Majhi, who expressed his people's fundamental right to live where they had always lived. The East India Company responded to the revolt with his immediate hanging. Today, governments borne of decolonization in India and of the policy of enforcing quotas on minority populations have taken over this exploitation process by dealing with the last tribal populations; so-called development policies constitute the main relay. Historic racist ideas that have developed about the Lepchas, describing them as forest elves, chaste, shy, obedient and cooperative, have also helped to forge these ideas of a territory not actually 'worked', not actually 'appropriated' in liberal terms, by hunters who are "idle and indolent living in a pre-Neolithic era".[29] In this respect, maintaining a 'native reserve' remains a permanent opportunity for governments to create new 'territories' for settlement, in terms of development and resource management, all concepts that fundamentally oppose the philosophy and the collective imaginary of inhabitants who not long ago could only describe their occupation in local existential terms. Thanks to their exemplary political struggle for the survival of the planet, these terms have become political arguments on a global scale.

REFERENCES

Balikci, A. 2004. 'Buddhism and Shamanism in Village Sikkim', Ph.D. thesis, London, School of Oriental and African Studies (SOAS), University of London.

– 2009. *Lamas, Shamans and Ancestors: Village Religion in Sikkim*, Leiden, Brill.

Bhasin, V. 1989. *Ecology, Culture and Change: Tribals of Sikkim Himalayas*, New Delhi, Inter-India Publications.

Clastres, P. 1980. 'Archéologie de la violence: la guerre dans les sociétés primitives', in *Recherches d'anthropologie politique,* Paris, Editions du Seuil.

Descola, P. 2005. *Par-delà nature et culture*, Paris, Gallimard.

Foning, A. R. 1987. *Lepcha, My Vanishing Tribe*, New Delhi, Sterling Publishers.

Hooker, J. D. 1980 (1849). *Himalayan Journals. Notes of A Naturalist in Bengal, the Sikkim and Nepal Himalayas, the Khasia Mountains, &c.*, vol. I (3rd Indian reprint), New Delhi, Today & Tomorrow's printers & publishers.

Klafkowski, P. 1980. 'Athing Joseph Rongong's Manuscript of the Rong (Lepcha) Story of Tashe Thing—a preliminary communication', in *Ethnologische Zeitschrift Zürich*, II, 1980, Bern, Verlag Peter Lang, pp. 137-142.

Lepcha, S. T. 1998. *Lepcha Folksong, Dance and Drama*, Kalimpong, Mani Printing Press, 1998.

Locke, J. 1982. *Traité du gouvernement civil* (new edition), Paris, Flammarion.

Plaisier, H. 2007. *A Grammar of Lepcha*, Leiden, Brill.

Risley, H. H. 1989 (1894). *The Gazetteer of Sikkim*, Calcutta, The Bengal Secretariat Press.

Sahlins, M. 1976 (for the French translation). *Âge de pierre, âge d'abondance, L'économie des sociétés primitives*, Paris, Gallimard.

Siiger, H. 1967. *The Lepchas. Culture and Religion of a Himalayan People*, part I & II, Copenhagen, The National Museum of Denmark.

Steinmann, B. 1992. '*Doila* tamang et rites de *mdos :* image des ancêtres à travers les textes rituels', in *Tibetan Studies, Proceedings of the 5th IATS, Narita 1989*, Narita, pp. 751-772.

– 2000. 'La cardamome au Sikkim : de la dette au pays des épices', in Hubert, A. and Ph. Le Failler (eds.), *Les plantes du plaisir et de la convivialité en Asie*, Paris, L'Harmattan, coll. 'Recherches Asiatiques', pp. 437-444.

– 2007. '*Khangchendzonga: Sacred Summit* by Pema Wangchuk and Mita Zulca. Little Kingdom, Gangtok, Sikkim, 2007, Book Review', in *Bulletin of Tibetology*, vol. 43, no. 1 and 2: 93-102.

Wangchuk, P. and M. Zulca. 2007. *Khangchendzonga: Sacred Summit*, Gangtok, Little Kingdom.

Younghusband, Sir F. 1910. *India and Tibet,* Delhi, LPP (repr. London, John Murray).

Notes

1. "The complex of Lepcha mythology is referred to as *lung ten sung*, 'oral traditions, folklore, legends, fables, fairy-tales' [...]. Traditional Lepcha narratives contain views and statements on fundamental matters of life and are aimed at the survival of the traditional values of the Lepcha community that they reflect" (Plaisier 2007: 37). The versions of the myths given here belong to this category of sung or orally recounted stories. They were collected in Nepali, between 1993 and 2004: they are possibly from stories that have already been heard or read by the inhabitants of Dzongu in various publications. Each storyteller has their own repertory. The Lepchas are very fond of stories and prove to be insatiable storytellers.
2. Dzongu, a reserved and restricted area in the Northern District is the most isolated area that suffers from political isolation which is accentuated by its geographical position. See Bhasin (1989), who carried out intensive fieldwork in the Dzongu area in the 1980s, when the area was totally out of bounds to foreign visitors. See also Balikci (2004, 2009).
3. We came across this account in Sakyong in 2003, in the extreme north-west of Dzongu reserve, where there are 17 very isolated dwelling units and three primary school teachers (two Lepchas and a Brahman) who, while sympathizing with the inhabitants, bitterly complained of how the government had abandoned them. These accounts have been transcribed and published by Shri Sonam Tsering Lepcha (1998), musician,

dancer and radio reporter (Kalimpong, Gangtok).

4. See last paragraph.
5. Descola 2005: 183-184.
6. Dawa Lepcha, *The Lepchas and their 'Soam'* (cane bridge), a film by D. T. Lepcha, A Kursong Entertainment Presentation.
7. Hooker 1980 (1849).
8. Younghusband 1910: 106-107.
9. Risley 1989 (1894): 93.
10. See Steinmann 2000.
11. According to Bhasin (1989): "Hee Gyathang in Lower Dzongu is the biggest revenue block in the whole reserve: when Gorer studied it in 1938, there were only 60 houses. In 1980, there were 220 houses, out of which 118 were of labourers other than Lepchas, and when I went for the first time to Dzongu, in 2001, there were more than 300 houses, grouped together in Kyong or small groups of three or four houses; or isolated in the forests".
12. See Klafkowski 1980, II: "The story of Tashe Thing forms the central part of the Rong (Lepcha) religious mythology. Although European scholars agree that the story is of Tibetan origin, the elders of the Lepcha tribe believe it is a part of their own heritage".
13. Foning (1987: 28-43) had widely commented on the importance of these 'thunder stones', supposedly prehistoric hewn stones, in a book now famous in Sikkim. This book can even be considered to be the driving force behind the Lepcha identity revival movement.
14. Wangchuk and Zulca 2007: 30, Steinmann 2007: 99-102.
15. Accounts collected in 2004 in the village of Tingvong (Haut Dzongu).
16. Descola 2005: 193-197. Based on Viveiros de Castro's 'perspectivism', Philippe Descola presents his theory as "an alternative to the socio-centric thesis of projecting social categories on the natural world [...] Perspectivism is allegedly an ethno-epistemological corollary of animism [...] The emphasis laid on the discontinuity of forms by animic ontologies must therefore be taken as a sign of a heterogeneity of life forms incarnated in the body as a focus of the perspective: a subject that is activated or '*agenté*' by the point of view".
17. The Tamangs in Nepal stage a drama which is acted out to celebrate the gods of the clan, around yeast, a symbol of the incestuous relationship of the first marriages between brothers and sisters. This myth has thus spread through the Himalayas, with interesting variations from the Lephas' point of view. See Steinmann 1992.
18. See Siiger 1967 I & II, on the origin of the *Mun*, shamanesses created underground by Itmu (the Earth) and the Lepchas' myths of creation.
19. The association Survival International regularly publishes information bulletins on the consequences of the destruction of sites inhabited by indigenous populations, and in particular the Indians of the Upper Amazon, hunted down in their own reserves. See the various bulletins on www.survivalfrance.org
20. Siiger 1967 I: 172-176.
21. Hooker 1980 (1849): 30-31, also gives a version of this story which describes a unique couple who were saved at Tendong.
22. The revival of the Tendong Hlo Rum Faat, which was celebrated at state level on 22 August 1993 in Gangtok's Community Hall marked the beginning of long and repeated negotiations between village delegates, the Prime Minister of Sikkim and

the Government in New Delhi, to protect and improve the miserable conditions of the Lepchas in Dzongu Reserve: figuring among their demands was the abolition of the Tribal Order Act of 1978. Many demonstrations were held in Gangtok, Mangan, Namchi, Jorethang and Ranipul between 1993 and 2001.

23. See the numerous publications and opuscules edited by K. P. Tamsang, one of the pioneers in Lepcha studies, for instance: 'The unknown and untold reality about the Lepchas', and 'A Study of the original Lepcha names of places of Sikkim and Darjeeling and their stories'. Kalimpong and Darjeeling, Deep Printers.
24. This association regularly edits a bulletin of the same name.
25. Balikci (2004: 189-195) has extensively related the different phases of the CCS struggle.
26. www.weepingsikkim.blogspot.com
27. Clastres (1980: 189) for example undertakes an analysis of the concept of territory as a constitutive dimension of all primitive societies necessarily implying war, that is to say the radical exclusion of the Other. This would in effect be related to the issue of ethnocentrism.
28. Locke 1982 § 32.
29. See Sahlins 1976.

2.1.3.

From Remote Area to Thoroughfare of Globalisation: Shifting Territorialisations of Development and Border Peasantry in Nepal

Ben Campbell

A new road connection through Nepal's Rasuwa District to the Tibetan border is due to be completed by 2012. This could restore the Trisuli-Bhote Kosi Valley to the major position it once held in the cartography of routes available to Transhimalayan travellers prior to the era of the internal combustion engine. The route to Tibet through this valley is the historical twin of the route taken by the Arniko highway to join Kathmandu to Lhasa. Even in winter horses could pass this way. What is this re-opening of contiguous regions for international trade likely to imply for the local population? What relevant ethnographic and analytical frameworks can help in understanding the reorientations of people's territorial locatedness at this juncture?

In this chapter I start from the ways in which building the road connection has been legitimised by the Asian Development Bank (ADB) as an endeavour to relieve poverty in this area of northern Nepal. This narrative of generous salvation from hardship, neglect and isolation, uses certain kinds of modern assumptions about how people will generically respond to extensions of the transport infrastructure. How people will eventually respond cannot be precisely known, but in the road project plans there is very little perception of local actors, other than an alert to certain risk-types, such as women and girls who might be trafficked, and a formal procedural compliance to have incorporated an Indigenous Peoples Development Plan.[1] My point is to evaluate certain elements of territorial thinking by examining points of difference between national development discourses about infrastructure and exclusion on the one hand, and ethnographically encountered accounts of people's relationships between livelihoods, place and movement, on the other hand.

This leads back to mythico-originary narratives of settlement. It considers the effects of existing roads, and brings into focus experiences of contemporary journeyings for employment. However, the actual accounts that will be discussed are not taken from the normally gathered and collated conversations of ethnographic practice. They are from interviews and conversations delivered to camera during the process of film-making in 2007. What difference does this make? In the three key interviews I discuss, there is a further yet only partly acknowledged audience being addressed, and a sense of putting the conversations on public record. One must bear in mind the means of communication in discussions about people's relations to territory, and the conscious possibilities for these viewpoints to travel and be heard by such means. In other words, there is an awareness that the camera can transmit to known and unknown others, wherever located. This adds to a concern that emerges in the implementation of the road project: that the poor of the district, who are targeted as beneficiaries, are not being listened to, and will be rendered mute by the noise of impending traffic. Their engagement with the project's impact on their livelihoods reveals variable degrees of alertness to the complex re-territorialising of opportunities and inequalities anticipated in the road's construction.

The perspectives that are revealed about place relation offer a way of thinking about the politics of place-making, and the emplacing of people in schemes of control and change. Against these schemes of modernist state territorialisation, the ongoing practices of fluid interactions bring people into a number of other socialities of place. Rasuwa's Tamang communities have notably come up against various strategies of boundary-making in projects of national unification and development, and have even memorialised them in ritual performance.

In terms of subsistence, the Tamang communities' pastoral movements have been contained in ever smaller circuits of transhumance over the last century. Even so, the Tamang village economy has maintained subsistence strategies based on as extensive an ecological range as possible, rather than conform to the models prescribed by development for intensification in livestock keeping or the designs for turning the district into a tree-fruit production zone. Over the 1980s, the World Bank funded Integrated Rural Development Program for Rasuwa-Nuwakot perceived Rasuwa as a temperate zone that should specialise in a nationally conceived model of economic efficiency. Other boundary-making effects came in the form of national park restrictions on the movement of livestock, and the use of swidden farming was prohibited. This, in effect, consisted in a territorial 'cleansing', to prevent a number of important livelihood activities from taking place within park areas.

The arrival of the road comes with a fanfare of planning optimism that the poor of the region will find economic opportunities to benefit from closer proximity to markets, but it is somewhat stuck in an old idea of releasing potential among remote agricultural producers. The standard gaze of development sees such people through an economic lens as poor peasants, as deficient producers, awaiting intervention through modern ways and means to become proper market actors, released from inappropriately traditional ties to land (Kearney 1996). Nor does the plan consider the poor as politically or culturally located, or in terms of anything but instances of generic human behavioural patterns.

> [B]y removing physical barriers through road networks and improving the policy and institutional capacity, the Project will improve the overall performance of the road sector—leading to reduction in poverty. The proposed roads under the Project pass through poverty-stricken areas and link trade, production, and marketing centers and other locations of economic activities and socioeconomic services. Therefore, the Project will contribute toward poverty reduction. (ADB 2006:57)

By contrast, I argue that a strategic historical vantage from the border region has given local people a sensibility and regard to the effectiveness and hold of the states on both sides *having reached their limits*, and therefore constituting a zone that is mostly neglected. This does not enter the explicit calculus of the road project's impact assessments, nor does the reality that the transition in political culture among people of the area occurs in gradualised and incremental ways rather than absolutely, with finite limits produced at a border line. A further component to the argument is that because of this historical neglect, and the protection of the area as a reserve of manual labour (Campbell 1997), the people have been used to seasonal labour migration. Instead of the road bringing new income opportunities via *in situ* production, the post-peasant realities of global labour markets has already taken the majority of the rural male workforce abroad.

Revealing a place made habitable for humans

The Asian Development Bank's plan for the road is to link up existing Tamang villages on route to the Tibetan border. The new road will upgrade the existing dirt road that has run since the mid 1980s from Trisuli Bazaar to Syabru Besi, and the mining works at Somdang on Ganesh Himal. Yet even before the dirt road, Tamang villages were concentrated in their settlement density along the main track, *mul bato* heading north. Let us go into one of these villages and hear some accounts of its relationship to the dynamics of residence and movement, past and present.

Purba Yelbo Ghale has been the head *Loben lama* of Tengu village since his father died about 30 years ago. I do not think he has ever seen TV more than a few times in the local town, or watched a film on DVD in one of the few houses in the village with a set (he complains now of poor sight), but a big impression was made on him by the film my brother and I made of the *bombo* (shaman) of the village on pilgrimage to Gosainkund in 1990. The lama had a bad foot at the time, so does not feature in that film. But after the death of his uncle, the *bombo*, Purba Yelbo saw that the film outlived the people filmed. He saw how the *bombo*'s grandson had learned from watching his grandfather on the video, and realised that he could speak to as yet unborn descendants of his own by means of film. Film could effect something of a similar character to ritual, which seeks to move out of ordinary time to enter the mythical time 'when people spoke with gods'. His clan was among the very first group that settled in this place, reportedly six generations before him. The older lineage branch of this founding clan provides the *lhaben* officiants of the territorial deities *Shyibda*, and *Kalleri Miktung*. He himself will sometimes perform solitary rites at the vegetarian *Shyibda* cairns on full moon days.

It was not a problem to find a theme for filming that was common to his interest of wanting to record some of his knowledge for future generations, and my wanting to record some passages of Ur-territoriality to set against the issue of the coming road. This was to be a walk through the ruins of the old village and along the old main path northwards, with an accompanying narrative track of myth—culminating in the taming of the place by Buddhist victory over demons. The old village site had been abandoned in the mid 1990s, after villagers resettled on the dirt road that was constructed 15 minutes walk uphill. The move from the old site had been made over a period of ten years. Initially just a few huts were built by villagers. Some people occupied a few more substantial houses that a powerful family from another village to the north had speculatively constructed. When road maintenance work was entrusted to a village team by the Department of Roads, this increased the number who built houses by the road. Most households had potato and wheat fields in the vicinity of the road, so the relocation did not necessarily lead to a reduced focus on farming livelihoods. Most years potatoes do provide good cash income. At the time of the move, there were roughly forty households. This relatively small size perhaps explains the move of the entire community. Larger villages have not followed suit, having slightly more prosperous households and substantial architecture, or being just too numerous to make a collective shift uphill possible. As to where they relocated near the road, there was not an exact re-mapping of the same old micro-neighbourhoods. Many families built where they already had fields, avoiding problems of land-ownership and disputed claims, which resulted from occupancy of the speculatively built houses mentioned above. There is more of a correlation between clan identity and field property in the distribution of houses that are located away from the immediate roadside cluster at Rishyang. For locals, this quasi-urban cluster has the feeling of a bazaar (actually there is just a handful of shops among the houses), as compared to the quieter social life round the bend in the road.

From the old village ruins, the walk northwards would take us past the line of old *purkhang* (small memorial *chorten* for the dead), beyond the nearby set of fields through a patch of forest, to the second grouping of fields, and other points of noteworthy events, on to the site where Guru Rimpoche (Padmasambhava) meditated in a cave, and killed a demon on route to Tibet. In the film of this guided walk, an animate landscape emerges from the evocation of actions and stories borne in the physicality and the enchantment of the past-in-place. Mythical and historical human life events are perceived as having literally *taken place*—that is, are legible in the presence of shapes and forms. So it was as we descended a steep path from the road to the old village, our first stopping point was to look across at a rock where lamas used to receive initiation. (These days new sites for initiation have been used that are above the dirt road—maintaining the structural relation of initiation site as above village habitation). Purba Yelbo began intoning the chant he had learned at initiation. He broke off after a couple of minutes, mentioning that we would never complete our walk if he did not edit the length of all possible narrative accompaniment. Further down, among the ruined walls of the old village, he helped me recall who had lived where, when I had first arrived to do fieldwork in 1989.

Before leaving the old village houses, he made it clear that, in proper fashion, we should have begun our tour of places of significance at Tengu U, the 'cave' (in Nepali *Thare orar*), but in fact only an overhanging rock, that was a famous night-stopping place for travellers between north and south. Purba Yelbo had heard from his father that it was at this cave that a Tibetan, Lama Maneba, had died, having first declared all land from this point to the north as belonging to the Tibetan 'insiders' (Campbell 1997). It was significant enough a place in the national itineraries of Nepal in the eighteenth century to receive a mention in Kirkpatrick, who noted from reports he had gathered at Nuwakot, that between Grang and Bokhajunda, there "is a celebrated cave, under a considerable eminence, called Thara-ooral, or the cave of Thara. It is also known by the name of Bhumakagoopa. The Trisoolgunga passes below" (1811: 312).

On leaving the ruins of the old village, one large flat rock beside the track attracted Purba Yelbo's attention. It was here, he told us, that one of the first clan settlers came upon a whole crowd of *mang* spirits. Taken by surprise, they dispersed at the sight of the on-comer and were chased off by a dog. Passing through scrub-forest, round a bend on the ridge, and arriving at Membarding with a vista north of the length of the valley, we met our first villager, Pangbang, a man with some Buddhist knowledge, who has lived and worked with Tibetan families in Kathmandu, and in tourist hotels. Purba Yelbo explained the purpose of our walk

> "I walk along. They film.
> I explain what it is we see here and there.
> What ever comes out of this, will be seen by how ever many children and grand children, even after I am dead."

Turning to the camera, he explained that it was from this ridge-point (*gang*) that Guru Rimpoche spied a demon (*simbu*) way over at the far side of the village fields, where a stream plunged down across the path. It was a man-eating demon that regularly ate people walking that way. Above this point was a very large, peculiarly shaped rock, like a step pyramid, in which there was the cave where Guru Rimpoche meditated, in order to overcome the demon. On the approach to this area, known as *Sangdormo*, Purba Yelbo pointed out an apparently random sequence of stones connected to the Guru Rimpoche story.

First he showed a rock that was formed by Guru Rimpoche having sat on it, leaving the imprint of his posterior, and in a notch, of his books that he carried strapped to his back. Then we came across a long stone tapering to a point. This was a plough used by the first farmers (he added that a man stepping over this rock would risk becoming impotent—'ploughing' and 'intercourse' being frequently linked semiotically). A large, long boulder to the side of the path was said to be a huge foot-operated threshing tool (*kwindilang*). Further along were some rocks that Purba Yelbo said were dogs that belonged to the first farmers, for protecting their crops. These rocks had been broken up according to an order in VS 2025

(1968) to widen the track at this point, and Purba Yelbo himself had been among the work team. A long flattish rock with marks like cross-hatching on it was a crop-drying mat (*phyó*). Beyond this, up some steps were the remains of other rocks that were ploughing oxen that had also been destroyed in the path-widening episode. Finally, we arrived at the flat stone under which Guru Rimpoche had lit a fire on which he had boiled the head of the defeated demon in a cooking pot. He had meditated for three years, three months and three days to build up his powers to kill the demon. On removing this obstacle to people's coming and goings, settlement and safe cultivation, Guru Rimpoche continued on his way north into Tibet (for another, colourful place-founding account of Guru Rimpoche and demonic dismemberment see Ramble 2008: 188-190).

At one point Purba Yelbo looked at the camera and declared that all this knowledge about the stones and their meanings came from listening to village elders, and to Tibetan lamas who he remembers used to pass through the village. Moved to comment on the longevity of this knowledge, given changing times and the re-orientation of his community to new pathways of connection and meaning, he added, "There is nobody else now in this village who can tell the stories of these stones."

In contrast to the notion of culture imposing designs on pre-existent territory, the stones do not at all appear as designed artifice. They are almost like emergent, mythical hypertexts, which only with the gift of knowledge from those with vision, are capable of conjuring up activity, craft, purposeful community, companion species, and the threat and violence that went with their establishment in this place. Humans came to supplant previous spirit-creature occupants, but in various seasonal rites people acknowledge the non-human sovereign Lords of Place whose protection is needed to avoid disease, natural calamity and mishap (Höfer 1997, esp. part II). Marie Lecomte-Tilouine (2009) writes about capacities for stones to be brought into meaningful signification elsewhere in rural Nepal. Stones and features of landscape bear a potential of immanence, and of vision revealed to 'the chosen'. Greater interest is evinced in manifest divine actions than in human representational capacity, requiring an alertness to the possibility of visual cues of divine agency.

The myth of Guru Rimpoche making a place safe for human habitation is inscribed in these rocks, which perhaps by no small coincidence mark the northern limit to the main concentration of village fields, where a waterfall in a gully presents a boundary to define village limits. The stones cement the idea of Guru Rimpoche's own journey that leaves behind a string of places made safe for humans and their livelihood activities. It is a place made into one, where the threat from enemies to human pursuits has been violently overcome, but the memory of them and the continuing possibility of their re-emergence still persist. Their presence can be read by the remaining signs of mythical time, beside what used to be the main passage for leaving the village to the north. Their continued subjugation is re-enacted every *baisakh purni*, when alongside the *lhaben*'s sacrifice of a goat to the deity *Kalleri Miktung*, the lamas perform a symbolic sacrifice with a ritual dagger. Elements of alternative versions to Guru Rimpoche's defeat of the demon were told me by the village *bombo*s, suggesting that beheading demons led to the transformation of their blood into the proliferation of leeches in the monsoon period.

In this section, notions of territory are emergent from encounters and conflict of different beings and presences in a place formed by their actions as a founding narrative of links in the lines of history and landscape. The political state does not belong in this particular narrative. In other ritual contexts of local territory the *lu* and *shyibda* are the local rulers, dispensing permissions and punishment. At *baisakh purni*, sacrificers and lamas enact simultaneous versions of village-based sovereignty and citizenry. No two rites or mythical episodes necessarily match. *Dasain* provides an occasion for territorial enclosure of the village and marking hierarchy by central state-endowed authority, but it is founded on a sinful state-required act of sacrifice that necessitates mourning for the buffalo (Campbell 1995). In sum, there is a pluriverse of incompatible territorial powers in co-residence, and in motion. Among these Guru Rimpoche provides one of the strongest extra-local impulses to set the local in wider networks of 'nomadic' connectivity.

From myth to history

In the early 1990s the villagers of Tengu had decided not to be by-passed by the arrival of modern transport communication through the valley, and had rebuilt their homes on the new roadside. The road itself provided employment for some, and the few shops obviously were crucially situated there for attracting passing customers, but there was no other direct economic rationale for the relocation, other than maximising the opportunity for villagers to stay in touch with other settlements in the valley, and keeping the village community in the line of sight of anyone known or unknown, travelling the road. The lesson I want to draw from this is that the relocation displayed a different starting point and approach to domestic and community viability than the logic and experience of development, of livelihood improvement, that the villagers were offered in the form of agricultural and livestock extension advice at the time.

After cautious hesitancy, the villagers saw that the heart of community life, and the young in particular, were drawn to what the new road offered, by way of connection, even if this was not directly 'economic'.[2] This kind of response was not, I expect, foreseen by the road planners, who were merely heading for the mine, via the district capital, irrespective of the community territories through which the route passed. In hindsight the move is explicable in terms of a livelihood rationale that has a very different notion of the importance of territory to be found in either mineral extractive projects, or in rural development plans of the times. The logic of development standardly proceeds by intensifying production through the application of new methods and technology on given, bounded territorial resources to increase the returns to inputs of labour or capital, and to produce items or services of value for markets that can be reliably known and predicted. The livelihood logic of this mountain border zone had other characteristics.

For the Tamang communities occupying agriculturally unpromising terrain, livelihood was primarily dependant on livestock keeping practices that required extensive movement to access seasonal fodder. Only with possibilities of grazing *across territorial boundaries* would flocks and herds from different villages manage to compensate for the inadequate array of resources available to them within specific village territorial domains. In the film, Lama Purba Yelbo tells how as a teenager he went five times with village flocks of sheep and goats to pasture across the Tibet border in the valley of Lende, before the Chinese occupation. Since the middle of the twentieth century, the major political and developmental changes in Rasuwa have reduced the territorial room for manoeuvre of subsistence production to ever smaller scales. With the closure of the Tibetan border and consequently no further access to pastures, the creation of the Langtang National Park restricted movement of livestock to within the territorial limits of the panchayat or *gaon vikas samiti* where livestock keepers held landed property.

Within the border zone region, villages occupied marginally different ecologies of niche diversity, varying according to location up or down the north-south valley, and all their livelihoods depended on *distributed* possibilities for access and exchange amongst them. Facilitating the movement of people across territories were clan affiliations that were continuous across even the linguistic divide between Tamang-speaking and Tibetan-speaking communities of 'Border People' (*Sépa*). (It is through these *Sépa* villages that the new road connection remains to be made.) In other words, there was a contiguous gradualism in the passage northwards, with language, clans, ritual, religious architecture, food, and kinship just somewhat rearranged rather than abruptly confronted. So it was in the economy of the border region. The standard target of development interventions—the peasant producer intensifying with technology to make finite land resources yield crops more efficiently to satisfy demand in integrated markets—was not a salient reality. More relevant would be to think of people occupying a number of different sets of entourages (of which village landed territory is but one) in the intersections of which the reproduction of livelihoods through networks of relational proximity, are managed by language, clan, co-operative reciprocity, ritual friendship, religious congregation and patronage, and traditions of hospitality. The narrative of Guru Rimpoche's mythical journey even provides a thread to stitch together the transitions en route.

As the testimony of Purba Yelbo further makes clear, the actual livelihood realities of these communities regularly consisted of forced labour demanded from each household by the *mukhiya*, frequent occurrences of famine, when families resorted to bartering bamboo produce for grain, and poverty was widespread in the region, reflected in people wearing patched-up rags for clothing. Wages had to be sought externally in portering and manual labour, mostly with bazaar merchants, but some Tamang *mukhiya* employed villagers on sheep trading expeditions to Dzongka. Such conditions of survival were in part a political outcome of the area having been protected from recruitment to Gurkha regiments in British and Indian armies (Campbell 1997). The communities of Tamang porters along the route to Kyirong were vital to the Ranas' trade interests.

In the normal run of things, people pursued livelihoods without too much reference to the needs of the centre, and the state normally kept only a minimal presence.[3] The occasions when state territory was disputed, and armies turned up to face each other, left a powerful trace in the memory of village history. Placing themselves at the centre of their worlds, but on the borders of others', the Tamang communities use the occurrence of the border wars after the Gorkhalis' unification of Nepal to proffer their own hospitable disposition each way, and to manage dharmically inspired compassion to the needs of wayfarers, military and mendicants. During the *tseene cheeba* performance of dancing kings and warriors, in Bharku village (featured in the film), the village ancestors place a rope on the ground in the middle of the road, in anticipation of the armies arriving from Nepal and China. This somehow encapsulates the arbitrariness of territorial aggression to the people whose own centre is the dividing line of sovereignty for polities for whom such a place is already socially a foreign community: *Bhote* to Nepalis, *Rongpa* to the Tibetans. With stereotypes confronting each other, and making threats via the stereotypes with which the Tamang risk being tarred, ('beggars for flour' vs 'arrogant and aggressive'), the two sides of the drama make visible state territorial presence as a performance of an arbitrary kind.

In the bazaar

For a perspective on the new road connection and its territorial effects for contemporary livelihoods, the film includes an interview with a woman originally from the same village as Purba Yelbo who moved to Dhunche, the capital of Rasuwa District, soon after the dirt road was completed in the mid 1980s. She settled down with a Sherpa who moved in from the Solu Khumbu area, and the two of them ran a shop, earning enough money to send their children to school. She therefore represents someone who has consciously taken up a residential and occupational position to be where change was taking place for her district. At the same time she has maintained a clientele for her shop that consists of many of her kinspeople from the village. She might well offer villagers credit, and they would seek her out to buy their produce, such as *raksi* (local alcohol), or fresh milk. She is a granddaughter of the Tengu *mukhiya* who was a successful sheep trader for the Dasain trade, and she demonstrates a capacity for strategic reflection on the consequences of new communications and trade opportunities in the valley.

In conversation with her about the road, I tried to gauge the success of the Asian Development Bank's efforts to persuade the people of the district that the road will relieve poverty, and bring them new development opportunities. As the principle road among three identified for construction in the Banks' Connectivity Sector Project it aims to "(i) support economic growth, particularly in rural communities; and (ii) help reduce the poverty of isolated people, mainly in hilly areas" (ADB 2006: ii).

After discussing aspects of the effects of the road on shops in the bazaar, we went on to talk about what was happening more generally with those looking for work. I started the interview with a question: whether she thought the road will benefit shopkeepers such as herself in Dhunche. For example, all kinds of new cheap goods might start coming from China.

Her answer came in a torrent, explaining how the people of Dhunche were not going to benefit. Rumours were circulating that the route the road will take will deviate from the original plan that it should pass along the contour of the existing dirt road, and instead it will run beside the river in the valley bottom, a thousand metres below.

> This place will be poor from now. It will not be good for business. What comes from above will be taken straight down. What comes from below will go straight up. Kyirong will be reached directly, and so will Kathmandu. There will be no trade for us. It will make us poor.

As if under imminent threat she declared "We'll be poor. There won't be enough to eat. How will our children be educated?" She put this scenario in context by describing how in her own life, she started off poor, then with her move to Dhunche some business was slowly built up. All this will be taken away if the road goes 'below'.

I suggested that they might relocate their shop, having the history of Tengu in mind, which, as previously described, had completely moved uphill after the dirt road by-passed the village. She replied:

> How are we going to move? We haven't any money. With no money how can you move? If we had money maybe we would. These households will go broke, or may be they will move away. Everything's ruined here. The road is being taken below by China. It is the place 'Big Pasture' that has ruined it most. [Where landslides block the road almost every year in the monsoon]. Having to pay to carry our goods across the landslide, half the money is eaten up. There's no profit. There's not enough for us to eat. We have to give so much money when someone dies, and when someone gets married.

Thinking how there might still be opportunities she could take advantage of with the new cross-border connection, I asked if she could not make business by selling produce from the area. In her response, she made clear to me just how much she had self-consciously left behind the position of being an actual cultivating villager, and what a marginal kind of activity being in a position of having to sell agricultural produce would be from her point of view now. To actually choose to make a living from selling farmed produce, as a decision selected from among other alternatives, was characterised by her as a precarious strategy, unless you were to be already cultivating as a normal villager would, and perhaps have a few kilos in surplus.

> We have no land. Others have land, not us. Without land what are we to do? Even if you sell radishes, how much will you get? Five, ten, twenty rupees. The competition is too tough.

The strident and almost desperate tone in which the shopkeeper gave her estimation of the effects of the new road can be recognised as part of a collective response of the community of the district capital's bazaar to the rumour of the modified route. The performance on camera was made with strong conviction, and a powerful scenario was drawn of the impact that a decision to re-route the traffic would make. This is not to diminish the genuineness of the sentiment behind the answers she gave, but her comments revolve around the possibility that the road could actually make people like herself poorer, in spite of the project's claim to do the opposite. Given the context of the interview taking place in front of a camera, this comes as if from the collective voice of the generation who had witnessed the first signs of development in the district. They had adjusted their networks of kinship, residence, politics, education and trade to completely new conditions as the district bazaar and other enterprises grew in size, before the onset of the Peoples War (which particularly hit tourism on the Langtang trek). In identifying the swifter alternative route passing directly up and down the valley beneath Dhunche, as almost an existential threat, it is the Tamangs' mediating positional ontology in this border district which is at stake. The Tamang middle-ground, in which regional tactics of flexible cultural affiliations and self-presentation have operated to some advantage, was now to be faced with a rapidity of movement between urban centres, and other road networks to the north and south, that could render their mid-way-betwixt transitional locatedness into being merely a point of observation on others' movements between destination points.

The transformation to be considered is one from the region only being accessible to 'wayfaring' twenty years ago, and that is now about to be thrown into full-scale 'transport' (Ingold 2007).

> Unlike wayfaring, [...] transport is destination-oriented. It is not so much a development *along* a way of life as a carrying *across*, from location to location, of people and goods in such a way as to leave their basic natures unaffected. (Ingold 2007: 77)

According to the Asian Development Bank's plan, the road will allow goods to find markets, and boost the economy of poor northern districts. It sees the lack of road infrastructure as a major constraint to growth, and as contributing to "the close nexus between poverty and excluded development". It intends that greater benefits will become available to disadvantaged groups in the conflict-affected areas, and specifically identifies tourism as "an instrument of poverty reduction". (Technical Assistance Note, point 5. June 2004).[4] When I talked with one of the local Tamang entrepreneurs who has invested in the tourism infrastructure at the existing roadhead in Syabru Besi, he told me there was little chance the rural poor were likely to benefit from the road. Those who are not entirely yoked to their hard labours of ploughing and carrying wood, have noticed that it is Nepalis from the south who have come in to build up businesses and dominate the prospects for trade in the future. He was firmly of the opinion that a young person with any sense would see better opportunities for an income out of the area.

The report states the second expectation of the project is to: "improve people's mobility so that they can access employment opportunities outside the community" (ADB 2006: 57).

This might be thought of as anticipating the outflow of the labour force, but there is no consideration given to class and gender differentiation among those who leave for work. School leavers might look elsewhere in urban locations for jobs, but if the road is meant to improve economic activity in the agricultural sector, the labour force will be deficient, if the trend for work abroad continues.

A remarkable telephone call was by chance filmed back in Tengu, at the village shop with the one and only telephone. A young wife was talking to her husband in Malaysia in an increasingly agitated tone. She reprimanded her husband so that the congregated village public would all be able to hear, rehearsing the shame in which she would have to humbly face her husbands' clan sisters and daughters, when they came for ritual greetings at the time of *tihar*. This made sure there was no doubt in the village information network that she had no money to live on or to keep up customary festive appearances, and that she had let her husband know the situation.

Woman on telephone:

> "If you [decide to] come back, come. If you [decide to] stay, stay.
> What good is [*tihar*] *tika* to me? You're just telling lies and lies.
> Which country has your money wandered off to and gone to sleep?
> If your money had come for dasain, that would be fine. You said it would come for *tihar*.
> I don't have a single coin here.
> 'I can send' you say. You can't [i.e. be honest, and say you cannot send money]."

This conversation makes questionable ADB's automatic assumption that road connections will intrinsically improve economic activity. The direct benefits calculated in terms of 'present value' of the Galchi-Syabrubesi road are Nrs 732.9 million (ADB 2006 report: 54). Based on this figure an assumed poverty impact ratio of 0.44 is added, but without any justification or argument being made to support this effect. Although analyses of road building elsewhere in Nepal have shown impacts on livelihoods before and after to be more positive overall than had been expected in reducing socio-economic inequalities (Blaikie *et al.* 2002: 1264), it is very hard to pinpoint causative processes, and even harder to predict outcomes, given the number of variables at work, and the difficulties of factoring in social benefits, environmental costs, gendered wellbeing, and cultural rights and resilience.

By improving road connections it might be expected that the situation of the woman would become even more common among the households of the district needing cash.[5]

On gender and development, the ADB makes an excessively confident prediction:

> The Project will have no negative gender impacts. The Project will benefit both men and women by providing them with direct construction employment opportunities, easier access to markets, better delivery of services, improved social facilities, and improved access to better quality health care facilities and higher education. (ADB 2006: 58).

There is a warning about HIV-AIDS at the end of the section, but no reference is made in the document to 'male out-migration', or to the preponderance of female-headed households. The project arrives with little evidence of anyone having understood the dynamics of socio-economic change in rural Nepal, and assumes as if by law of nature that roads will increase economic activity to the benefit of the poor. As Scott says of high modernist planners in general, they routinely "ignore the radical contingency of the future" (Scott 1998: 343), and populate their before and after scenarios not with different kinds of people but with standardised subjects of development. It is as if by making an appropriate national map of connection, a territory for national economic welfare will then emerge, when in fact the global directions of the new nexus between cash-needing households and wage offering employment takes mostly men abroad. This is reminiscent of the blind faith in road building, with scant regard for the conditions the road is formally intended to address, that Ferguson (1991) describes in the case of Lesotho. The idea of building roads for the purpose of increasing production had not been matched to the fact that local producers had migrated en masse for waged work in South African mines decades previously. On top of this, roads bring in cheaper food than can be produced locally.

Back in the conversation with the woman shopkeeper in Dhunche, I had asked one of her friends, holding a baby, whether the arrival of the new road would make it more likely for the young in Dhunche to stay or leave. She said:

> If this road is built below here, the young will fear poverty and leave for anywhere-wherever.

Even so, the actual wage-earning experience elsewhere is not wholly positive.

> "They go but they come back again. They are not able to earn.
> They come back empty handed. They are not able to go to 'big countries'.
> They go to 'small countries', and come back with nothing.
> It costs them, one lakh, two lakhs. The people of Dhunche are poor.
> They stay one year, two years, and return with debts owed to friends.
> They are not able to earn. If they have some education they can earn.
> Some have been beaten [by their employers]. They don't understand the language, and end up being beaten. That is how it has been for many who come back."

In other words, far from the road being seen as bringing opportunity, it is expected to redirect the flows of value currently accessible, and comprehensively pull the rug from underneath the advantageous location that the district capital once represented. To play with Ingold's opposition between transport and wayfaring, the road will introduce a *transport* system that will send the young people into a condition of rootless *wayfaring* in search of work far away.

Conclusion

In the last decades Rasuwa District has been characterised as a remote area. It has low levels of education and considerable conditions of poverty as compared to other districts, which have longer histories of remittance economy connection. The population is 80 per cent Tamang. Prior to the Chinese occupation of Tibet, the Bhote Kosi valley through Rasuwa was the main route into Tibet along with the one through Kuti.

In this chapter 'territorialisation' has been used to analyse relationships between people, places, and the modes of representation that are available to locate identities, collective difference, relative claims of belonging and entitlement, and the power relations that build on such representations. What is of particular concern in this paper is the relationship between different languages of territorialisation, and the possibility for them to take account of each other. Looking back over foundational discourses of human settlement, and adjustments over the centuries to changes in cross-border relations and conflict, the place-specific histories of livelihood and interdependence form links through trade and non-exclusive rights along contiguous strings of communities stitched together by the thread of Buddhism's placation of territory. My argument is that by contrast the road project is legitimised by a language of planned transformation, which assumes to know about, and be attentive to, the communities who will be affected, and in whose name the road construction is being justified.

Official documents about the project display how a distinctly 'productionist' (Kearney 1996) view of these communities' interests and motivations is adopted. This presents the poor of the area as lacking the conditions of infrastructural connection which will enable them to develop their potential as economic actors - as rural producers who can be brought into developed modernity through better national integration with markets. The ideology of development in Nepal 'placed' its villagers as subjects in need of change for their own benefit and for national transformation (Pigg 1992). In 'remote' underdeveloped areas like Rasuwa the models for change imagined modern economic actors to emerge into the designs of national economic planning: adopting new agricultural and livestock keeping practices, or responding to the opportunities of tourism, so that their village economies would connect with national markets, or through education join the literate classes who would modernise the nation. The aspired-to horizon of modern developed nationals displaced attention from actually existing practices of livelihood and social and cultural networks orientated to multiple and mobile sets of locatedness. Instead, models of development were premised on emergent agricultural cash-croppers, spreading practices of production intensification, and an image of producers grounded through property and kinship, who would improve their quality of life through the provision of skills and infrastructure as the cogs and wheels of the mechanics of national economic integration.

This projection from poverty into a future of connectivity is ahistorical, neglecting the existing bases on which infra-regional and cross-border relations are likely to be configured from the point of view of the local population's cultural and economic links to the north and south. It also conveys a distinctly national territorialisation of effects, in that it imagines that the energies of the people of this remote area will be channelled according to the project's designs into circulation within Nepal. It neglects the existing condition of trans-national labour migration that has already taken much of its labour force out of the district and the country. Nor does it (could it) consider how the emerging dialogues about a post-conflict constitutional federalism could be affected by the road.

The generic poor of development tend to be simply presented as of-and-in-place. Road projects imagine their effect as opening up obstructions to communication of people and goods. This is a distinct kind of territorial imagination. It needs to be set alongside other territorial imaginations, including those of oral history, and current migration experiences.

The 'people of the middle ground' is the conscious ontology of the Tamangs being peripheral to state imaginaries from both sides. Perceived in inferior glances from above and below, as derogatory *Bhote* or *Rongba* from Nepal and Tibet respectively. If state-territorial, exclusive boundedness always co-exists with nomadic boundary-crossing practices, the prospect of the road will run rough-shod over the ambiguous points of their overlay. In the malleability between levels of cultural-economic spheres, there have been dispositions for actions of livelihood afforded by historical knowledge of potential value in the movement of people and things from one place to another, across the place-specific configurations of value, access and socialities of return.

In remaining peripheries, as in many of the areas of South and Central Asia where 'cultural diversity' is recognisable, this is enabled by certain skills of mountain living, with some physical advantages e.g. water, and neglect or autonomy from states (Brower and Johnston 2007: 20). Roads will change this. Such specific conditions in which concrete needs and their provision can remain invisible from the ways of 'state seeing' are often a feature of places that have remained peripheries and not been drafted in to central sightlines and designs:

> the premodern state was [...] partially blind; it knew precious little about its subjects, their wealth, their landholdings and yields, their location, their very identity. (Scott 1998: 2)

State territoriality is of necessity arbitrarily performed to make places separate. In the Bharku dance, Tamang village authorities witness the battles for division on their home ground, and the enactment of contrasted, stereotypes of the meat-eating Tibetan warriors and the officious Nepalis defending their grain stores against beggars from Tibet—in-between which the Tamangs position themselves in terms of contextual alignments, facing one way or another, according to the power dynamics of a given interaction, and creatively hybridising the possibilities of who they can be.

Harvey and Knox, writing of contemporary border road planning between Peru and Brazil, contrast the planners' forthright rendering of places, goals, scales, objects, and agents within frameworks of *compatibility*, and the aspects of social life in border regions which are innovative, diverse, and intently engaged with the art of addressing the difficulties of making conversations, negotiating across differences, and achieving mostly provisional compatibilities. There is then a disjuncture between visualisations of the future through the provision of connectivity by making places simpler and speedier to get to, by privileging technical solutions, and the modes of operation of people caught up in these designs. From their perspective "it is clear that smooth, speedy integration involves erasures, removals, and disappearances that draw forth active responses" (Harvey and Knox 2008: 89).

Questions then need to be asked about whether these territorial discourses can speak to each other, or whether they remain mutually remote and isolated, lacking *discursive* connectivity. As the voices heard in the film announce—the road intended to relieve poverty will do the opposite of the stated goal and make them poorer.

REFERENCES

ADB. 1999. *Policy on Indigenous People*, available at: http://www.adb.org/Documents/Policies/Indigenous_Peoples/ADB-1998-Policy-on-IP.pdf

– 2006. *Proposed Asian Development Fund Grant and Technical Assistance Grant. The Kingdom of Nepal: Road Connectivity Sector I Project*, available at: http://www.adb.org/Documents/RRPs/NEP/37266-NEP-RRP.PDF

Blaikie, P., J. Cameron and D. Seddon. 2002. 'Understanding 20 Years of Change in West-Central Nepal: Continuity and Change in Lives and Ideas', *World Development*, vol. 30, no. 7: 1255-1270.

Brower, B. and B. R. Johnston (eds.). 2007. *Disappearing Peoples. Indigenous Groups and Ethnic Minorities in South and Central Asia*, Oxford, Berg.

Campbell, B. 1995. 'Dismembering the Body Politic: contestations of legitimacy in Tamang celebrations of Dasain', *Kailash*, XVII (3 and 4): 133-146.

– 1997. 'The Heavy Loads of Tamang Identity', in Gellner, D., J. Pfaff-Czarnecka and J. Whelpton (eds.), *Nationalism and Ethnicity in a Hindu State*, Chur, Harwood Academic Publishers, pp. 205-235.

Ferguson, J. 1990. *The Anti-Politics Machine: 'development', depoliticization, and bureaucratic power in Lesotho*, Cambridge, Cambridge University Press.

Harvey, P. and H. Knox. 2008. 'Otherwise Engaged: Culture, deviance and the quest for connectivity Through Road Construction', *Journal of Cultural Economy* 1(1): 79-92.

Höfer, A. 1997. *Tamang Ritual Texts II. Ethnographic studies in the oral tradition and folk-religion of an ethnic minority in Nepal*, Stuttgart, Franz Steiner Verlag.

Ingold, T. 2007. *Lines: A Brief History*, London, Routledge.

Kearney, M. 1996. *Reconceptualizing the Peasantry*, Boulder, Westview.

Kirkpatrick, W. 1986 (1811). *An account of the kingdom of Nepal*, New Delhi, Asian Educational Services.

Lecomte-Tilouine, M. 2009 (2003). 'The Nepalese Landscape: Exegis and Appropriation of the Country', in Smadja, J. (ed), *Reading Himalayan Landscapes over Time. Environmental Perception, Knowledge and Practice in Nepal and Ladakh*, Collection Sciences Sociales, no. 14, Pondicherry, Institut Français de Pondichéry, pp. 161-192

Pigg, S. 1992. 'Constructing Social Categories through Place: Social representations and development in Nepal', *Comparative Studies in Society and History*, 34(3): 491-513.

Ramble, C. 2008. *The Navel of the Demoness*, Oxford University Press.

Scott, J. 1998. *Seeing Like the State: how certain schemes to improve the human condition have failed*, New Haven, Yale University Press.

Tilman, H. 1952. *Nepal Himalaya*, Cambridge, Cambridge University Press.

Notes

1. While fully recognizing the sovereignty of the borrowing country, the Bank accepts that it has a responsibility to ensure (i) equality of opportunity for national minorities and (ii) that its operations and assistance to developing member countries do not negatively affect the welfare and interests of national minorities (ADB 1999: 2). The ADB (1999) defines Indigenous Peoples by two significant characteristics (p. 5) (i) descent from population groups present in a given area, most often before modern states or territories were created and before modern borders were defined and (ii) maintenance of cultural and social identities; and social, economic, cultural, and political institutions separate from mainstream or dominant societies and cultures.
2. I recall hearing many of the older generation say they would never abandon their old houses. They did eventually.
3. Tilman visited the border post of Rasuwa Garhi in 1949 and mentioned 5,000 man-loads of salt per year passing through, adding that at that time trade was 'blessedly free' of customs duties (Tilman 1952: 55).
4. http://www.adb.org/documents/tars/nep/tar-nep-37266.pdf
5. "The nature of the rural household itself (and possibly even 'the village') has been transformed, if not fragmented, by the rise in individual migration. The household is also now, in structural ways (as opposed to visual), much less 'rural.' Many rural households have become a sort of spatially disparate extended family. The results are: increasing feminization of rural life (with the potential for greater exploitation and oppression of women, but also offering possibilities for women of significant improvements in their relative position), and the demographic re-structuring of households and villages (as the economically active men seek employment elsewhere and women are left with the children and the elderly). The risks for those 'left behind' are great, if monetary remittances from younger, mainly male migrants decrease as the moral economy of the household and the village disintegrates." (Blaikie *et al.* 2002: 1268)

2. 2. MIGRANTS, NEW SETTLERS AND BONDS TO TERRITORY

2.2.1.

Redefining Belonging and Bonds to Territory: Multiple Forms of Mobility and Itineraries among the Tamangs of Central Nepal

Blandine Ripert

Labour migration is not a recent phenomenon in the Himalayas, and has been the subject of numerous studies showing important variations in different areas of the chain, especially in Nepal (Gil 2003, Gurung 1987, Seddon *et al*. 2001, Thieme *et al*. 2007, Thapa 1990, Sagant 1978, Kollmair *et al.* 2006, Von der Heide *et al.* 2001). In general, mobility has been shown to vary between nomadism and seasonal migration, between temporary and permanent displacement, between closer and more distant destinations, and between shorter and longer durations; one must equally consider the purpose in migrating (Acharya 2000, Bruslé 2006, Bruslé 2008, Krengel 1997), and take into account the migrations of individuals over the course of their lifetime. Less attention has been given to the effects mobility has had on bonds to territory and on how these bonds relate to the group of origin. We shall attempt to examine these effects using the example of a Tamang group of central Nepal (classed as Western Tamangs, as distinct from the markedly different Eastern Tamangs), dwelling in the districts of Nuwakot and Dhading (Fig. 1). An ethnic group whose language belongs to the Tibeto-Burman family, they principally inhabit the mountainsides northwest of Kathmandu, a few days' walk from Trisuli Bazaar and Dhading Besi. The villages in this area are effectively mono-ethnic (with only a few families of an Indo-Nepalese sub-caste of blacksmiths also living here), in contrast to the lower valleys.

Territorial unity is centred on the villages and typically corresponds to the mountain slope, whose clearly demarcated boundaries are delimited by the ridges and the rivers, and it is characterised by an agro-silvo-pastoral system of land farming (Toffin, Meyer and Garine 1986). The village territory thus comprises a distinct entity with recognizable boundaries. The integrity of the territorial unit is particularly manifest during collective rituals in which all village families participate (Holmberg 1989); associated with these rituals are certain protective spirits, guardians of the soil, which ensure the fertility of the land and the prosperity of the community (Höfer 1972).[1]

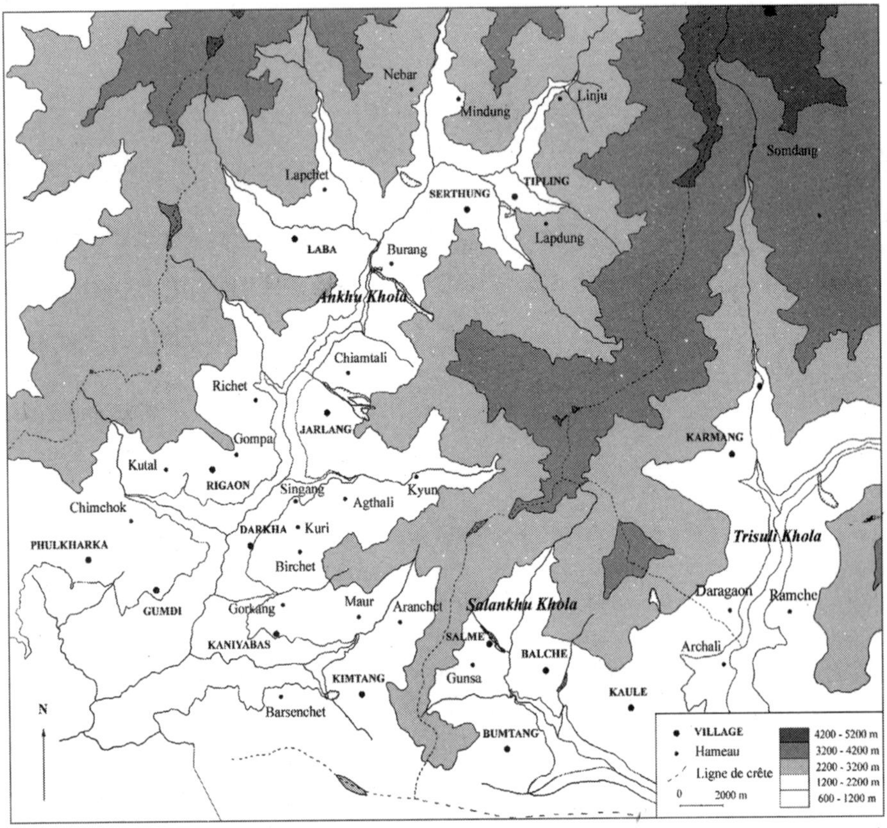

Figure 1. Nuwakot and Dhading districts in Nepal

Labour migration has always played a vital role in the Tamang villages' economy, owing both to the products (most often food, particularly cereals) brought back by migrants and to the drop in demographic pressure on village land that occurs during migration periods (Toffin 1974). If this has varied over the years, in recent times it has come to play a dominant role, profoundly altering both village activities and, so it would seem, the group's bonds with localities and sense of belonging. By examining the mobility of these villagers over the past forty years (1970-2008) in terms of destination, duration and type of occupation, and considering the migration routes followed during migrants' lifetimes, I will try to show how the emergence of new types of mobility seems to have altered their perception of group belonging and of village territoriality, leading, in some cases, to renewed assertions of identity.[2] After making a historical comparison of migration and its links to Tamang adherence at group level, I will then complete the general picture by focussing more particularly on some individual cases. I will then attempt to place this analysis within a more global perspective.

Like the majority of Nepalese ethnic groups, the Western Tamangs have been familiar for a long time with diverse forms of mobility, whether it be their settlement migration on the southern slopes of the Himalayas that dates from the seventeenth century, at which time, prior to their settling, their mobility was most probably of a nomadic type; or their periodic displacement of mobile herder huts (sheltering both livestock and herdsmen) over their cultivated territory according to the agricultural calendar; or their seasonal, temporary or occasionally permanent migrations to labour markets. In recent years, these latest migrations have become increasingly diversified, while acquiring an importance they no doubt rarely had in the group's past, although the paucity of information concerning the more distant past makes it difficult to confirm this with any degree of certainty. If labour migrations have long supplemented what the local economy was unable to provide—cereals for year-round sustenance—and were formerly conducted during periods of reduced agricultural labour intensity, today they involve such faraway destinations and such longer stays, and are so much more frequent, that they take precedence over agricultural activity with no regard for the seasonal planting calendar. From simple porterage activities, road construction or agricultural work in the south of Nepal or northern India, migrant labour has now become greatly diversified due to new opportunities, resulting from increased labour mobility, to include agro-industry and non-agricultural sectors, both in South Asia and in other parts of the world.

Migration was long viewed as a means of keeping the village alive. Thus, recourse to external sources served the interests of the group's economic and social preservation within the village territory. There was a shift in this migratory focus in the mid-1990s, when a combination of changes in the region's agro-silvo-pastoral system[3] enabled numerous villages to achieve self-sufficiency in food production. This has not, however, stemmed the flow of migration: the Tamangs in the area have continued to migrate, but now with the aim of raising their living standards in the village. Income from migrations is no longer used to feed one's family but to obtain consumer products on the market or to invest in agricultural projects which in the past were beyond the means of even the most prosperous. The obligation of the well-off to spend any surplus on village social and religious ceremonies permitted a certain amount of income redistribution and considerable political influence for the wealthy.[4] If the economic consequences of these changes have been clearly visible in the arrival of new products, materials and objects transported from local bazaars to the villages, the social ones are equally evident.

In short, three periods in the development of migrations from the area can be identified.

From the 1970s to the mid-1980s, migrations were seasonal, the majority taking place in winter, the period of reduced agricultural activity in village fields. They were for the most part collective and involved relatively proximate destinations such as southern Nepal or the north-eastern states of India.

From the end of the 1980s to the mid-1990s, migrations were more towards the west of the Himalayan range (Himachal Pradesh, Uttaranchal and Kashmir), while the North-East was closed to migration from the outside as a result of political tensions. Southeast Asia, undergoing a full economic boom at the time, exerted a strong pull on Tamangs up until the 1997 economic crisis, which dealt a severe blow to what were then known as the 'four dragons.' During this period migrant activities took on a more diversified character, ranging from the transport of heavy freight and subsequently of more valuable products (clothing, electronic goods, gold, and more rarely, drugs), to the construction of bunkers and the transport of arms, food and mail from north-western India towards the Pakistan border for use by the Indian army.

From the end of the 1990s to the 2000s, migration has been to even more distant destinations in the Persian Gulf (particularly Saudi Arabia, Kuwait and Qatar) and deep into South-East Asia (Malaysia, Indonesia). These migrations are no longer linked to the agricultural calendar; they are most often temporary, and may extend over a period of several years to ensure that the cost of the airline ticket and the recruitment agency fees are recovered. They involve physical labour (porterage, building construction, road construction, etc.) under difficult conditions (Bruslé 2009). Concurrently, since the 2000s new mobility towards the regional cities of Nepal and the capital Kathmandu has emerged, particularly among the younger Tamangs, and has been partly linked to Maoist guerrilla activity.

The three maps below illustrate the three different stages in migration that have been identified in the course of the last forty years (Fig 2.3.4). They clearly indicate the progressive broadening of the scope of destinations depending on the opening and closing of labour markets that have been sending the Tamangs further and further away for longer periods of time.

These migrations are part of a surge in the globalisation of labour inciting the Tamangs to move, forcing them to take greater risks, particularly economic risks. In such a context, unsuccessful migration may plunge an entire family into unprecedented levels of debt.[5] Thus, as regional labour markets close, the Tamangs are now projecting an international dimension onto their villages, are leading them out of isolation and are living in very different contexts from what was previously the case. As a consequence, at family level one observes a fragmentation of living space: a young son leaves to study at a regional secondary school in Trisuli or Dhading, making frequent trips to and from his village, which remains his place of reference; his father sets out almost every summer to work for six months in Kashmir or Himachal Pradesh in India, along with a group from the village, sometimes taking along a younger son; the elder son goes to the Persian Gulf for a few years; one or two daughters get married in neighbouring villages, or in a settlement lying outside the home village; the mother attempts to carry on the agricultural work with the help of the remaining children, if there are any, or by hiring village help when the agricultural tasks prove to be beyond her abilities. Thus, the living space shared by the family, the territory, is fragmented into various widely separated locations and different social and economic contexts. This fragmentation of living space can actually occur within an individual's itinerary.

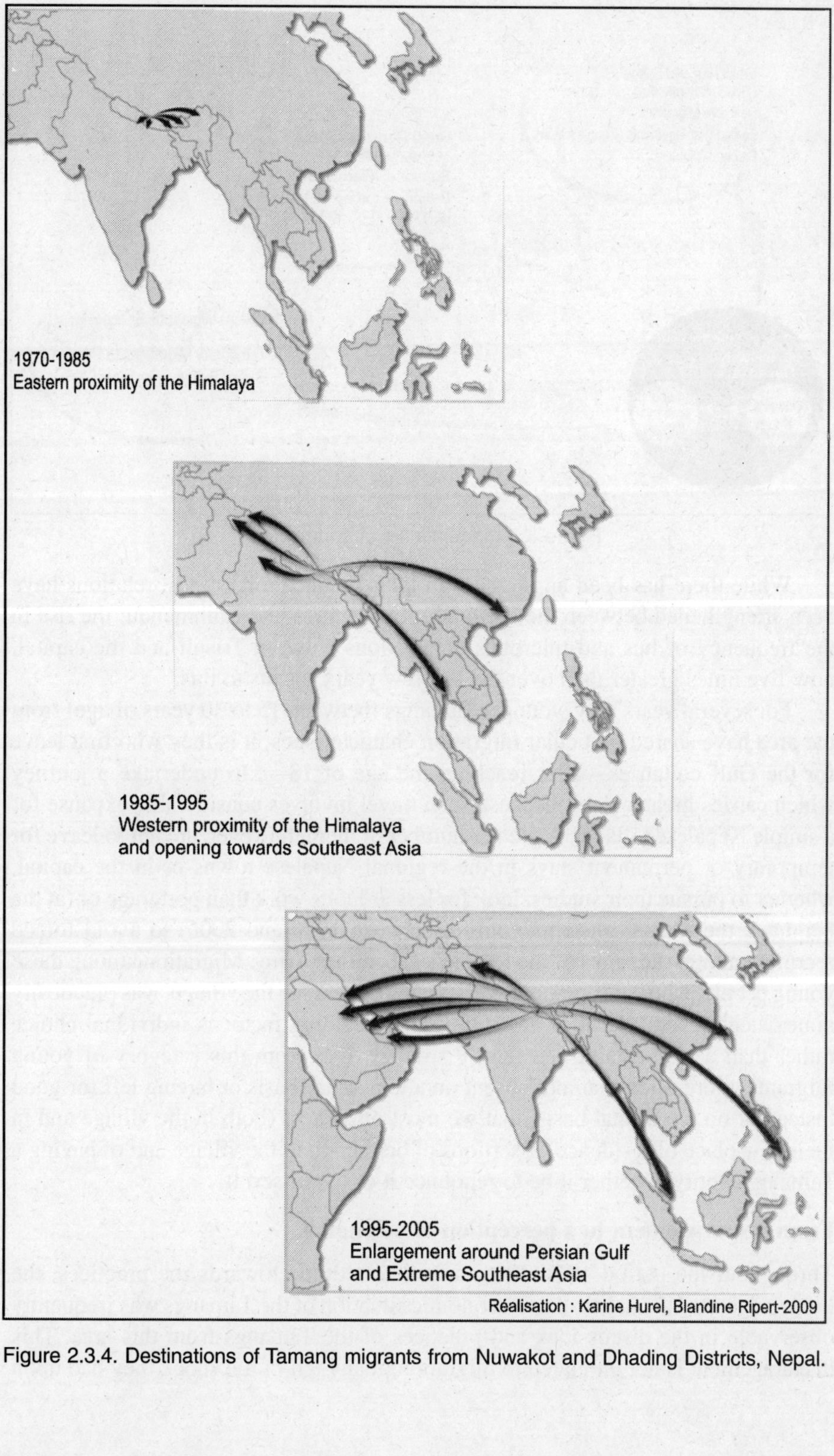

Figure 2.3.4. Destinations of Tamang migrants from Nuwakot and Dhading Districts, Nepal.

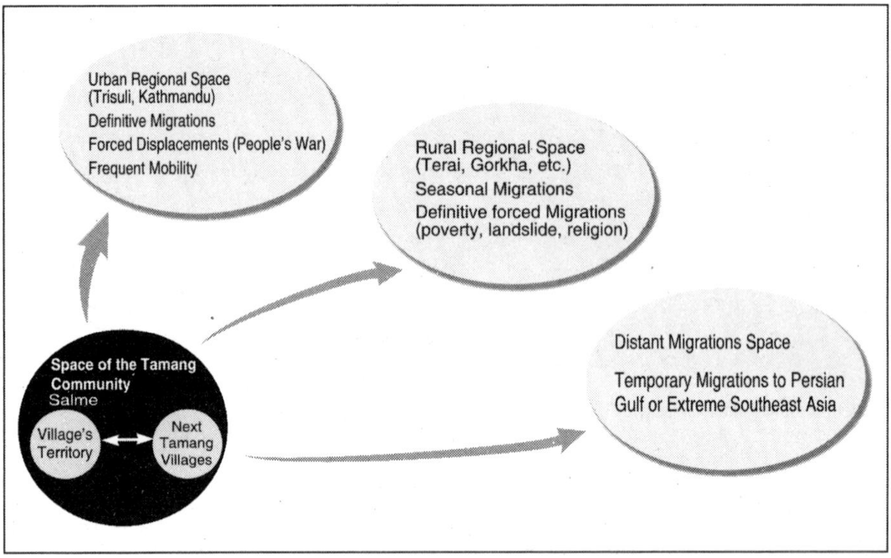

Figure 5. Fragmentation of living space

While there has been an increase in long-distance migrations, relations have been strengthened between the regional urban centres and Kathmandu: the rise in the frequency of bus and microbus connections between Trisuli and the capital, now five times greater than over the last few years, attests to this.

For several years now, younger Tamangs (between 15 to 30 years of age) from the area have shared particular migration characteristics: it is they who first leave for the Gulf countries—after reaching the age of 18—, to undertake a journey which carries greater risks, because such travel involves considerable expense for a simple Nepalese villager. A greater number of them, however, are apt to leave for temporary or permanent stays in the regional Nepalese towns or in the capital, whether to pursue their studies, look for less arduous work than porterage or (at the height of the Maoist guerrilla conflict between 1996 and 2006) to avoid forced recruitment into the ranks of the People's Liberation Army. Migration among these young people, who are sometimes very young on leaving the village, was practically nonexistent several decades ago. Often, the deciding factor is individual choice rather than a well-established family strategy. It is from this category of young migrants, more often than not absent on a temporary basis or having left for good instead of on a seasonal basis, that we most often hear (both in the village and in their new place of residence) assertions of belonging to the village and of having a Tamang identity, whether it be to renounce it or to reassert it.

From disparagement to a perception of belonging

Throughout the period 1970-2000, a critical attitude towards the practices, the knowledge, and the political and economic situation of the Tamangs was frequently observable in the discussions and practices of the Tamangs from this area. This disparagement is not unrelated to migration: many tell of the shock they had upon discovering other places in the world and other people, while perceiving the disparity in economic and living standards between themselves and others, to a level of which they had previously no conception. Fascinated by what they perceive as great wealth, modernity and the general facility of living abroad, many disdain life in the village with its economic poverty, which they tend to associate with overall cultural, social and political. The formal schooling of the young has contributed, at least initially, to such cultural disparagement by promoting 'Nepalisation' and national integration, especially in the post-1991 period.[6] This younger generation of Tamangs had held no part in the democratisation movement of 1990 or in the popular identity movements of the ethnic groups (known as *janajati*) that preceded and followed it, which had reached their height at the time of the promulgation of the 1991 constitution to assert the rights and equality of minority groups long dominated by central government.

Aside from these claims, a new consciousness of a Tamang identity has materialised in discussions among younger Tamangs over the last five years or so. This attitude seems to emerge with some consistency after a few years spent in the city, far from the village. After an initial phase of disparaging and belittling Tamang culture, during which these young people unceasingly extol the splendours of the city that has become their new living environment, a longing to return to their origins frequently begins to operate, after their bonds to the village have weakened and memories of it have somewhat faded. The village, now the scene of an idealized childhood, is depicted as a place whose loss they regret, where the air was fresh, the water pure, the food healthy and family ties strong, in contrast with the polluted city and the loneliness that they eventually encounter in their urban life.

Along with these attitudes, these young Tamangs migrating to the cities have adopted new practices over the last five years. Some note down songs heard in their village, others study and recount the different Tamang festivals celebrated in the region, such as Tsechu; others investigate local history by interviewing the oldest members of their village, re-transcribe and often publish local works of poetry, or assist in compiling dictionaries and manuals for learning the Tamang language.

These literary activities involving the recording of their oral Tamang culture have not, for the most part, been linked to any engagement of a more political nature, as in organisations such as *Tamang Gedung Shang* that assert Tamang identity. Of particular relevance is the fact that some of these young Tamangs have converted to Christianity (Ripert 1997, 2001, 2004); I observed that the reaffirmation of Tamang culture has occurred more slowly among these Christians. After their conversion, they distance themselves from a great number of Tamang cultural practices considered by Protestant missionaries as pagan, and in particular from all the festivals and ceremonies which are mostly of a religious nature, whether Buddhist, shamanist or part of local religious cults. As a result, these young converts, now urban dwellers, are subsequently faced with the dilemma of sorting from their Tamang culture what they can possibly re-subscribe to without entering into conflict with their religious affiliation, the more so as they are sometimes barely familiar with their own culture. Indeed, some have spent many years away from the village in Christian-run schools, as is the case, for example, of numerous Rigaon (right bank of Ankhu Khola, see Map 1) youths educated in the Terai. If none of the young Christian Tamangs from this area belong to *Tamang Gedung Sang*, this is precisely because the association maintains that Tamang identity is inseparable from Buddhism, thus excluding Christians, whether they regard themselves as Tamang or not.

Individual itineraries

Such generalities at group level must be nuanced and supplemented by a closer analysis at individual level. Some examples of migration routes, together with the connections to group belonging and village territory that have been maintained, should throw some light on the relationship between territory, mobility and identity. The following examples were selected for their being both distinctive and representative.

Surje Man Tamang,[7] now about 50 years old, never attended school. Nevertheless, he has often left for seasonal migration in winter, principally to northern India, where he has worked as porter, agricultural labourer, and sawyer. His wages have enabled him to provide for the needs of his family (mother, brother, sister-in-law and nephew) while bringing in the money long needed to purchase the cereals lacking for some months of the year, and new tools, as well as for projects on improving the agricultural yield. A large part of the money has also been used to care for his ailing mother, and later to pay for her burial costs. Now that thanks to improved harvests they are able to feed themselves all year-round, his money has been used to build a new house in the village on the ruins of the previous one and to purchase livestock. The Maoist conflict forced him to leave the village, due to longstanding local conflicts rather than to any recent political tension (Ripert 2006). In 2004, he, his wife and their young child sought refuge in Trisuli Bazaar, where they encountered great difficulties, renting a very small room, often only for a few months before finding a new place to live. Unable to find work, he attempted to survive in the town selling alcohol he made himself from millet purchased on the market. He was highly dependant on the local support network among the Tamangs, eating every day at the house of a family from his village, depending on whoever was able to feed him.

Since moving to Trisuli, he has proudly worn a Tamang turban,[8] something I never saw him wear while in the village.[9] He declares himself proud to be a Tamang, asserting his 'Tamangness' while talking about the mountains and wonders of his village, in contrast to previous years when he would lament the village's underdevelopment compared to what he had observed during his migrations to distant India. His enforced stay in Trisuli Bazaar, not far from his home village but dominated by Indo-Nepalese castes, appears to have awoken in him an awareness of an otherness that he is now intent on valorising, while his greatest hope is to be able to return to his village to live on the territory that remains his point of reference. The mere sight of that territory, visible from the ridges that dominate the town, fills him with great nostalgia.

Birendra, a young Tamang about 30 years of age, belongs to the generation of the 'educated young' (Ripert 2009). He attended the village primary school and continued his education for a few years in the local town, reaching a level equivalent to the lower high school certificate, before eventually moving away to Kathmandu. After working as a dishwasher in various restaurants, he has now become a pop singer in one of the city's nightclubs. He had dreamt about this ever since he lived in the village in the 1990s: he showed talent singing along to the Nepalese popular music he heard on the radio, and he was impatient to leave for the city to prove himself. At the time he viewed life in the village very critically and would not have sung Tamang songs for all the money in the world. Today, after a few years in Kathmandu that have thoroughly urbanised him, he has become curious about the Tamang origins he rejected for so long. Whenever he visits his family in the village he makes recordings of Tamang songs, which he has started writing down. Meanwhile, his best friend has recently completed a written history of the village after interviewing the community's oldest members. Along with friends from the village who now live in the capital he plans to organise and celebrate the Tamang festival of Tsechu there, which in recent years has rarely even been celebrated in the village. They plan to organise the festival in Balaju neighbourhood (northern part of Kathmandu where buses arrive from the north), where many Tamangs from Dhading and Nuwakot districts live today. If Birenda's involvement in this sort of cultural revival would have been unlikely five years ago, ten years ago it would have been inconceivable. His education, combined with his long stay in the city, was decisive in this return to his origins and in the use of his writing capabilities in order to record songs. Interestingly, in 2006 I learned that some Tamang families from the village of Salme, now living in Trisuli, were performing—or planning to perform—'Rimdo' in their rented living space, a ritual traditionally performed by lamas every two years in the villages of this area in order to drive away malevolent influences and spirits from the territory. If this report were to be confirmed, it would mean the displacement of a local ritual, traditionally linked to a territory, and now "relocated" to a new place of residence. Although this practice would have to be observed and carefully analysed, it might provide a way of reasserting the memory of a territory displaced in terms of space.

The case of Dilman is more complex and less representative of what we have observed. A man of about 50 and one of the few educated persons of his generation, he served for a long time as headmaster of the village primary school. If he has migrated, it is certainly not in order to carry heavy loads or to work in the fields of northern India. His migrations have been of a short duration, mainly to Southeast Asia, from where he brought goods (more or less legally) for a Tibetan employer in Kathmandu. He says the great benefits he derived were mostly from being able to discover a developed urban world. In the past he has also travelled to North-East India with some other villagers to follow a training course in tea cultivation, which he has successfully begun to grow in his own fields. Most of the money earned during his migrations was invested in the construction of a private school in his village, to which he invited paid instructors he had met in North-East India to teach

English and other subjects. The Maoists, who were present in the area during the guerrilla war, punished this private initiative and his social status by destroying his house, confiscating his livestock and expelling him and his family. Since then he has been living in Trisuli, teaching at a public school, while his wife runs a small restaurant on the main road. Since moving to Trisuli, Dilman has deliberately distanced himself from his Tamang origins and even tends to conceal them. In the town he has woven a new network of relationships from which Tamangs are excluded; he says he associates with them as little as possible (a statement confirmed by other people). He says that he wears the Nepalese *topi* (the national cap) to look more like a Chetri (a dominant Indo-Nepalese sub-caste), and takes care over his clothes so that nothing reveals his origins. He speaks Nepalese to his wife and children now, instead of Tamang, and does not plan on returning to the village after peace has been restored, because his links with the village territory have weakened since settling in town. While it is always possible that his attitude may change in a few years from now, at present there is nothing to indicate that this will be the case.

These diverse examples show how individual itineraries, where migration strategies are not always accommodated in a broader family context, differ both with regard to type—temporary, definitive, forced, successful or unsuccessful—and as to whether or not a return to the territory of origin is envisaged.

Generally speaking, our observations have shown that an individual's sense of belonging to the Tamang group appears to emerge in their discourse when they are in a situation of long-term migration, as long as either a return to the territory is planned, or strong bonds are maintained with the territory and its inhabitants. The level of education, religious affiliation, migration routes, varying from one individual to another, sometimes totally unrelated to an overall family strategy, lend nuance to the manner in which group belonging can be achieved, which, in the end, provides us with a wide variety of situations.

REFERENCES

Acharya, M. 2000. *Labour Market Development and Poverty (with Focus on Opportunities for Women) in Nepal*, Kathmandu, Thanka Prasad Acharya Foundation in co-operation with Friedrigh-Ebert Shiftung.

Bruslé, T. 2006. 'Aller et venir pour survivre ou s'enrichir. Circulations de travail, logiques migratoires et construction du monde des Népalais en Inde', Ph.D. Thesis in Geography, Poitiers, Université de Poitiers.

– 2008. 'How to choose a working destination? Nepalese migrants to Uttarakhand (India)', *Mountain Research and Development*, 28(3/4): 240-247.

– 2009. 'Who's in a labour camp? A socio-economic analysis of Nepalese Migrants in Qatar', *European Bulletin of Himalayan Research*: 32-49.

Dobremez, J.-F. (ed.). 1986. *Les collines du Népal Central, écosystème, structures sociales et systèmes agraires*, 2 volumes, Paris, INRA.

Gill, G.J. 2003. *Seasonal Labour Migration in Rural Nepal: A Preliminary Overview*, Working Paper 218, London, Overseas Development Institute.

Gurung, H. 1987. 'Factors Influencing Migration in Nepal', *Journal of Development and Administrative Studies*, 19: 94-124.

Höfer, A. 1972. 'Note sur le culte du terroir chez les Tamang du Népal', *Langues et Techniques, Nature et Société II - Approche ethnologique, approche naturaliste*, Paris, Thomas, Bernot Eds., Klincksieck, pp. 147-156.

Holmberg, D.H. 1989. *Order in paradox: Myth, ritual and exchange among Nepal's Tamang*. Ithaca, Cornell University Press.

Kollmair M, S. Manandhar, B. Subedi and S. Thieme. 2006. 'New figures for old stories: Migration and remittances in Nepal'. *Migration Letters* 3(2):151–160.

Krengel, M. 1997. 'Migration and the danger of loss: Some aspects of cultural identity in Kumaon/Indian Himalaya' in Stellrecht, I. and M. Winiger (ed.), *Perspectives on History and Change in the Karakorum, Hindukush and Himalaya*, Cologne, Rüdiger Köppe Verlag, pp. 171-188.

Ripert, B. 1997. 'Christianisme et pouvoirs locaux dans une vallée tamang du Népal Central', *Archives de Sciences Sociales des Religions*, 99: 69-86.

– 2000. 'Dynamiques spatiales et transformations de la société. L'exemple des Tamang au Népal central', Ph.D. thesis, Université Paris 7-Denis Diderot.

– 2001. 'Innovations among the poor: conversions to Christianity in Central Nepal', in Amaladass, A. (ed.), *Profiles of poverty and networks of power in India*, Madurai, DACA Publications, pp. 131-142.

– 2004. 'Le 'lama pasteurisé', vers de nouvelles figures d'autorité religieuse chez les Tamang de l'Ouest', in Bouillier, V. and C. Servan-Schreiber (ed.), *De l'Arabie à l'Himalaya, Chemins croisés en hommages à Marc Gaborieau*, Paris, Maisonneuve & Larose, pp 453-465.

– 2006. 'Si loin', *EspacesTemps.net*, Mensuelles, 5.4.2006, available at: http://espacestemps.net/document1937.html

– 2009. 'Parcelling of Land, Privatisation along with Collective Management of Space and Resources on the Salme Mountainside', in Smadja, J. (ed), *Reading Himalayan Landscapes over Time. Environmental Perception, Knowledge and Practice in Nepal and Ladakh*, Collection Sciences Sociales, no. 14, Pondicherry, Institut Français de Pondichéry, pp. 467-492.

Sagant, P. 1978. 'Ampleur et profondeur historique des migrations saisonnières népalaises', *L'Ethnographie*, Paris, CNRS, Gabalda: 93-119.

Seddon, D., G. Gurung and J.Adhikari. 2001. *The New Lahures: Foreign Employment and Remittance Economy of Nepal*, Kathmandu, Nepal Institute of Development Studies.

Smadja, J. 2009.'A Reading of the Salme Tamangs' Territory and Landscape', in Smadja, J. (ed), *Reading Himalayan Landscapes over Time. Environmental Perception, Knowledge and Practice in Nepal and Ladakh*, Collection Sciences Sociales, no. 14, Pondicherry,Institut Français de Pondichéry, pp. 199-239 and 563-589 + map.

Thieme S., R. Bhattarai, G. Gurung, M. Kollmair, S. Manandhar and U. Müller-Böker. 2005. 'Addressing the needs of Nepalese migrant workers in Nepal and in Delhi, India', *Mountain Research and Development*, 25:109–114.

Thapa, P. 1990. *Nepal: Socio-Economic Change and Rural Migration*, Delhi, Vikas Publishing House Pvt. Ltd.

Toffin, G. 1974. 'Les populations de la haute vallée de l'Ankhu Khola', *Objet et Mondes*, 14(4):325-336.

Toffin, G., E. Meyer and I. de Garine. 1986. 'Salmé, Ethnologie et démographie d'un village tamang', in Dobremez (ed.), *Les collines du Népal Central, Ecosystème et activités dans un village népalais*, Paris, INRA, pp. 55-120.

Von der Heide, S. and T. Hoffman (eds.). 2001. *Aspects of Migration and Mobility in Nepal*, Kathmandu, Ratna Pustak Bhandar.

Notes:

1. Joëlle Smadja (2009) has indeed shown in her research on place names how the territory on the Salme slope is marked with names, and sometime stones or trees, referring to these spirits and divinities, responsible for its territorial integrity.
2. The data used in this work have been taken from the Programme Versant (CNRS-INRA) for the years 1970-1980 (see for example Dobremez [ed.] 1986) and from my own research in the field in the decades that followed. This research was conducted during an 18-month period of fieldwork carried out between 1994 and 2000 as part of a doctorate in geography, combining semi-directive interviews, participant observation, and, for migrations, a survey of all migrants from the village of Salme, the principle investigation site, as well as less exhaustive investigations in the neighbouring Ankhu Khola Valley, mainly in Tipling, Rigaon and Jharlang. The most recent data are from a stay in 2006 and from written correspondence since then.
3. See Ripert 2009 for more detailed information on the agricultural innovations responsible for these changes, which are not necessarily representative of Nepal in its entirety.
4. The purchase of cattle would also offer some people a certain degree of security.
5. Most often, the cost of migration towards Gulf countries is covered temporarily through loans to local notables at very high rates of interest. For most migrants, the sums at stake are so high that they would be obliged to work for decades in Nepal to reimburse them. Furthermore, migration is also often considered an option for reimbursing previous loans.
6. The school curriculum has been entirely revised since the democratisation of 1991 to take more account of Nepalese cultural diversity, though, in fact, more as part of a national integration process, rather than to valorise identities (see changes in the history and geography curriculum discussed in Ripert 2000).
7. Names have been changed.
8. This is a red turban, locally weaved according to a special design with a colourful square pattern and traditionally offered by sisters during the *Dasain* festival.
9. Few Tamangs do in this region and they are very rare below the age of 50.

2.2.2.

Urban Fringes: Squatter and Slum Settlements in the Kathmandu Valley (Nepal)

Gérard Toffin

Introduction

Over the last decades, migrations and population displacements have produced new peripheral spaces throughout the world, on the margins of national states and of urban territories. Among these sites are refugee camps, slums, squatter settlements, resettled enclaves, and so forth. At best, migrants live in buildings or camps provided by their employers. Yet the key features of most of these spaces are the non-permanent and transitory conditions, the vulnerability, and the poverty of the populations. For the sake of analysis, they can be called *outplaces*, i.e. neither belonging to the urban territory nor to its outside space. Their uncertainty has a serious impact on education, economic conditions, and the exercise of citizenship rights (Agier 2008). More often than not the people settled there are hardly integrated into global all-encompassing society and are considered urban or national pariahs. They are implicated in national conflictual causes, and are easily manipulated by political leaders and organisations. In South Asia, these spaces are principally multicaste, multiethnic and multilingual. They mix people from different geographical origins and stand in sharp contrast to the previous pre-industrial territories based mainly on kinship, ethnic group and caste hierarchy. A new social fabric is emerging from these settlements, characterised by: new collective identities; achieved status as far as leaders are concerned; social bonds based on a common neighbourhood and shared impoverished economic conditions; and lastly a vital role played by associative life. This article intends to provide a case study of such *outplaces* in the urban geography of the Kathmandu Valley, Nepal. I will focus on slum settlements along riverbanks, and address the various political and sociological issues which are central to the populations of these urban fringes. The data were collected over the last years in Kathmandu Metropolitan City, amidst growing traffic jams and thick clouds of car exhaust fumes.[1]

The anarchic urbanization of the Kathmandu Valley (1970-2010)

The Kathmandu valley, which encompasses a surface area of only about 600 sq km, has undergone unprecedented and dramatic changes over the last four decades. The massive increase in its population (from 500,000 in 1970 to above 3 million in 2010) and its subsequent overall urbanization have to a large extent reduced the open spaces available and agricultural fields which formerly surrounded the three major historic cities in the basin: Kathmandu, Lalitpur, and Bhaktapur. Similarly, the cities' urban fabric has become denser. The height of new buildings is continually on the rise. It has been estimated that within the next twenty years, the entire Valley will be covered with constructions, leaving far behind the former traditional divide between cities and clustered villages which prevailed in the area for centuries (Toffin 2007). The growth of settlements has been (and still is) generally spontaneous. Little planning intervention has been enforced by the government and municipal authorities. Urban areas continue to grow haphazardly, with no appropriate infrastructure such as water supply and sewerage systems, despite warnings by environmentalists.[2]

Such steady urbanization has produced a considerably negative impact, including traffic congestion, atmospheric pollution, and a total collapse of the former fragile ecological equilibrium between man and his environment. The situation is aggravated by the high elevation of the Valley (1350 m), which accentuates vehicle emissions, and its bowl-shaped topography, which restricts air movement. The rivers have also undergone tremendous pressure from the increase in demographic growth and in economic activities. They are now highly polluted by the discharge of untreated sewage and the widespread dumping of solid waste. They often resemble open sewers. For those who were there in the late 1960s or early 1970s, the Valley has changed beyond any recognition. Its local architecture of dazzling beauty and its exceptional landscape made up of green rice-fields covering rural areas have nearly disappeared or are on the wane. Even the view of the Himalayan peaks in the foreground is now barely visible due to a haze of pollution. The overall state of deterioration is so serious that Unesco is threatening to declassify some sites, especially the Pashupatinath area along the Bagmati River, which were designated as major items on the World Heritage List in 1979.

Demographic growth (more than 4 per cent per year) includes both natural growth and immigration from different regions of Nepal, and even from Northen India. The Kathmandu Valley, which is Nepal's political, cultural, industrial, and hospital centre, has become the favourite destination for rural people migrating from the hills. The concentration of political and economic power, as well as of tourist centres, with their employment activities and numerous opportunities, has favoured urbanization. Due to the Maoist insurgency (1996-2006), there has been a huge influx of internally displaced people in recent years in search of security, employment, government aid, and shelter. The population of Kathmandu, which in 1971 amounted to 150,000 inhabitants (105,000 in 1952), had already reached 671,000 in 2001, and is most probably more than one million today (admittedly within a larger administrative territory) (*Kathmandu Valley. Environment Outlook*, 2007). The rate of growth between 1991 and 2002 was 4,67 per cent per year. The density of inner city areas is high compare to the Valley as a whole. In 2001, they were 11,099 persons per sq km in Kathmandu City, 6,808 in Lalitpur City and 5,700 in Bhaktapur City (*Kathmandu Valley* 2007). The Valley's average population density was 1,837 persons per sq km.

This random urbanization began to gain ground in the Kathmandu Valley in the late 1950s. However, the main turning point in this process came from the 1970s onwards. Even in the 1970s, approximately 90 per cent of the entire population lived in rural areas. The economy was dominated by the agricultural sector, which accounted for 71 per cent of the gross national product. The mushrooming of housing developments from this time onwards has resulted in the conversion of a large section of prime agricultural land. Between 1984 and 2000, the amount of agricultural land in the region dropped from 64 per cent to 42 per cent, that is an annual decline of 7.4 per cent (*Kathmandu Valley* 2007). If this trend continues, by 2025, there will be no agricultural fields left in this once fertile Valley. The national urban population is 12 per cent, yet the Valley's share of this urban population is 54 per cent.

Economic and human pressure on territory has brought about an incredible increase in land prices. Between 1990 and 2000, these prices shot up 40 times in most sectors of Kathmandu Metropolitan City. Within the immediate suburbs of Kathmandu, near the Ring Road, plots of land suitable for building are now sold for more than 70 euros per square meter (15 to 30 *lâkh* of Nepalese rupees per *ânâ*,[3] that is 2.4 to 4.5 *crore* a *ropanî*, or 19.2 to 36 *crore* per *acre*).[4] In the new suburb of Koteshwar, a short distance from the airport, the price is 320 *lâkh* (= 3.2 *crore*) of Nepalese rupees a *ropanî*. In central areas of the capital, it fluctuates between 5 and 6 *crores* a *ropanî* and can even reach higher prices in exclusive, much sought after places. There seems to be no stopping the upward surge, even if the recent (1995-1996) introduction of a 10 to 1 per cent by government tax on land sales has slightly curbed the boom. The growing scarcity of land in sought-after places tends to lead to a steady price increase. Interestingly enough, the soaring of property prices in Kathmandu Valley is quite similar to the one in large Indian cities, such as Calcutta or Delhi (Toffin 2007: 18-20).

The price of land has also increased tremendously in rural areas: in Pyangaon village (Lalitpur District), the cost of one *ropanî* of irrigated land with no access to any roads or pathways amounts to 6 euros per square meter in 2007 (3 *lâkh* NPR, Nepalese rupees, per *ropanî*). This corresponds to 24 *lâkh* NPR per acre (0.4 hectare), whereas in the same locality, three years earlier the price was 2 to 4 euros per square meter (1 to 2 *lâkh* a *ropanî*). In the more central villages of Harisiddhi and Sonaguthi, a plot of land bordering a pathway costs around 5 *lâkh* a *ropanî*. The price increases considerably when the field is situated along a road: 17 to 24 euros per square meter in 2005 (8 to 12 *lâkh* a *ropanî*). In Sainbu Bhainsepati (Khokana), the price of a well situated building plot is about 8 *lâkh* of Nepalese rupees for one *ânâ* (2010). The cost of house construction is much less.

The three major towns in the Valley were the capitals of Hindu city-kingdoms which existed during the late Malla period (sixteenth-eighteenth centuries). In terms of urbanism, they still lag far behind the modern megacities of South-Asia, such as Mombai, or of South-East Asia, such as Bangkok or Singapore. Yet they are becoming more and more cosmopolitan in character, with their multi-faith, multi-ethnic and transnational inhabitants. A Westernized educated middle class has emerged, easily identifiable by their activities and expenditures, new dress code, and specific values oriented toward modernity (Liechty 2008). In addition, traditional territories have been dramatically transformed and reconfigured. Let us take for example two major trends involving new social and spatial hierarchies within urban areas: 1) migration towards the periphery by people formerly living in the heart of cities, which maps a new social geography of the region, is very different from that of the past. The houses built in these suburbs are of a different type, made of a mixture of concrete and bricks, with a flat roof, and giving onto a garden; 2) the traditional opposition between village and city is in the process of becoming blurred. Villages are gradually being swallowed up by cities and the construction of houses in the peri-urban areas along new and old roads has already joined up with former separate settlements. Wealthier peasants are progressively abandoning farming to take up other non-manual activities, for instance trade or government employment. All these changes have prompted major sociological breaks and a major decline in the local rural economy.

No effective policy has been implemented to regulate such urbanization. A number of plans and reports have been drafted by various foreign agencies, such as the 'Physical Development Plan for Kathmandu Valley' conducted in 1969. Experts proposed several recommendations and devices. Yet none have been seriously implemented. The future outlook seems even gloomier than it has been over recent decades. Two mega-projects, which are presently under discussion in government offices and local bodies, will obviously further accelerate this unrestrained urbanization. The first one of these projects, run by the Ministry for Physical Planning and Works, intends to found a new town south of the Kathmandu Valley, to provide full facilities for around 150,000 to 200,000 people. The project, called 'Harisiddhi New City Project' (HNCP), is located around Harisiddhi and concerns the Village Development Committee of Harisiddhi, Imadol, Thaiba and Siddhipur, south of the Ring Road, in Lalitpur district. It aims at turning 12,240 *ropanîs* of land, mostly given over to agriculture, into a residential town, with its own markets and central business district. To quote their own terms, "the population problem of the capital city will thus be solved for several decades" (*The Himalayan*, Sept. 2005). According to the proposal, the city area is expected to have a cumulative road network of 195 km. The cost of its completion is estimated at 3.6 billion Nepalese rupees. The second project concerns another Ring Road, much longer (72 km) than the first one, encircling a larger zone. This Outer Ring Road, as it is called, is expected to regulate the flow of traffic and to provide better access to rural settlements. It is heartily supported by the rural areas concerned. However, it will increase mass housing (in an area prone to major earthquakes), air pollution due to vehicle exhaust fumes and it will cause a general degradation of the environment, as was the case with the first circular highway built thanks to a Chinese cooperation. Some people are already lobbying to bring the road alignment closer to their settlement to reap the benefits from the expected rise in land prices (Shrestha and Shrestha 2008: 53). The preliminary estimate shows a cost of eight billion Nepalese Rupees, excluding the cost of land. Both projects will probably worsen the already degraded situation.

Squatter communities (slums) in Kathmandu Metropolitan City

The growth of the urban population has contributed to a surge in squatter communities. Such settlements have emerged in various parts of the Kathmandu Valley (Hada 2001). About 75 settlements have been identified so far, 65 of which are located in Kathmandu Metropolitan City. The majority are established along riverbanks, which traditionally formed the borderline between cities. The rivers concerned are mainly the Vishnumati, which flows north-south, west of Kathmandu city, and the Bagmati which borders Kathmandu to the south. These two waterways converge in the heart of the capital. A smaller group of squatter settlements are located in a non-riparian environment, in the Kathmandu suburbs, often on the periphery of former independent settlements. This is the case near Bauddha (Bodnath), Chabahil, Maharganj, and Guhyeshvari. In Nepali, all these areas, riparian and inland, are called *sukumbâsî bastî* or *bastî*, 'settlement', and the squatters living there *sukumbâsî*, a word applied to displaced and landless persons, as well as to families illegally occupying land or a recently deforested area. As a matter of fact, this word has become synonymous in Nepal with an excluded person, with no means of subsistence.[5] Legally, a *sukumbâsî* is a person who can prove that nobody in his family over the last three generations held a land title, *lâl purjâ*.

For the whole Kathmandu Metropolitan City, which covers an area of 50 sq km and concentrates ninety per cent of these shelters within the Kathmandu Valley, *sukumbâsî* represent a population of about 15,000 persons (*New Beginnings* 2005, Lumanti NGO). The figure is relatively low (between 1 and 2 per cent of the city's urban population) compared to mega-cities of India, such as Mumbai, Delhi or Kolkata. It is probably due to the lack of wasteland in the Kathmandu municipality. Nevertheless, the number of squatters is growing steadily. In 1985, their population represented 2,134 persons for the municipality of Kathmandu. For the same area, the figure reached 4,295 in 1990, 11,862 in 2000 and about 15,000 in 2005 (*New Beginnings* 2005). It is therefore an acute problem that municipal authorities have to face and deal with accordingly. This rise in numbers has not slowed its pace since the end of the civil war and the abolition of the monarchy. Since the winter of 2007-2008 and spring 2008, two large *bastu* have appeared (or rather reappeared in one case): the first, made up of about 300 households, in Thapathali (Photo 1), near the bridge linking Kathmandu to Lalitpur, and the second, larger (about 500 households), in Balkhu. Both these riparian settlements have been set up on the banks of the Bagmati River. The first, which has been given the name of Naya Paurakhi Gaon (from *paurakhi*: 'valorous'), already existed—though to a lesser extent—in the 1990s, but was razed in 2001 at the time of a SAARC Conference. It is located in an area given over to the United Nation to commemorate the fiftieth anniversary of the Organization, and scheduled to be transformed into a UN Riverside Park.

Photo 1: Slums along the Bagmati River at Thapathali (2008)

The oldest of these settlements dates back to the 1950s. However, most of them have sprung up over the last two decades. Some of the better known areas are Sankhamul (Photo 2), on the right bank of the Bagmati, near Naya Baneswar, Sinamangal near the Airport, on the Bagmati, Balaju (Photo 3), in the north, along the Vishnumati, Khadi Pakha (KMC, no. 5), Tripureshwar (Bansighat), Tankeshwar (KMC, no. 13), Ramhity (KMC, no. 6), Kumaristhan (KMC, no. 16), etc. In 2008, a significant percentage of Kathmandu's riparian corridor was lined with permanent *sukumbâsî* housing. The banks of these rivers do not belong to the Municipality but to the state. This is one of the reasons for the concentration of squatters in these areas. Each cluster is made up of 50 to 300 families (*dhuri*), each household occupying a shelter.

These new territories, seldom studied up to now, are inhabited mostly by Nepalese families who come from various districts of Nepal in search of employment, better facilities than in their native places, and safer environment. The Nepalese living there form a very composite group of people, reflecting the diversity of the country's population. Most residents (48%) come from the hills and belong to various ethnic groups, *janajâti*, of Nepal: the Rais, Limbus, Tamangs, etc. However, 28 per cent of residents also belong to Hindu high castes—a figure that is not in keeping with the discourse of the *janajâti* organizations and their common rhetoric on social exclusion in Nepal—and 13 per cent to the Newars. Some squatters come from the Tarai plains. As far as most clusters are concerned, it is said that they migrated from the 75 districts of Nepal. Nepali is the language for communication. Leadership roles tend to reflect a person's length of stay in the community, as well as age, respectability and financial prosperity. Interestingly

Photo 2: Slums along the Bagmati River at Sankhamul (2007)

enough, six per cent of these squatters are Christians, a high percentage compared to the mean national figure (about 1%). Some dwellers (between 15 per cent and 20 per cent) are transient Indian workers from Northern India (Uttar Pradesh and Bihar for the most part), very often in the Valley on a seasonal basis. In addition, these Indian people establish temporary camps on riverbeds during the dry season, from November until May.

Photo 3: Slums in Balaju (north of Kathmandu) (2007)

The squatter areas represent a new form of urban periphery inhabited by marginalized people driven out to the city's traditional limits by their impoverished economic conditions and displaced status, just as the untouchables were (and still are to a large extent) relegated to the outskirts of the city by caste rules (Tanako 1997). These two forms of exclusion cumulate their effects and accentuate social and economic marginalization. Their shanties are merely one-storey dwellings with very limited space (only one or two rooms). The lanes separating the long rows of shelters are very narrow. Corrugated iron sheets and sometimes plastic tarpaulins are held down by stones to cover the roof. The walls are generally made of poor quality brick. Among the very poorest dwellers, the walls are made of bamboo and mud.[6] Nearly 10 per cent of the shelters house grocery shops or teashops. Most dwelling-places have electricity, sometimes even television. However, a third of them only have private latrines. Water facilities are extremely shoddy: water comes from public taps or hand pumps connected to tube wells and dug wells. There is no solid waste management. Besides, these riverbanks are prone to natural disasters such as seasonal landslides and flooding. Squatters pile up sandbags to protect themselves from the rising level of the river. This is one of the reasons why inland camps are much sought after. The people settled there benefit from better living conditions and their houses are of a more solid structure than in riparian settlements. Yet in both cases, from a legal point of view, the future of the inhabitants' tenure is under constant threat.

These squatter communities often include permanent residents, with second and even third generations sharing the same shelter. Some shanties are partly or totally rented to newcomers or fresh immigrants deprived of all belongings.[7] Interestingly enough, a common process of creating a new *bastî* is to move one part of a settlement (in particular those living in a more precarious way, newly arrived relatives and tenants) and establish a new *bastu* some distance from the first one. A camp, in other terms, foretells the possibility of new beginnings in other parts of the town. More often than not, this move is made in a coordinated manner by several families. Such an internal development process contributes to the mushrooming of encroachments.

Except for inland squatter settlements, which are often scattered among nearby permanent dwellings possessing land titles, these camps are unequivocally place-bound. Although there is no common place of worship or common religious processions during festivals (as is the case elsewhere in Nepalese local communities), riparian *sukumbâsî* camps tend to generate a new sense of belonging among its habitants. Setting up a school specifically in the *bastî* and having a committee to represent its migrant and refugee population before the urban municipality help to build an identity based on local ties. The site itself has its own collective memory based on the recollections of the first settlers. In spite of their multicultural and multiethnic character, these spaces thus pave the way for new common ground and new forms of commonality. In many ways, these *outplaces* have been reterritorialized.

The squatter population comprises a large number of unemployed persons: 41.9 per cent (*New Beginnings* 2005, Lumanti).[8] Those who have a job often work as servants or have their own small business. Some women earn money in spinning wool. Significantly, 50 per cent of this Nepalese squatter population have no citizenship card and 60 per cent no electoral card either. This situation reflects low citizenship consciousness and profound marginalization. Similarly, the small size of households (approximately 5 members) reveals fragmentation of extended families into small nuclear units.

These settlements contribute—albeit partly—to the general degradation of rivers, full of plastic, excrement, sewage and refuse.[9] One of the consequences is the sharp decline in the use of the river for ritual and daily activities. It must be remembered that in the Hindu religion rivers are deified and considered sacred. They are viewed as a means of purification and liberation. They have the power to give progeny or to cure disease, and they are the focus of important pilgrimage practices. "River banks are by tradition a particularly appropriate place to practice alms giving, make ancestor offerings, and perform Vedic sacrifices" (Feldhaus 1995: 72). As far as Kathmandu is concerned, the Vishnumati was traditionally the sacred river for the original inhabitants of the city, the Newars. Depending on the exact place where they lived, the location of their ward, and of their caste, people used to burn their dead at riversides and perform a number of rituals on these spots, including bathing (Toffin 2007). These days, dead bodies are generally burnt elsewhere, in particular in a riparian place called Teku, while funeral parties bathe at nearby taps and merely sprinkle a little river water on their heads, instead of bathing. Ashes are still sprinkled over the river but the thin trickle of water running through mounds of refuse and building debris no longer seems to be powerful enough to send the souls to heaven. In addition, few ailing people are brought to the river, to breathe their last breath on a carved stone, their feet dangling in the flowing water, while the last rites are performed.

The politics of slum and squatter settlements: *sukumbâsî, svabâsî, hukumbâsî*

Squatters are the object of fear, anxiety, suspicion and misconceptions among most of the population. The word *sukumbâsî* itself carries negative connotations. The Nepalese belonging to the urban middle-class in particular view them as dangerous social outsiders, even invaders, and river polluters. Their shanty-encroachments on rivers are seen as obstacles to restoring the original riparian landscape and ecology. It is also said that these illegal settlers are puppets in the hands of Maoists, communists and other left-wing parties. They supposedly represent a 'clientele' ready to be summoned at any time to participate in demonstrations and rallies organised by these activists. It is thus believed that *sukumbâsî*s formed the backbone of the huge demonstrations that succeeded in overthrowing King Gyanendra's direct rule and in abolishing the monarchy in April 2006. In addition, the idea prevails that these people are fake indigents and are helped unjustifiably by local bodies and foreign agencies. A word has been coined, *hukumbâsî (*probably derived from the term *hukum*, meaning 'order'), to designate this category of person. A *hukumbâsî* is someone who pretends to be a *sukumbâsî* in order to obtain a land title, *lâl purjâ*, as well as other advantages granted to the underprivileged (Yamamoto 2007: 141). The term is used outside the Kathmandu Valley as well as in other squatter settlement contexts. This discourse renders illegitimate the squatters' claim to be relocated and arouses strong suspicion about them.

The *hukumbâsî* issue has even poisoned relations between squatters themselves. To take just one example, the newly established *sukumbâsî* mentioned above (Thapathali and Balkhu) are seen by older squatters as an operation launched by fake *sukumbâsî*. In 2008, I met the leaders of several internal squatter organizations who openly consider the people living there as *hukumbâsî*. They stressed that most of the people living in squatter settlements are not 'genuine *sukumbâsî*', *bastabik sukumbâsî*, but in fact possess some family land documents in their home district. Some settlers, it is said, are actually tenants who sublet the shelter (or a part of it) where they live. As we saw above, this is indeed often the case. Such internal conflict provokes distrust among squatters and a lack of solidarity.

For their part, defenders of housing rights and the landless poor, such as the Lumanti[10] Support Group for Shelter NGO set up in 1993 to fight against urban poverty and marginalized housing, and other non-government associations, play down these issues and lay emphasis on the poverty and the marginalization of most squatters. These agencies point out that river restoration projects threaten the security of thousand of landless migrants settled in riparian zones. They assert that a solution to their problems is needed before any action can be taken on the urban riverscape. Members of the Lumanti association also underline the fact that *sukambâsî* play a relatively minor role in the river degradation process. They quite justly maintain that sand extraction from the riverbed, used to make cement for construction projects, has much more damaging effects on river morphology and the riverbanks than any squatter intervention. On the whole, housing rights activists are fighting for better sanitary conditions and schooling, but most of all against eviction. They propose to inscribe housing rights in the future constitution, in the same manner as a citizen's other fundamental rights are recognized. Their credo is that political parties are not seriously interested in solving *sukumbâsî* problems. That is why, so they argue, urgent action is needed by civil society organisations. However, they do recognize the difficulty of their undertaking and are embarrassed when accused of indirectly encouraging more squatting in the Kathmandu Valley as in other regions of Nepal.

Environmental and squatter issues are therefore a bone of contention and political issues. Anne Rademacher (2005: 128-133) has rightly demonstrated in her work how these urban fringes are embedded in three competing "narratives": the one of the state and the squatter, plus what she calls the "cultural heritage narrative". Whatever the case may be, urban fringes are the object of vehement political debates. Two local squatter associations founded in the year 2000 play an important role in mobilizing people and fighting for better living conditions: *Nepâl Basobas Bastî Samrakchan Samâj* ('Squatters' Federation') and *Nepâl Mahilâ Ekatâ Samâj* ('The Nepal Women's Unity Society'). Both try to provide adequate schooling for children and to facilitate a micro-credit programme in the squatter settlements to promote local initiatives and entrepreneurism. These associations were formed mainly to prevent any possible evictions and to develop mutual cooperation. When asked specifically about the aim of her women's group, Bimala Lama, the president of Mahilâ Samâj, explained: "The work of women at home and in domestic affairs is not recognized by men. Women need their own association to fight against the administration". Both were founded in the year 2000 and cover the entire Nepalese territory. Besides, a number of *sukumbâsî*s in the Kathmandu Valley belong to various left-wing political parties and to the NEFIN Federation of 'autochthonous people', *âdivâsî/ janajâti,* which encompasses all the ethnic groups of Nepal, including in the Tarai plains.

NGOs and local bodies (Nagar Pâlikâ) in charge of these much discussed and politicized areas make a distinction between squatter settlements and slums (*A Situation Analysis* 2001: 12-13). The people ascribed to the first category of settlement live on marginal government-owned land and, for the most part, come from outside the Kathmandu Valley. They do not possess any property title (*lâl purjâ*) for their shelter. On the other hands, those living in slums have been in the Valley for a long period of time and are sometimes even considered the original dwellers in the region. They mostly belong to low Newar castes, such as the Dyola fishermen or Shahi butchers. These castes, among the lowest of Newar society, used to dwell in rudimentary houses on the outskirts of historic cities. Slum dwellers are not *sukumbâsî* per se: some have land documents, others do not. Their houses are small, dilapidated, and have poor sanitary conditions. Whole sectors of the overpopulated centre of old Kathmandu City belong or could belong to this category of housing.[11]

To differentiate themselves from squatters, a new category of persons emerged in the early 2000s: the *svabâsî*, which can be translated as 'self-settlement dwellers'. As opposed to *sukumbâsî*, which refers to families that have moved from one place to another and are economic or political refugees, with no local roots, the *svabâsî*s are poor people, mainly belonging to low castes and having lived in the Kathmandu Valley for a very long time. They are assimilated more or less to 'autochthonous people', that is *âdivâsî*s. Dinesh Shahi, the president of Jhigu Manka Samâj organization set up in 2000, explained to me that his aim is to provide property titles to all 'self-settlement dwellers' who, for one reason or another, have no *lâl purjâ*. "We don't consider economic criteria, he asserted, we help all Newars, Tamangs or Parbatiyas".[12] By the same token, Dinesh expressed his enmity towards *sukumbâsî*, who are outsiders and do not show any *ijjat* (honour) by illegally occupying land belonging to the government. A conflict seems to be emerging here between the old code of honour, very much attached to hierarchy, and new values enhancing the concept of dignity, which is based on egalitarian premises.

The Vishnumati Link Road and the Kirtipur relocation project

In 1980, a government programme was instigated to build a 2.8-km long road following the course of the River Vishnumati through the heart of the capital, to link Kalimati in the south to Sorakhkutte in the north, thus joining two sections of the Ring Road. This project aimed at improving the traffic flow in Kathmandu City. In 1992, the Norwegian Institute of Technology conducted a study at the request of His Majesty's Government. In order to build this new road (Vishnumati Link Road), 142 houses and shelters built illegally along the right bank of the river were to be demolished. They were scattered over five wards (*tol*): Dhukhal, Chagal, Kushibahil, Tankeshwar and Dhaukel, and were inhabited mostly by Newar low castes, butchers, Pode fishermen, and Hala Hulu, formerly classified as impure castes according to the old Hindu national legal code, involved as they were in occupation considered low and ritually defiling. This population originally came from the other side of the River Vishnumati, where these low-status groups traditionally settled, at the boundary with the historic core of Kathmandu City (due to successive divisions of fathers' houses by the sons). A small number of Tamangs also lived in this area. The oldest squat dates back to 1952 (Tankeshwar), and the most recent (Chagal) to 2000. None of these squatters are formal title holders (*lâl purjâ*).

The families concerned did not oppose the construction of the road. It was clear that the banks of the river were a filthy place to live and that the sewage-filled Vishnumati needed to be rehabilitated. The squatters merely demanded compensation for the loss of their homes and in many cases, their work space. The Lumanti NGO took action among the squatters to defend their rights and mediate with the government and municipalities. *Lumanti* members at that time mostly came from the Newar community of the Kathmandu Valley and chiefly operated in the region. At the time of the Vishnumati Project, this non-government association helped people to submit applications for compensation and organised meetings with the municipality's ward offices. Squatter associations also played an important role in mobilizing people against eviction.

In January 2002, the government published notices warning residents about the move. *Lumanti* worked with residents, the Mayor of Kathmandu, Keshab Sthapit, and various government departments to try to delay the eviction and secure an agreement. A Memorandum of Understanding was finally signed. According to this agreement, residents who were identified as *sukumbâsî*, 'genuine squatters', i.e. residents possessing no land elsewhere, and who could not afford housing on their own income, would be paid 2,000 Nepalese rupees a month for three months' rent once they had moved. After three months, they would be provided with alternative housing. After extensive discussions among community members, a list of about 50 households, the most vulnerable among the people concerned, was drawn up. In April 2002, bulldozers moved into the area and demolished any structure still left standing.

People resettled wherever they could find shelter, mostly in the nearby area. The money for the three months' rent was paid. However, in June 2002, the national government dissolved all elected local government bodies. The Mayor was forced to stand down. For more than a year, nothing happened. The situation became difficult for those who had been displaced. Rent payments had ceased, with no alternatives being offered. Lumanti decided to seek help for buying land and building houses, which could then be sold to the families on a low-cost credit system. *Action Aid*, a United Kingdom development charity organization working in Nepal since 1982, expressed its interest and started to explore options for funding. In September 2003, Sthapit was reinstated as Mayor of Kathmandu and new negotiations took place. Finally, Lumanti and the Kathmandu Metropolitan City succeeded in establishing an "Urban Community Support Fund" with several national and international development agencies. The objective was to buy land and to provide low-interest loans to the families concerned. The municipality, which at first was reluctant to launch a relocation programme, played a very positive role in this affair. In 2003, six *ropanîs* of farm land (3000 sq m) were purchased for 30 *lâkhs* rupees by the Fund at Paliphal, beneath the hill-top settlement of Kirtipur, about 45 minutes by bus from Kathmandu City. In collaboration with the displaced community, a total of 44 low-cost two-storey houses (of a total surface area of about 70 sq m for each house) made of bricks and concrete with corrugated iron for the roof, were built on the site, with an adjacent open space and water facilities.

The Kirtipur Housing Project was inaugurated on 24 December 2005 in the presence of the Mayor of Kathmandu. The houses were immediately occupied. The respective families will have to pay about 350,000 or 320,000 rupees (depending on where their house is located, at the front or back of the settlement) over the next 15 years to the Fund to obtain full ownership of their houses. They actually pay between 500-100 Nepalese rupees per month. This rehabilitation and resettlement project, based on a partnership between the urban poor and local government, is the first of its kind in Kathmandu and probably in the whole of Nepal. It has proved to be a complete success and can be taken as an example for the future. Unfortunately, the price of land is so high at present, especially in the Kathmandu Valley, that it is difficult to launch a new project of this kind. Incidentally, the road on the bank of the Vishnumati was opened in 2009.

I made several visits to Paliphal between 2007 and 2008 and conducted interviews with people who had undergone ressetlement. All together, 43 houses are currently occupied, with one used as a commonal building for meetings. Noticeably, a kind of community solidarity has been forged through the squatters' common struggle against the state to establish their right to live on public land or to be relocated. In spite of some internal dissensions which occurs at some stage, such committed cohesion has supplanted older forms of belonging (caste and kinship networks) that prevailed in their father or forefathers' days. The majority of the population is made up of Newar butchers and Parbatiya Vishvokarma blacksmiths, two low castes. Besides, the dwellers of the settlement comprise four Newar Dyala fishermen, two Tamang families, two Newar Napit barbers, two Dalits from the Tarai (Pariyar), two Newar Maharjan farmers, one Rai, one Parbatiya Brahman, etc. Three houses have been sold to poor Newar farming families from Kirtipur, to maintain good relations with the Newar community of the neighbouring city. The same credit advantages have been granted to these families. Contrary to what has sometimes been said and casually asserted, all resettled families still live in their Paliphal houses. Nobody has sublet their house to tenants. Most male inhabitants have found jobs outside Paliphal, mainly in Kathmandu, or have set up a small business. The people I interviewed all expressed their satisfaction at having been resettled in such good conditions and are perfectly aware of how lucky they are. Everybody has private toilets, though water has to be fetched from a common tap operated by a motor pomp. Relations between families seem excellent and the local committee has established a series of rules regarding the consumption of alcohol, quarrels, and the noise level which seems to be respected by most inhabitants. House-dwellers feel totally at home. There is even a sense of pride in having been relocated there. The beneficiaries of the programme will receive their house property titles when the rent has been paid in full.

Conclusion

The frenzied urbanisation of the Kathmandu Valley has thus created zones of uncertainty, poverty, and unemployment that are the subject of delicate political issues. For instance, in these shanty-enclaves there is a growing concentration of people that can be easily mobilized by populist and skilled politicians to rally their causes. They also provide a convenient source of 'vote banks' for political parties which encourage settlers to enrol on the electoral lists. That is why, so it is said, they are not evicted. For a large part of the urbanite population, *sukumbâsî* squatters have become a figure of otherness, localised on riparian urban margins, an image of a 'social other' who does not share the same values as other urbanites and is a threat to urban sites, cultural integrity and ecology. They are looked upon as an undesirable population. In other words, a study of these urban fringes sheds light on the broader political context and reveals a nascent class conflict between, on the one hand, the poor, and, on the other hand, a middle-class that has taken advantage of the economic changes and which does not recognize the rights of these illegal settlers. In many ways, it is a valuable key to understanding contemporary urban entities.

Furthermore, the building boom over the last decades has considerably degraded the environment and has produced a highly dangerous situation given the seismology of the region. The probability of a major earthquake occurring in the near future is unfortunately very high, with an expected loss of thousands of lives. Obviously, the failure of the state and municipalities to manage these problems is related to the political crisis that Nepal has been undergoing for two decades. State-development bodies are notoriously incapable of enforcing any regulations, despite the plethora of proposals circulated in reports, conferences and housing policy statements, and they are well-known for their ineffectiveness. The Kathmandu Municipality is even incapable of banning existing practices of sand extraction from rivers which is causing serious damage. Interestingly enough, any attempt at addressing these ongoing problems is passed on to NGOs, which in turn tend to despise politicians for their inefficiency. Such overall degradation of the ecological environment and of the cityscape seriously challenges current methods to develop the former Himalayan Kingdom.

In addition, it must be borne in mind that slums and squatter settlements are not a phenomenon restricted to the Kathmandu valley. It concerns many other regions of Nepal situated in the lowlands. In 1998, the National Planning Commission estimated that seven per cent of city dwellers throughout the country lived in squatter settlements (Pradhan and Perera 2005). This figure is still on the rise. The eviction of illegal residents had already taken place before 2008 in places such as Nepalgung and Dangadhi. Since the end of the civil war and the promulgation of a democratic federal republic (May 2008), the marginal squatter population has continued to grow. As a matter of fact, peace has not yet totally been restored in several parts of Nepal, and a large number of hill people who settled in the Tarai plains long ago have migrated to the Valley for safety reasons. For an increasingly large number of Nepalese, the Kathmandu Valley is not seen only as a source of employment, but also as a refuge from outside threats and the uncertainties of the current political regime. Today, the fate of many Nepalese people seems to be either migration abroad, or the option of settling in the Valley which, for most, is still regarded as a desirable place to live.

REFERENCES

Adhikari, J. 2007. 'Urbanization and Change in Pokhara, Nepal', in Ishii, H., D. N. Gellner and K. Nawa (eds.), *Nepalis Inside and Outside Nepal*, Delhi, Manohar, pp. 17-55.

Agier, M. 2008. 'Quel temps aujourd'hui en ces lieux incertains ?', *L'Homme*, 185-186: 105-120.

Dupont, V. 2007. 'La place des slums dans les villes indiennes', *Revue Urbanisme* (Dossier : Villes indiennes), 355, July-August 2007: 1-11.

Feldhaus, A. 1995. *Water and Womanhood. Religious meanings of Rivers in Maharashtra*, New York, Oxford University Press.

Hada, J. 2001. *Housing and Squatter Settlements. In City Diagnostic Report for City Development Strategy*, Kathmandu Metropolitan City and the Word Bank.

The Himalayan, daily newspaper, Kathmandu.

Kathmandu Valley. Environment Outlook. 2007. Kathmandu, International Centre for Integrated Mountain Development (ICIMOD).

Liechty, N. 2008 (2003). *Suitably Modern: Making Middle Class Culture in a New Consumer Society*, Kathmandu, Martin Chaudhuri.

New Beginnings: Housing the Urban Poor. A case study of Kirtipur Housing project. 2005. Kathmandu, Lumanti, Support Group for Shelter.

Pradhan, P. and R. Perera. 2005. *Urban Growth and its impact on the Livelihoods of Kathmandu Valley, Nepal*. Urban Management Programme, Asia Occasional Paper no. 63, Pathumthani (Thailand), Asian Institute of Technology.

The Physical Development Plan for the Kathmandu Valley. 1969. Kathmandu, HMG, Department of Housing and Physical Planning, Ministry of Public Works.

Rademacher, M. 2005. 'Culturing Urban Ecology: Development, Statemaking, and River Restoration in Kathmandu', Ph.D. thesis, Yale University.

Shrestha, B. J. and S. Shrestha. 2008. 'Proposed Outer Ring Road for the Kathmandu Valley', *Spaces* (Kathmandu), 4, 6: 52-58.

A Situation Analysis of Urban Poor Communities in Kathmandu and Lalitpur. 2001. Lumanti Support Group for Shelter.

Tanako, M. 1997. *Conditions of Low Income Squatters in Kathmandu*, Kathmandu, Lumanti.

Toffin, G. 2007. *Newar Society. City, Village and Periphery*, Kathmandu, Himal Books/ Social Science Baha.

Yamamoto, Y. 2007. 'The Sukumbâsî transformation from Communitas to Community: the birth and death of 'Proto-Charisma' among squatters in Pokhara, Nepal', in Ishii, H., D. N. Gellner and K. Nawa (eds.), *Nepalis Inside and Outside Nepal*, Delhi, Manohar, pp. 115-158.

Notes:

1. I am grateful to Sama Vajra, a leading member of Lumanti NGO (Tahachal, Kathmandu), who acquainted me with these slums and their politics in contemporary Nepal. Sama introduced me to some leaders of the local federations concerned, with whom Lumanti is currently working. I am particularly indebted to Bimla Lama, Krishna Pariyar, Nani Hera, Dinesh Shahi, and Arya Ram Kumari, for their help and the interviews they granted me. I would also like to express my thanks to Anuj Rimal, who accompanied me to these settlements during his spare time, to Amita Baviskar for her comments and bibliographical advise on my draft paper presented at the Villejuif workshop (December 2007), to Tristan Ruslé for his remarks on an earlier version of this chapter, and to Rajendra Pradhan.
2. For the Pokhara Valley, see Adhikari (2007).
3. 1 *ânâ*: 1/16 *ropanî* (= 31,79 sq m). One *ropanî* = 508 sq m.
4. 1 *lâkh*: 100,000 rupees. One *crore*: 100 *lâkh*.
5. Slums are sometimes referred to as *picharâ bastî*, 'backward settlement', meaning that they are backward in terms of housing, water facilities, schooling and so on.
6. The shacks belonging to Indian migrants are much more rudimentary than Nepalese ones.
7. It is difficult to make an estimate of the number of fully or partly rented shelters. It varies from one settlement to another. Yet in most cases, the percentage does not seem to exceed 25 per cent.
8. This figure needs to be viewed with caution, as most squatters work in the informal sector of the economy.
9. Noticeably, rich and large private houses have also been built on public land close to a temple and sometimes on an exposed riverbed. These wealthy riverbank encroachments are yet again evidence of the blatant failure of the 'democratic' days between 1990 and 2001.
10. *Lumanti* means 'memory' in Newari.
11. The distinction between slums and squatter settlements also exists in India (Dupont 2007).
12. The office issuing such documents is the Malpot Karyalaya.

2.2.3.

From a Green Happy Homeland to a Cramped 'Island': Discussing the Territories of a Nomadic Group Inhabiting the South-Eastern Edge of Ladakh

Pascale Dollfus

In Ladakh (Fig. 1),[1] territory occupies a central place in building an identity and in how one refers to groups. Ethnonyms are not based on the names of clans or ancestors, as is often the case elsewhere in the Himalayas; they are built on place names to which one adds the nominal particle—*pa* which means belonging to and may be translated by 'Those from'.[2] Men name the territory, but the latter provides them with an identity, expressing mutual dependence, which ties men to the land. To define oneself as when referring to the Other, one's geographical location comes first, with reference to 'one's native land', *pha yul* (fatherland) or *skyes yul* (birthplace) taking precedent over one's place of residence whenever these two differ. Interesting enough, this is even the case for peripatetic communities such as itinerant musicians and entertainers,[3]or religious mendicants. According to the context, the place where one is speaking and the knowledge available to the interlocutor, the place thus named, referred to as *yul,* will indicate a more or less vast entity: nation, province, region, valley, hamlet and locality. Thus, a person will introduce themselves as being from quarter A, from hamlet B, from village C, from valley D, from region E, and even from nation F—in this case India—if in addition they are abroad at the time. (Contrary to other populations on the Indian sub-continent, a large number of whom leave their country to work abroad, especially in the Persian Gulf countries, Ladakhis do not emigrate and rare are those who have left India.) Take for example, a man born in Kharnak: he would say that he is from Kharnak (Kharnakpa), from Changthang (Changthangpa), from Ladakh (Ladakhpa), and eventually from India (Gyagarpa).[4]

Defining a *yul*

Yul is "a place constructed by men, where Culture is imposed on Nature and religion on barbarity" so A. W. Macdonald says.[5] Similar to the French word 'pays', *yul* can mean a village, a country or a province; a land, a region, or realm. It denotes any territory inhabited by a community and constituting a defined geographic reality. *Yul* implies inhabitants (*yul pa*), so when there are none, this has to be specified through the addition of *stong pa*, an adjective meaning 'empty'.

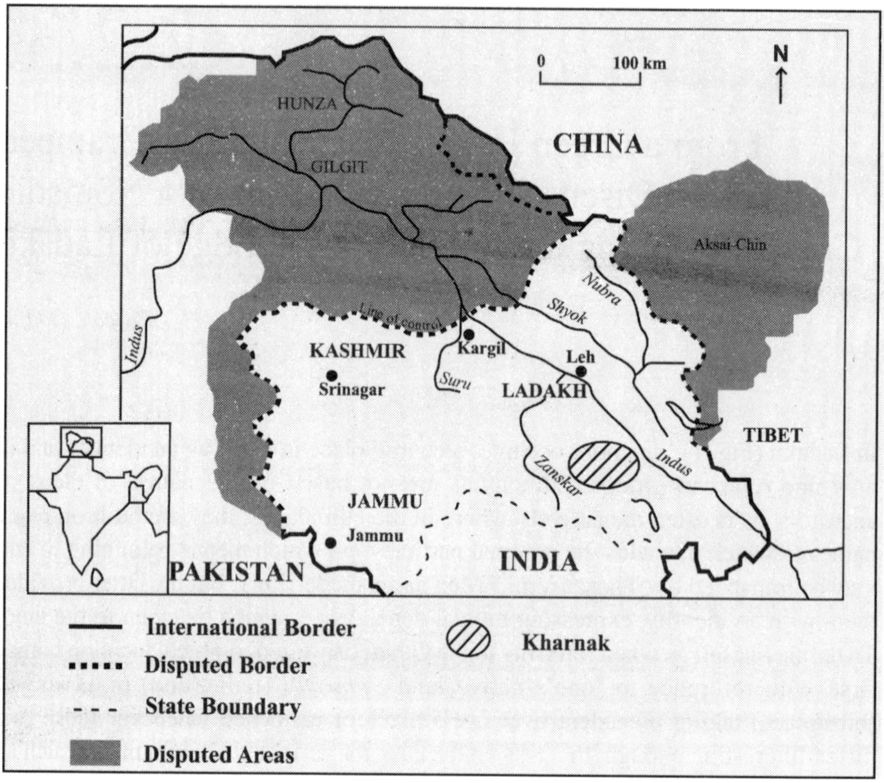

Figure 1. Map of Ladakh region locating Kharnak

As Ravina Aggarwal (2004: 61-62) points out, for its inhabitants, "*yul* is both an imagined community and a social reality, an abstract theory and a contextual reference point for various locales. In its narrowest territorial sense, *yul* indicates a place where one's house is located. In a wider context, it can mean a whole region or nation (for example, *rgya-gar-yul*, the Ladakhi term for India). It can be the land of one's birthplace (*skyes-yul*), the land of one's fathers (*pha-yul*), or simply a place where one dwells." In the same way, *yul pa* denotes a villager, citizen, native, local, and, with a possessive, a person from the same village, region or country, a compatriot; *yul khrims* means local customs, and *yul skad* a local dialect.

For sedentary farmers, *yul* in its primary sense means village. It is an uninterrupted territory (often the watershed of a body of water which keeps it alive), organized along cardinal lines and situated in a dramatic physical landscape in which 'up'(*g.yen*) is opposed to 'down'(*thur*). As Maria Phylactou (1989: 55) remarks, a basic expression of this preoccupation with vertical space emerges in expressions of greetings; encountering anyone on the road, at the bazaar, on the mountainside, leads to a stereotyped exchange, with one person asking the other where they are going, and to the invariable reply: either going up, *g.yen la*, or down, *thur la*.

The territory thus defined is conceived as a closed world. It is clearly delineated, marked by religious buildings constructed by the piety of the inhabitants: chortens (reliquaries), cairns, or low stone prayer walls. These sacred landmarks show the way and draw borders between an inside, which is organized and controllable, and an outside, which is largely untamed and potentially hostile. Therefore, when a villager leaves home, he takes with him prayer flags, previously blessed by a lama. Then at every new pass, he hangs them on branches stuck into heaps of stones, as an offering to the local deities of whom he is not the 'protégé'.

This precisely demarcated space is provided with an ideal—when not geographic—centre, i.e. often a fort around which it is organised (Fig. 2). In the past, peasants' houses were grouped under the protection of the fort within fortification walls. Outside lay fields, meadows and summer residences. Further away on the periphery, for fear of contamination, people suffering from leprosy were placed in quarantine in the *thang*, a barren and stony plain.

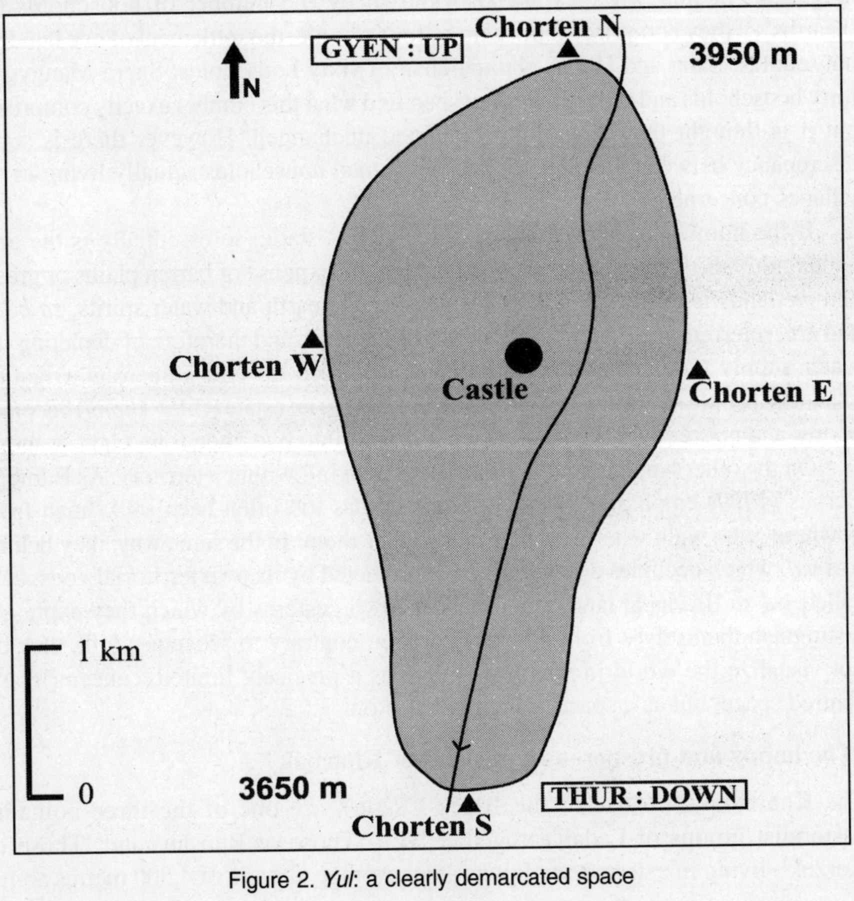

Figure 2. *Yul*: a clearly demarcated space

A *yul* is defined through the set of interconnected soils (*terroirs*) which make it up: *zhing* (field), *spang* (grass- and wetlands), *thang* (barren plain) and *ri* (mountain).[6] Within its limits, only its members, *yul pa,* have the right to use both rare resources such as water and the vegetation, and supposedly endless resources such as sand and stones. Land is continually being used over time and in the space marked out within a property. A village is further defined by its durability, its permanence through generations. It is a fixed group of households according to which all social, political and agricultural activities in the community are organized, and which are united in a web of cross cutting alliances.[7] Traditionally, the older generation moves into a smaller, dependent dwelling with their unmarried children, leaving the main dwelling to their eldest son (or daughter) and his/her family. In this case, the primary unit is known as the *khang chen,* 'big house', while the offshoot dwelling is referred to as the *khang chung*, 'small house'.[8] Settlement in the village as a whole is divided into administrative sections known as *bcu tshogs* ('a group of approx. 10 [houses]'), *srang tshogs,* ('a group of lanes or alleys'),[9] or *'khor 'khor po.*[10] Each includes an average of ten to twelve houses, which work together on matters of village administration, ritual and agricultural tasks such as irrigation. In fact, villages are also known by the number of households (or 'hearths')[11] they comprise. Temisgam is, for example, the settlement of one hundred households, Sabu and Hemis-shukpa-chan of sixty households, Sgera-Mangyu of forty households and so forth. It is not specified what this number exactly comprises, but it is thought that it has long remained unchanged. However there is some discrepancy between this number and the named households actually living in the villages concerned.

If the number of households is perceived as static, so essentially is the area under cultivation. Expanding cropped land at the expense of barren plains or grass- and wetlands is to run the risk of displeasing the earth and water spirits, *sa bdag* and *klu*, referred to as the natural owners of the land, and therefore of depleting the water supply (streams and springs). When divided up into numerous irrigation channels, people say, water resources run out, just as an estate (*zhing khang*) becomes poorer and poorer over the years when it is split into two, then four, eight or more.

On the other hand, nomadic pastoralists also fall within a territory. As Edmond Bernus (1999) emphasises—they are not, as has too often been said, 'men from nowhere' who walk wherever their fancy takes them. In the same way, they belong to a *yul*, which occupies a physical space protected by its own territorial gods, here called *yul sa* (lit. local land)[12] and with its own customs by which they expressly distinguish themselves from others. However, contrary to sedentary folk, they do not visualize the world in which they live as a precisely limited, concentric and centred space, but as a space, which radiates out.

"The happy and prosperous homeland of Kharnak"

The Kharnakpa, 'Those of the Black Castle', are one of the three nomadic pastoralist groups of Ladakh, together with 'Those of Rupshu' and 'Those of Korzok', living in extreme conditions at an average altitude of 4,500 metres above sea level in a remote area on the south-eastern edge of Ladakh. The black castle in question—a few dark-coloured crumbling walls—stands in ruins high on a spur above the eponymous valley where shepherds sometimes graze their flocks and several families used to grow barley on patches of land. The Kharnakpa raise goats, sheep and yaks on natural pastureland, producing for their own subsistence but also for commercial purposes. They do not venture to unknown places, but they move with the herds and their whole family to known and named places, following the same route season after season (Photo 1). The availability of grass takes precedence over water when contemplating a move and choosing a place to camp. In the past, men travelled for weeks to Zangskar and Lahaul with large flocks of sheep and goats loaded with salt and wool, which they exchanged for barley, tobacco and other basic necessities.[13] Nowadays they sell livestock products, such as *pashmina* wool locally called *le na*[14]—the winter undercoat of a variety of domestic goat—and culled animals for meat, mostly to Leh traders (primarily Muslim), and they purchase foodstuffs, such as tea and food grains which are part of their staple diet, as well as many of the supplies and equipment they use (clothes, kitchen equipment, torches and radios …).

Photo 1. Moving camp. Yaks, loaded with luggage, make the move to a knew camp site. Photograph P. Dollfus, June 2007.

The "happy and prosperous homeland of Kharnak",[15] as its inhabitants call it, is located westwards from the Leh-Manali highway. It roughly covers the Zara River watershed to the east and the Kharnak River watershed to the west. It is oriented north-west/south-east and may be crossed in a three-day walk. Here distances are calculated in hours. As Henry Strachey (1853: 14) points out, "it is geographically the N.E. corner of Zangskar". However it is generally referred to as part of Changthang or the 'Northern plain' (Tib. *byang thang*), the huge high-altitude massif that Ladakh shares with Tibet. In reality, the area is less of a plain (or a plateau) and more of an endless stretch of rugged ranges, deep canyons, and ragged bluffs. On the north and the south it is hemmed in by high mountain ranges impassable except by a few high passes. To the west, a narrow gorge that can be traversed only by incessant fording and wading of the Kharnak River in the dry season marks the border with Zangskar. At one point it goes through a natural archway of rocks so low that a horseman has to lower his head to get through, and at another point through a rocky gorge aptly called 'Horse way no saddle way', which is almost narrow enough to scrape the saddle off a horse's back, whilst the bottom is filled with dangerous accumulations of snow or water. On the contrary, to the east, there is no topographical accident or physical obstacle to materialize the boundary; it merges with the territory of their closest neighbours, the Rusphu nomadic pastoralists, via a large stony plain.

The nomadic pastoralists of Kharnak usually break and make camp about six times a year to gain access to resources such as grass and water, making only short moves. Their territory with its indistinct limits, as long as these are not challenged, is thus split up in time and space between several places—main campsites and secondary or satellite camps—occupied as the seasons come and go by the whole group or by only some members of it.[16] In fact, its inhabitants do not perceive it as a marked out entity, enclosed within continuous and well-defined limits. They see it as an open space, built around several locations, but which only have any meaning when linked together by a more or less regular move.

When called upon to draw up a map, the Kharnakpa look confused. They evade the question saying that they do not know how to draw. When asked again, they end up listing place names, the valleys duly marked out between which there are vast spaces with no name. Any areas with no resources are ignored, they are obscure—they do not exist—even when they are in the middle of the territory. People then mention Zara, Spangchen, Jagang, Dad and Samartse where they have their main campsites; Kharnak and Lung where they raise barley and peas; Selung, Sheyen, Sheyung, Kharlog and Lungthonpo where they go to cut grass in early September; the localities Sherol, Yabuk, and Hasamta, where shepherds lead their herds and set up secondary campsites, and many other places too.

The "happy homeland of Kharnak" is thus visualized as a set of itineraries punctuated by landmarks and locations along with the resources they have to offer. It evokes a territory made up of a combination of lines and surfaces, which brings to mind that of the Inuit huntsmen in Canada, as highlighted by Béatrice Collignon (1996), or "the reticular space" of Melanesian societies as described by Joël Bonnemaison (1986-1987) using as a case-study the Vanuatu archipelago.

Within this broad area, each named place refers to a practice, to a story. Tubak Golok, the rocky cliff 'Which Reverses Guns' (*tu bag mgo log*) alludes to the story of a hunter who once shot a mountain goat there. However, he did not succeed in killing it and was hit by the bullet in his own leg. Memele, 'The Grandfather's Pen'(*mes mes lhas*) denotes a square-shaped basin where an old man from the community used to herd and graze sheep and goats. Indeed, in this shepherd country, the valleys' names give the lion's share to vegetation or, on the contrary, highlight the absence of any kind of plant. For example, the place name Zara, locally translated either as 'The Stinging Nettle Enclosure'(*zva ra*) or 'the Food Enclosure'(*za ra*) refers to the general shape of the valley, which looks like a large pen, and to the nettles in abundance there in June; this wild edible plant was the only vegetable available in the past; Samartse, written *rtsva dmar rtse*, 'The Summit with Red Grasses', or *rtsva dmar sa,* 'The Soil With Red Grasses', brings to mind the reddish colour made from autumn onwards by the rushes growing on this wet land. On the other hand, Kamsang lung, 'The Completely Dry Valley' (*skam sang lung*) points out the extreme aridity of this glen lacking any water, river or spring.[17]

By aptly naming places, the space—a neutral stretch—becomes an environment filled with men's past and present, and therefore humanized, inhabited. The territory is thus 'taken over by the word'.[18] The appropriation also adopts other forms, such as the building of religious edifices, the setting up of traps and wolf pits, or else marking out pitches for tents and the like: pens and sheds, large stones around which nomads tie up their yak calves or to which they fix their looms, etc.

Contrary to sedentary farmers, nomadic pastoralists do not conceive their *yul* as a static and permanent entity set within frozen boundaries, but as a living body. At any time, the territory can be displaced, reconstructed, enlarged or reduced due to variations in water and grass resources; to an increase or decrease in the number of men or animals; to an economic and/or political crisis; or to a divine summons, as we will see in the following examples.

The extension eastwards: an answer to the increasing number of men and animals

According to local tradition, at the very beginning, there were only two households in the happy land of Kharnak, both living within the Kharnak River watershed but at different locations. One was known as Chukpo Lobde and settled at Dat where a dried up spring still bears its name. The other was named Togoche and resided at Samartse, a nearby tributary valley, five-hours' walk away. These two families did not like each other and did not co-operate at all. Once upon a time the country was hit by a winter of unheard harshness. For weeks on end the weather was freezing cold and there were abundant snowfalls. All the goats and sheep owned by the Chukpo Lobde family died of starvation. Bereft of animal herds, the head of the household and his son had no work. Eventually they decided to go to see what had happened to the Togoche family. They did not take 'the low road' along the river, but walked along 'the high road' up through the deserted glen of Kugshel. When they arrived at the top of the pass, they looked down on to the valley. Everywhere the snow was strewn with bones. Driven by hunger, wild dogs had greedily eaten all the small livestock. Not a single animal had survived. United in their sorrow, the two families decided to help each other. For days and nights people sang and danced with no respite. With all the dead animals lying everywhere, 'dead meat' was plentiful. On the other hand, it was difficult to find any barley to brew the beer without which no feast was quite complete. Each successive grain was a step in the right direction. Wedding celebrations were promptly organized, and a year later, children were born. They in turn had children and their children had children. One generation after the other, the number of households gradually increased. The herds grew accordingly. In search of more pastures to feed their flocks, the Kharnakpa ventured eastwards. They crossed the Yar pass that marks the limit between the Kharnak River watershed and the Zara River watershed, and they settled on the other side of the pass: first only at its foot, then little by little, in the surrounding areas. Some of the former main encampments were left and replaced by new ones, others became secondary camps. This was the case, for example, of Omalung, 'The Valley of Milk' *(o ma lung)*. The place was left deserted for Lungmoche, 'The Large Valley' *(lung mo che)*, a campsite abandoned—in turn—some years later for that of Jagang, 'Full of Slates' *(g.ya' sgang)*, and of Zara, 'The Stinging Nettle Enclosure' located further east. Nowadays, people only stop overnight there when moving from the summer camps to the winter camps and vice versa.

New land was conquered with no fighting or bloodshed. All these valleys were uninhabited, 'empty'. One simply had to go out and take them. There was no frontier *(sa mtshams)*, no army *(dmag mi)*, and no *gurkha*, as local people called the migrants working for the Public Works Department, whatever country they belonged to. Everyone came and went as they pleased. At the time, nomads from Rupshu tended their herds very far up along the Indus River and spent winter in faraway Kagzhung beyond the present border between India and Tibet [China]. There were no Tibetan refugees. There were only a few *rkyang* (Tibetan wild ass, *Equus hemionus kiang*), thus no forage competition with livestock. Grass was plentiful.

According to Mes Sonam Dorje born in 1925 (Tiger Year, *stag lo*), a respected elder of the community and an expert in local history, Kharnak numbered one hundred households at its peak. In those days, its territory was much bigger and included pastureland near the Tso Kar Lake and stopping places on the way to the Tibetan salt lakes in Ruthog. However, when war with the Chinese (Aksai Chin conflict in 1962) broke out, everything changed. Soldiers came and settled nearby in Shangdong, 'The Wolf Pit' *(sbyang dong)*. With the help of 'outsiders' *(phyi pa)*, they built roads and barracks here and there.

A case of conflict: The Indo-Chinese war of 1961 and its consequences on territorial restructuring

As was the case for many Ladakhis,[19] the Sino-Indian Border conflict, triggered by the construction of a road through Aksai Chin, which China regarded as a strategic link between the Chinese administered territories of Tibet and Sinkiang, had a major impact on the lives of Kharnak shepherds. Many military camps were established in Leh, Ladakh's district headquarters, and at strategic points. Roads were built to supply them with men, munitions, fuel and food. At the time, the Leh-Manali highway that runs along the eastern side of the current Kharnak territory was built. The military population soon grew larger than the civil one. India closed its borders across which many exchanges used to be made. Ladakh nomadic pastoralists were no longer authorized to go to the western Tibetan lakes where they previously collected salt, taking with them hundreds of sheep and goats. They lost the benefit of this lucrative trade along with winter pastures. Since the 1960s, the influx of Tibetan refugees into Ladakh as a consequence of the Chinese occupation of Tibet, then the loss of land due to the Indo-Chinese war in 1961 has affected nomads' migratory patterns. Apart from changes in the route, the pattern in which they moved has also altered slightly. The increase in population and herds, followed by a decrease in pastureland, has put a strain on grass and water resources, and so relations between the different groups of pastoralists having to share them have become tense, often leading to blows, and territorial redistribution. According to Monisha Ahmed (1996: 85, note 42), "When the Tibetans first came, problems arose over the route of migration they followed. It appears that the Tibetans would go wherever they pleased and not follow the route laid out by the Rupshupa. [...] Finally Rupshu took the Tibetans to court, and in an agreement drawn up between the two parties on 5 October 1985 it was stated that: 'In future the movement of livestock from one pasture to another and timing for grazing will be decided by the local Nambardar [chief], which will be conveyed to the Tibetan Refugees through their Nambardar. Whosoever, among the Tibetan Refugees violates the above mentioned regulations after thorough enquiry if he is found guilty of such violation shall be ousted from that place."

Moreover, with the closing of the borders in the early 1960s, an alternative source for supplying salt had to be found. It is in this context that the Tso Kar Lake, where up till then the salt was only thought to be fit for animals, became a much sought after salt mine and a bone of contention.[20] In fact, the issue was not simply the salt, but much more, i.e. the land and grass. The vicinity of Tso Kar yields some of the best grazing land, and it was this along with the salt that Rupshu was trying to protect from a takeover by Kharnak.[21] The former offered as proof the fact that their traditional camping grounds were located in the vicinity of the lake as well as their monastery, Thugje. Besides, old land records in the archival and revenue offices in Leh, dating back to 1908, demonstrate Rupshu's ownership of Tso Kar. On the other hand, the latter, with the backing of Hemis Gompa, the largest and one of the wealthiest monasteries in Ladakh, argued that they previously owned land by the lake, but lost it two or three generations ago, when a violent row broke out between the two groups. They went to battle against each other with knives and slingshots. One man was killed, and others were badly injured. The police were called in, and Rupshu took the case before the court. New boundaries were fixed. Kharnak lost three glens at the northern end of the lake (Norbu Khar,[22] Tamalung and Rang), and in compensation obtained Spangchen, 'The Big Meadow' *(spang chen)*, a large and green valley irrigated by a Zara tributary and located on the western side of the Leh-Manali highway. They retained the right to access the lake, but only for salt extraction. This arbitration was quickly followed by other rows. Those of Rupshu reproached those of Kharnak for openly abusing them and for having come to the lake riding horses and leading hundreds of goats and sheep introduced as pack animals. Slingshots started flying again. A few men were badly hurt; many animals were killed or stolen. Once more, the police were called in and the case brought before the court. An agreement was drawn up for a period of five years, which stated that though only the Rupshupa had access to the lake, they had to give a portion of the salt to Kharnak in exchange for a small amount of money. When the agreement expired, the Kharnakpa reiterated their claim, and savage fighting broke out again.

As recounted by Monisha Ahmed (1996: 306-308), who narrates the story in detail, according to Rupshupa's point of view, "Tsering Samphel, a Member of the Congress Party at that time, who was sent out to Rusphu to survey the situation stated: 'There is little doubt that Tso Kar belongs to Rupshu. It is just that these Kharnak people are more influential and financially stronger because they have the support of Hemis that they are trying to take it away from them. In the courts it was agreed after the fight that Kharnak would pay a fine of Rs. 3,000 to each family in Rusphu, but when the agreement was written up it said that Kharnak would pay Rs. 3,000 to the whole of Rupshu!'

When the courts proved futile, Rusphu approached the LBA [Ladakh Buddhist Association] for an out-of-court settlement and a compromise was reached between the two places. The terms of the agreement state that while only Rupshu has access to the lake, they must give a portion of the salt to Kharnak. This was fixed at eight hundred *lug-sgals*[23] in exchange for which Kharnak would pay Rupshu twenty-five paise for each bag. Thus, as soon as the salt reaches Rupshu, a message is sent to Kharnak to come and collect their share of salt. Each tent in Rupshu contributes eight or nine *lug-sgals* towards Kharnak's share. Some men in Rupshu deride this agreement saying it was made only because their chief at that time had been weak: 'Look, we do so much hard work removing this salt and Kharnak gets it for free!'"

What has now become a major concern for Rupshu is to prevent trespassers from encroaching on Tso Kar. To deter intruders and 'salt thieves', five guards are posted at a time by the lake and its precincts for a period of four to five months, from June to October or November.

Alas! again in June 2000, fighting broke out. Having not received their share of salt the previous autumn, the Kharnakpa decided to go to the lake and help themselves. All the men aged from 15 to 65 years old were mobilized to take part in the expedition, and the group's 113 horses were requisitioned. At 9 p.m., posted in front of their tents, under the cover of darkness, the elders prevented their dogs from barking and from following the convoy. The women went with their men to the campsite entrance, offered prayers and juniper fumigation to gods to ask for protection of their fathers, husbands and sons. Next morning at noon, the expedition came back victorious. The guards had not woken. No slingshot had been fired and no knife had been brandished ...

Of course, the story the Rupshupa told was entirely different. Indeed nobody got injured, but the guards were attacked by armed men in superior numbers. They were bound with ropes and only recovered their freedom the next day when they were relieved from duty. In addition, if the people of Rupshu had not given the salt as promised in autumn, it was only because they had not extracted it in time due to unexpected heavy snow falls.[24]

To sum up, due to pressure from external events, scattering and flow have been replaced by grouping and boundaries. The nomadic pastoralists of Ladakh, who, in the past, followed their herds over a vast area along their own itineraries, occasionally making a chance encounter, now control exclusive and bounded territories, which leave little room for movement. In keeping with the example of "Those of the Black Fort", their cramped territory is now entirely located to the west of the Leh-Manali highway. From now on, Kharnak and Rupshu do not share any grazing land except for the Zara valley, which neither of them has wanted to abandon. Both groups have right of access, but at specific periods in the year. Those of Kharnak stay there in June, at the beginning of summer, camping in the middle of the valley, while those of Rupshu arrive in the second week of November. Usually, they spend one month at Zara Do, at the bottom of the valley *(mdo)*, followed by three weeks at Zara Phu near the high-altitude pasturelands (*phu).*

The weight of gods: "This is my mountain, this is my grass"

The gods also have their word to say about demarcating the territory. Yet, according to a local tradition, the boundaries between Kharnak and neighbouring Zangskar were fixed at the end of a period of rivalry between two territorial gods, Kalabukyong and Sharchog, the first ruling over Kharnak and the second over the east of Zangskar.

The story goes as follows:

Once, Kalabukyong (Tib. Ka la bu skyong) went to the Charcher pass. He stuck a long prayer flag at the top of the pass, illustrating, with this gesture, his hold over all the land below. Hearing about this, Sharchok (Tib. Shar phyogs) became very angry and made up his mind to have it out with Kalabukyong. He thus went to meet Kalabukyong and asked him to take down the prayer flag. He left his horse at the bottom of the Charcher pass. One can see its white shape imprinted in the rock at this location Tachok Lungba 'The Valley of the Excellent Horse' (*rta mchog lung ba*). When he reached the top of the pass, Kalabukyong had already left. Sharchok followed his prints and found him not very far from Rabrang. At this point two small hills face each other. Kalabukyong climbed to the top of one, Sharchok to the top of the other. Then indicating the land around them with a sweep of their arm, in a loud voice each of them claimed to own it: "This is my mountain, this is my grass" (*nga'i ri nga'i rtsa*). After many hours of relentless discussion, nobody was willing to back down, so the gods agreed to fix the border separating their respective properties half-way between these two hills. From that day on, this has been the locality 'My Mountain, My Grass' where Kharnak's *yul* stops and Zangskar's starts.

In the same way, some of the main campsites were not chosen according to grass or water availability, but by a divine summons. This is the case of Sangtha, a place located on the right bank of the Zara River on the way to the Morang pass opening on to the Tsarap valley and its agro-pastoral hamlets.

At the beginning of the nineteenth century, on the orders of Karma Tendar, a lama native of Kharnak, who had had a dream, Sangtha became one of the community's main sites. Stone houses replaced the shelters in rock cliffs roughly fitted out by herders. Chortens and low stone prayer walls were built. Then the holy man died. Later on, for a reason still unknown, Sangtha was left deserted. Only shepherds still went there with their herds. Such was the situation until autumn 1994. A year earlier, in 1993, the community had been heavily hit by misfortune. As early on as October, heavy and abundant snow, making travel perilous both for men and their herds, had blocked passes and covered pastures, causing the death of hundreds of animals. An appalling bout of epizootic decimated the large cattle, which were already weak from several weeks without feed. Finally, eleven healthy young people died suddenly, stricken by some sort of devastating evil.

In an attempt to avert such misfortunes, the inhabitants multiplied apotropaic rituals, but to no avail. Death was still on the prowl. When summoned to a meeting, the heads of the households therefore decided to send a messenger to Duwang Rinpoche (*sgrub dbang rin po che*), a great Tibetan master. This highly respected lama arrived at the end of the spring, as soon as the passes had been cleared. After a period of deep meditation, he delivered his point of view. The cause of all these ills was the goddess Tsheringma, the powerful leader of 'The five long lived sisters' (Tshe ring mched lnga), who is revered as a territorial god by the people of Kharnak. The goddess was thus venting her anger towards her devotees whom she accused of negligence. To put a stop to this adversity and to restore prosperity, the community had to immediately celebrate a great ritual to make up for this and to promise to worship her every month, in winter as well as in summer, at her main shrine, one hour's walk from Sangtha. Since then, when the nomads leave the Zara River watershed and their summer camps for their winter settlements in the Kharnak River watershed, a dozen families settle in Sangtha to honour the goddess.

Kharnakling, "The Island of Kharnak", a new territory?

Kharnak, like the rest of Ladakh generally, has been undergoing a process of continuous change for as long as we can trace back over time. Over the last 15 years, dozens of families have already been lured away by the prospects of city life; having access to proper schools and medical facilities, electricity, warm houses, stores, and entertainment. They have moved from Changthang reducing the mobile community by nearly eighty per cent. Most of them have sold their livestock and have settled in urban areas on the outskirts of Leh, especially in Choglamsar.[25] There, they have established a permanent urban settlement known as Kharnakling, 'The Island of Kharnak'. In June 2007, every family from Kharnak –whether settled or still nomadic– owned a plot of land for building, if not a one- or two-storey house erected in the middle of a high-walled courtyard. Deprived of any animals, except for a few horses kept for tourists, they are engaged in various skilled and unskilled jobs, including that of coolie, work with the army as contractors, and in small businesses. In Kharnakling, the migrants referred to as 'Those who have come down'(*'bab mkhan*),live as though in a waiting room; they have not made this place their own. As they emphasize, Kharnakling is not a *yul*; moreover there is no local territorial god *yul lha* ruling over this place and protecting its inhabitants.[26] It is only a 'colony', that is a settlement abroad established where a group of people from the same place or with the same occupation live together forming a distinct community within a larger city. In fact, nobody says that they are 'from Kharnakling', 'belong to Kharnakling', i.e. 'Kharnaklingpa'. Interesting enough, no prayer wall, chorten, heap of stones, or other votive edifice marks its gates as is traditionally the case in villages and even in campsites. Indeed the settlement is divided into three administrative sections *('khor)* but none of them is named or numbered in the same way as the quarters of Tibetan refugees, which circle them (Fig. 3). Contrary to village sectors *(bcu tshogs, srang tshogs)* previously mentioned, which are bound in their components and territories, these *'khor* are not fixed, either in their number or in their spatial dimensions. Indeed, their layout changes as newcomers settle and the population expands. Here, the primary unit and its offshoot dwellings are not built nearby and do not belong to the same sector, as is traditionally the case for villages. Furthermore, these three administrative sectors shelter a mixed population: people originally from Kharnak, but also from other nomadic places (Rupshu, Korzok) as well as remote villages.

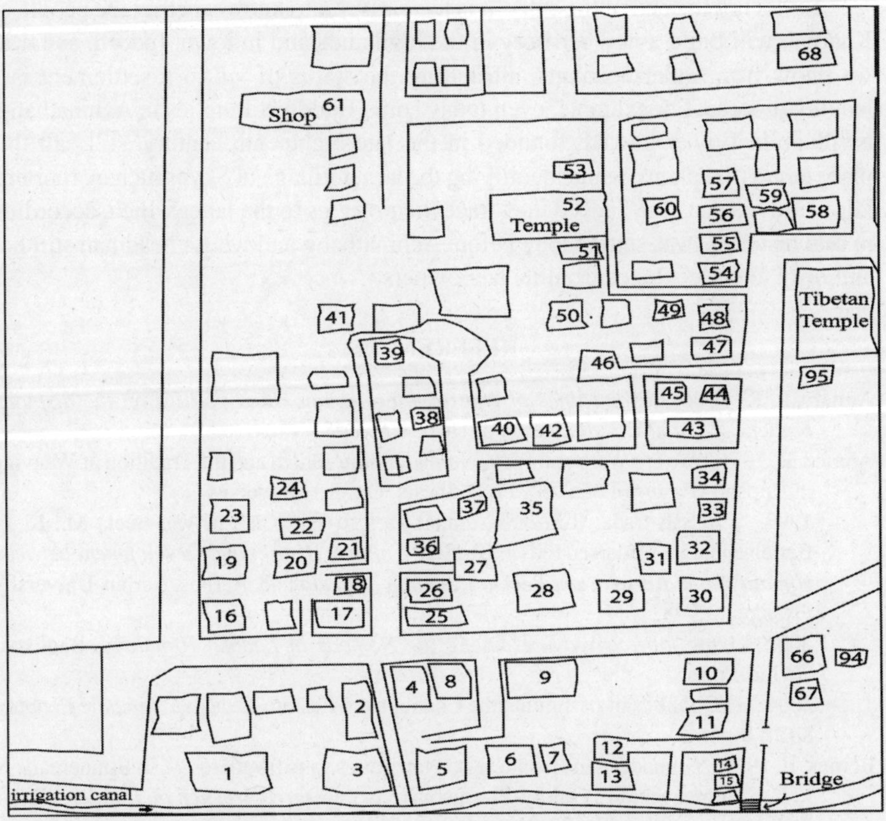

Figure 3. Settlement of Kharnakling. Map drawn by Angchuk from Kharnak.

For the youngest who have grown up here and know about life in Kharnak mainly through the tales recounted to them, the situation is slightly different, Kharnakling is more of a reference framework. They have woven a special intimate relationship with it. However, their discourse on Kharnakling does not take into account physical and topographical features. In fact, this young people's new territory does not stop at the limits of the settlement where they live, but stretches to other places in Ladakh, where they have the opportunity of going to spend some time for a sporting event, entertainment or shopping, to pursue their studies or to work: picnic spots at Shey on the banks of the Indus; the lush green meadow of the Peace garden at Choglamsar, where the birth anniversary celebrations of the 14th Dalai Lama are held; Old Bus Stand in Leh...

The same is true for the young Inuits whose new territory 'in town' also encompasses other villages within the same province. But the latter's territory also includes places situated outside the arctic region, such as Yellowknife, the nearest Canadian town, and even dreamed of places perceived only through the distorted image of television broadcasts: ice rinks for hockey matches or baseball and American football fields, the California of the television series "Baywatch" (Collignon 1996: 202).

This is not the end of the story. It is too early yet to know whether the people of Kharnak will build a new territory in the highlands and in town. Indeed, as far as we know from other accounts, attributing the status of *yul* to a settlement is a lengthy process. For example, even today some Buddhist families in Achinathang, a village in Lower Ladakh founded in the late eighteenth century, still call this place *thang* (the plain), while identifying the larger village of Skyurbuchan, fourteen kilometres east, as *yul*. Indeed they trace their origins to the latter which, according to oral histories, was settled long before Achinathang and whose headman still has authority to collect taxes from its landowners.[27]

REFERENCES

Aggarwal, R. 2004. *Beyond lines of control: Performance and politics on the disputed borders of Ladakh, India*,Durham, Duke University Press.

Ahmed,M. 1996. 'We are Warp and Weft. Nomadic Pastoralism and the Tradition of Weaving in Rupshu (Eastern Ladakh)', Ph.D. thesis, Oxford University.

– 1999. 'The Salt trade: Rupshu's Annual Trek to Tso Kar', in Van Beek, M., K. B. Bertelsen and P. Pedersen (eds.), *Ladakh: Culture, History and Development between Himalaya and Karakoram. Recent Research on Ladakh 8*, Aarhus, Aarhus University Press,pp. 32-48.

– 2002. *Living fabric. Weaving among the Nomads of Ladakh Himalaya*, Bangkok, Orchid Press.

– 2004. 'The politics of pashmina: the Changpas of Eastern Ladakh', *Nomadic Peoples,* 8 (2): 89-106.

Bernus, E. 1999. 'Nomades sans frontières ou territoires sans frontières ?', in Bonnemaison, J., L. Cambrezy and L. Quinty-Bourgeois (eds.), *Le territoire, lien ou frontière,* Paris, L'Harmattan, vol. 1, pp. 33-41.

Bonnemaison, J. 1986-1987. *Territoire, Histoire et Société dans l'archipel de Vanuatu (Mélanésie)*, 2 volumes, Paris, ORSTOM, coll. 'Travaux et Documents', no. 201.

Collignon, B. 1996. *Les Inuits. Ce qu'ils savent du territoire*, Paris, L'Harmattan.

Dollfus, P. 1989 (new edition 2005). *Lieu de neige et de genévriers. Organisation sociale et religieuse des communautés bouddhistes du Ladakh*, Paris, Editions du CNRS.

– 2004. 'Vers une sédentarisation univoque ? Le cas des éleveurs nomades de Kharnak, Ladakh, Himalaya occidental indien', *Nomadic peoples*, 8 (2): 200-213.

Dollfus, P.and V. Labbal. 2009a. 'Ladakhi Landscape Units', in Smadja, J. (ed), *Reading Himalayan Landscapes over Time. Environmental Perception, Knowledge and Practice in Nepal and Ladakh*, Collection Sciences Sociales, no. 14, Pondicherry,Institut Français de Pondichéry, pp.85-106.

– 2009b. 'A Foray into Ladakhi Place Names', in Smadja, J. (ed), *Reading Himalayan Landscapes over Time. Environmental Perception, Knowledge and Practice in Nepal and Ladakh*, Collection Sciences Sociales, no. 14, Pondicherry,Institut Français de Pondichéry, pp. 241-262.

Goodall, S. 2004a.'Rural-to-urban Migration and Urbanization in Leh, Ladakh. A Case Study of Three Nomadic Pastoral Communities', *Mountain Research and Development*, 24(3): 220-7.

– 2004b. 'Changpa nomadic pastoralists: differing responses to change in Ladakh, North-West India', *Nomadic Peoples*,8 (2): 191-9.

Gutschow, K. and P. Dolma. (n.d.).'The Changing Face of Obstetrics in Leh and Zangskar', unpublished.

Henrion-Dourcy, I. 2004. 'Ache Lhamo. Jeux et enjeux d'une tradition théâtrale tibétaine', Ph.D. thesis, Université Libre de Bruxelles and Ecole Pratique des Hautes Etudes.

Karmay, S. G. 1996. 'The Tibetan cult of mountain deities and its political significance', in Blondeau, A.-M. and E. Steinkellner (eds.), *Reflections of the Mountain. Essays on the History and Social Meaning of the Mountain Cult in Tibet and the Himalaya*,Vienna, Verlag der Österreichischen Akademie der Wissenschaften, pp. 59-76.

Macdonald, A. W. (ed.). 1982. *Les royaumes de l'Himalaya, histoire et civilisation : Le Ladakh, le Bhoutan, le Sikkim, le Népal*, Paris, Imprimerie Nationale.

Phylactou, M. 1989. 'Household organisation and marriage in Ladakh, Indian Himalaya', Ph.D. thesis, London School of Economics and Political Science, University of London.

Pirie, F. 2002. 'The fragile web of order: Conflict avoidance and dispute resolution in Ladakh', Ph.D. thesis, Oxford University, Wolfson College.

Riaboff, I. 1996. 'gZhon nu mdung lag, mountain god of Zanskar–A regional scale divinity and its cult's territorial ordering', in Blondeau, A.-M. and E. Steinkellner (eds.), *Reflections of the Mountain. Essays on the History and Social Meaning of the Mountain Cult in Tibet and the Himalaya*, Vienna, Verlag der Österreichischen Akademie der Wissenschaften, pp. 23-38.

Rizvi, J. 1999. *Trans-himalayan Caravans. Merchant Princes and Peasant Traders in Ladakh*, New Delhi, Oxford University Press.

Strachey, H. 1853. 'Physical geography of Western Tibet', *Journal of the Royal Geographical Society of London*, 23: 1-69.

Smadja, J. 2009. 'A Reading of the Salme Tamangs' Territory and Landscape', in Smadja, J. (ed.), *Reading Himalayan Landscapes over Time. Environmental Perception, Knowledge and Practice in Nepal and Ladakh*, Collection Sciences Sociales, no. 14, Pondicherry, Institut Français de Pondichéry, pp. 199-239 and 563-589 + map.

Notes

1. Bordering Tibet (China) on its northern and eastern edges, Ladakh is part of the eastern portion of India's most northerly state, Jammu-Kashmir, often referred to simply as Kashmir.
2. The nominal particle *-pa* is used to express membership to a country, religion or profession. It can be added to almost any number, place name or noun to make a single person or a group of people associated with whatever it is.
3. As for Tibet, Isabelle Henrion-Dourcy (2004: 216) notes: "The names of theatre troupes are invariably formed around the name of the actors' native village. [...] Even the name of itinerant troupes such as those found prior to 1959, was taken from a place name (founding member, main lama for the actors." (my translation)
4. Gyagar (Tib. *rgya dkar*), literally 'The White Extent', denotes India, especially all parts other than Ladakh.
5. My translation from A. W. Macdonald (1982: 44).
6. On these landscape units *(zhing, spang, thang* and *ri)*, see Dollfus and Labbal (2009a: 85 *et seq.*)
7. On the construction of village community, see Pirie (2002: 150).
8. *khang chung* in Zangskar; *khang bu, khang gu* in the Indus Valley.
9. One also finds *srang phyogs*, from *phyogs* (pronounced 'chhogs') meaning 'party, side'.
10. *khor* comes from the verb *'khor cas*, to roam, tour and is indicative of the rotating rounds of feasts and tasks between neighbours, see Aggarwal (2004: 79).
11. *phu mkhan*, lit. 'someone who blows (the fire)'.
12. *yul sa* has the same meaning as *yul lha* in Zangskar and in several Tibetan enclaves of northern Nepal (see Riaboff 1996: 26). As Samten Karmay (1996: 66) points out: "The expression [*yul lha*] is often translated by 'the god of the country' or 'dieu du pays'. However, the term *yul* in *yul lha* has, in my opinion, the connotation of 'local territory' in the sense of a defined locality and not simply 'country' as it can often mean"; and further on: "The process in which the term *yul sa*, 'local land', has come to mean 'deity of the local territory' as that of *yul lha*, further proves the fact that the concept of the *yul lha* type deity was originally connected with the territorial divisions of the polity of the early clanic society."(*id.*: 67)
13. On this trade, see Rizvi (1999: 69 *et seq.*).
14. Each winter, this fine, soft, and warm fibre grows beneath the coarse outer coat of Tibetan goats to enable them to survive temperatures dropping as low as minus 40°C. On *pashmina* trade, see Ahmed 2004.
15. *yul la g.yang chags skyid po'i mkhar nag.*
16. The main campsites, duly listed, are called **desa*, a term translated as 'the place of the group' (from *sde* 'the group; the tribe') or 'the place of dwelling' (from *sdod byes* 'to live, to stay, to inhabit'). They are occupied each year for one or several months by the whole community, and are divided between summer encampments *(dbyar ka *desa)* and winter settlements *(dgun ka *desa)*. Each household disposes of a fixed suitable place, which, even when vacant, cannot be occupied by others. On the other hand, secondary or satellite camps are characterised by a contingent occupation left to the individual's initiative. They are not referred to using a generic name.
17. To learn more about the local perception of territory and landscape through place names, both in Ladakh and Nepal, see Dollfus and Labbal 2009b, and Smadja 2009.

18. The expression "possédé-par-la-parole" comes from Le Moel and is quoted by Collignon (1996: 45).
19. For a broader picture of this period as seen by a sedentary farmer from Lower Ladakh, see Dollfus (1989: 67-68).
20. For the stories recounting how salt was 'found' at Tso Kar once the trade routes had been abandoned, see Ahmed (1999: 36-40).
21. Kharnak people began encroaching on Rupshu's land from the west a long time ago.
22. Written Nalbu Khar by Ahmed (2002: 47)
23. A *lug sgal* is a saddle-bag made for sheep and goats which can be filled with fifteen to twenty kilograms of salt.
24. The removal of salt generally takes place during September or October. It depends on the amount of rainfall during the summer and how quickly the water in the lake dries up.
25. On this rural-to-urban migration, see Dollfus 2004 and Goodall 2004 a and b.
26. With regard to this issue, the birth pollution attitude is a telltale sign. While Ladakhis have fewer qualms about offering a pregnant couple a room in the newly-built—squatter settlements, there are clear and consistent reservations about offering pregnant women rooms in more traditional houses and neighbourhoods. It emerges that people are anxious about the birth pollution they might be exposed to if their tenants go into precipitous labour and deliver at home or the pollution she would bring back upon her return from the hospital. Indeed, they are less concerned about the local deities they might offend in these new settlements, known as 'housing colonies'. See K. Gutschow and Dr. Padma Dolma, unpublished.
27. Aggarwal (2004: 62).

2.2.4.

From Home to Abroad, and Back Again: The Expanding Territories of Nepalese Migrants

Tristan Bruslé

Introduction

Everyone who has been to Nepal in recent years has noticed how foreign employment has become a major social phenomenon. Every day, more than 1600 international migrants fly from Kathmandu to the Gulf states and Malaysia. At least equal numbers of Nepalese head for India, though no accounts of these are kept. It is difficult to estimate the number of Nepalese living abroad: depending on the authors, it ranges from 1.3 to 3 million. Thieme (2006) asserts that "considering the estimations for India and the Gulf states, the percentage of the total population absent from Nepal would be between 6.5 and 14.7 per cent, compared to the officially recorded 3 per cent". From 10 to 24 per cent of the population over the age of 15 would therefore live out of Nepal (*Ibid.*).[1]

International labour migration has never been so high and so highly publicised. In 2011 funds sent home by migrants represented approximately four billion dollars, as much as 22 per cent of the GDP (World Bank 2008). Any political, development or economic actor realises that temporary migration, referred to as 'poverty exportation' by Harka Gurung,[2] is one of the only booming industries in the country. Migrants are like commuters who travel between their place of work and their place of origin: the common way of talking about temporary migration to India is 'to come and go' (*aune-jane*). In Qatar the same expression is used, although the temporality of the moves differ greatly. It shows, however, continuity between 'traditional' labour migration to India and new forms of long-distance migration.

On a more individual level, being a migrant means living in a foreign place, far from one's native village, in a very different environment. Migrants usually describe their exile experience as painful (*dukha*). A farmer spending at least half of his life abroad must attribute some importance to the place where he lives as a migrant. I would like to put forward the hypothesis that to understand Nepal's socio-spatial developments, investigations need to be carried out at other places where migrants stay, that is their 'migration territory'. The migration territory can be defined as places abroad which migrants feel familiar with, and where they recognise landmarks. The migration territory is meaningful and is appropriated so that migrants do not feel lost.

To be able to survive or 'to feel at ease', migrants create social, mental and material spaces where they feel comfortable. The concept of *topophilia*, which "can be defined broadly to include all of the human being's affective ties with the material environment" (Tuan 1990) will be used. Since these ties "differ greatly in intensity, subtlety, and mode of expression" (*Ibid.*), I will take into account both the material and affective dimension when migrants build a territory. Territory shall be defined as a portion of land that is appropriated, not in the legal sense but appropriated in the sense that people have special links with it, *topophilia* among others.[3] It is made up of places (on different scales) that are linked together. The territory includes two major dimensions, a material one and a mental one. According to the forms and degree of appropriation, there are different kinds of territory, which have different values, particularly regarding the feeling of belonging. The more appropriated it is and the greater the achievements, the more the territory represents an important layer of spatial identity.

My focus, therefore, will be on how a migration territory is built, how migrants appropriate it and how they regard it. Then, I will try to view the possible spatial consequences in Nepal. Thus links will be established between the two spaces migrants live in.

This paper is based on an eight months' fieldwork spent in India between 2001 and 2003 for my Ph.D. thesis,[4] and on a five-week field trip to Qatar in 2006 and in 2008. Emphasis will be laid on far western Nepalese migrants in India and when possible, comparisons will be made with Qatar.[5]

1. Living abroad: the appropriation of places, from public to private spaces

Whereas life in far western Nepal is spent typically in a rural environment, going to India and to Qatar means living in a very different world, in a context of both compulsion (*badhyata*) and habit (*calan*) that drive people away from their home. A strong 'culture of migration'[6] exists in far western Nepal, as in other parts of Nepal (Sharma 2007): going abroad for a few months or years has become normative.[7] Nonetheless, migrants in India feel that their situation is the opposite (*ulto*) of what it should be (Bruslé 2007). Whether in Qatar or in India, migrants discover new places they have to become familiar with in order to cope with the burden of exile and of the separation with their native places.

In towns in Uttarakhand, Delhi and Doha

The division between the mountains (*pahad*) and the plains (*madesh*) is one of the major vernacular spatial dimensions through which migrants divide the Indian space. These two categories embody different social worlds and different migration patterns, whereas Qatar is included in the broad *bidesh* category.[8] The statement by Pfaff (1993: 99) about Bajhangi migrants is valid with regard to our case study:

"members of the rural societies in remote mountain areas have great difficulties in coping with the increased complexity away from their local context. Depending on their class position, their disposition to handle the outside world varies".[9] Psychological factors and a person's level of self-confidence influence the representation of other places. The perceived distance to the social world migrants are about to enter may govern their choice of destination. And lastly, the 'culture of migration' and the tales told by returnees help would-be migrants to figure out what the world to be discovered is like.

Seasonal or temporary labour migration from far western Nepal to Uttarakhand is nothing new. For at least two or three generations, male peasants have headed to Kumaon and Garhwal during the agricultural off-season period (Dahal *et al.* 1977, Winkler 1979, Pfaff 1993) to work as porters, servants, road workers or miners. The niche assigned to the Nepalese is always at the bottom of the social ladder. Although many migrants work in the same town as their father does or used to, showing that even migration destinations are inherited, some of them also venture into towns where they have no particular contact. It is easy for any Nepalese migrant to go to Uttarakhand and roam the streets of any town seeking work as a coolie: there is no need for any social capital to become a porter. It is a totally different story regarding jobs through contractors who have a very bad reputation: belonging to a network guarantees working with a reliable and honest boss. Nevertheless, and contrary to what migration theories usually assert, belonging to a network is not compulsory. For board and lodging, *dharamsalla*[10] and cheap hotels are available, whereas working as a coolie does not require any introduction: being Nepalese and showing willing is sufficient. Beyond migration networks, the migration territory also exists as a potential space. Even migrants who have never been to a Uttarakhand town before dare to venture there because there is a common feeling and conviction that Uttarakhand is a familiar place where Nepalese people find a job easily. The migration territory in the Himalayan Indian state is an open one.

Contrary to Kumaon and Garhwal, further away destinations such as Delhi and Doha (Qatar) are relatively closed spaces with regard to their access. To reach these towns, there are solid psychological, financial and social barriers and not all migrants overcome these. People from Mid-Western and Far-Western Development regions usually do not have the social and financial resources to undertake such journeys to the Gulf states.[11] For others, the way of securing a job contract is paved with pitfalls. For first-time migrants, shady agents (*dalal*) and not-so-honest manpower agencies are the only paths available. Full of high expectations, many of them, either gullible or simply too trusting in anyone end up being cheated and robbed of their money. Enjoying solid social capital and having some knowledge of how the manpower agency market works are real advantages for migrants who want to step beyond the traditional pattern of migrating to India. For mountain dwellers, to face a very different world is an ordeal that not all are prepared to face. As a matter of fact, it happens that migrants who used to work in the hills of Uttarakhand go to the plains at least once to see what it is like. Most of them never repeat the experience as they find it difficult to adjust to (and understand) the completely unknown space they find themselves in. In Dehli, it is essential to belong to migrant networks that help men to find a place to live and a job as watchman (see Thieme 2006). Indeed, networks shape the way people perceive the city and ultimately determine their knowledge of the town.

Building a territory in public spaces

Although dominated when abroad, Nepalese migrants do manage to build individual and collective territories in the public spaces of the towns where they work and live. These territories, which are in no way established by legal appropriation but by appropriation through practices, reflect the subordination of migrants and their varying ability to leave a mark, other than a temporary one, on the towns where they work.

In Uttarakhand, coolie is the major ethnic niche for Nepalese workers. As coolies, they interact with local Indian people, particularly tourists and wholesalers. As their job does not entail a continuous day's work, once they have carried out an errand, coolies gather at particular places in town. These areas, which are always the same, are strategic places where Nepalese migrants know that would-be clients will come. These places, such as old and disused bus stations, can be seen as being appropriated and, as such, are part of the migration territory (Photo 1). Contrary to the private space that is often transformed, however slightly, such places within the public space are not altered. At night, when migrants go back to their rooms, there is no trace left of their passage through them. In Mussoorie for example, Nepalese migrants know their landmarks well, and these are places where they can stop for a while, have a sleep or simply chat with fellow compatriots. On a town scale, this kind of territory is invisible to others and as such can be considered as "heterotopias" or "other spaces" (Foucault 1984) only meaningful to a particular category of people, in our case Nepalese migrants. Due to their position of inferiority, the Nepalese choose not to flaunt themselves or draw attention to themselves. These places, situated within the town but in marginal areas, reflect the social position of migrants, who are a vital cog in the economic machinery yet destined to remain outside local society. The marginal position of Nepalese migrants is also reflected in the *ethnic* restaurants run by fellow compatriots (Photo 2). In Pithoragarh for example, about eight to ten such restaurants are scattered over the centre of town. These cheap food outlets serve typical Nepalese food and mainly cater for migrants from the far western region. They are a real part of the migrants' territory and serve as shelters where migrants exchange news about work, using the *pahadi* dialect, and where one can listen to popular Nepalese music. They contribute to the sentiment of 'feeling at ease' abroad, in a place where for a short while the status of foreigner is forgotten. Be it at work or during their leisure time, migrants make foreign places their own by moving from one point to another. They become familiar with ordinary places thanks to their repeatedly frequenting places year after year. However, the appropriation of towns is strongly linked to work, contrary to the village territory which is inherited and also linked to daily routine.

In small towns in Uttarakhand migrants quite easily get to know the geography of places, whereas in Delhi the places they really know, where they feel comfortable, are limited to the area where they work. When I asked migrants about their knowledge of the towns, the question made them feel awkward. For migrants, the notion of wandering through (*ghumnu*) unknown areas to discover new places or simply to escape from the daily routine is unusual. As one old man said: "what is the use of wandering? Only fools (*pagal*) wander". Given the non-stop working week of the Nepalese in the Indian capital, who work as security guards and car washers, they have very little time in which to get to know the town. Moreover, the point of discovering new places, which may especially be the case for younger migrants, is not the major preoccupation for them all, especially when one has "to pay even for a glass of water". On mental maps (See Gould and White 1992) drawn by migrants, their living space in the Indian capital can be seen as a bounded area. When asked to draw a map of Delhi, migrants working in East Delhi, almost only include places they know through their daily personal practices. The Delhi of migrants is made up of several large lanes and encompasses, for younger migrants, the cinema halls where they sometimes hang out. *Hauts lieux* or symbolic places well known to Indians as representing the Nation, are seldom drawn: Nepalese migrants do not really belong to the mainstream city.[12] Only a few of the migrants met had visited India Gate, Rashrapati Bhawan or Connaught Place. Whereas migrants express an interest in seeing the Palace of the former king in Kathmandu or the famous Pashupati temple at least once, visiting places in Delhi is not considered to be of any interest.

Although in 2008 the Nepalese were the first foreign community in Qatar, the traces they have left on the city are less visible than other South Asian nationals.[13] There are at least six Nepalese restaurants in Doha (Photo 3).[14] The oldest one, Nepali Bhansa Ghar, is a particular landmark, not for all Nepalese but for the ones working in the service sector around the city centre. All its customers are middle-class Nepalese and the decor reminds them of their country. In a way, there is a folklorisation of Nepalese culture, since carved wooden objects and symbols of Nepal hang on the walls. These kinds of restaurant, known by migrants as a place to eat 'identity food', are part of the migration territory. As in Uttarakhand, they are places of some importance where migrants dare to venture, contrary to other places such as shopping malls or exclusive shops where they seldom go, feeling that 'they are not fit for these places'. This idea has been reinforced by some private and public institutions that claim that bachelor migrants are *persona non grata* in certain places on Fridays.

This need for migrants to meet up, when possible, is particularly well illustrated in Doha. Near the central area of 'National', an immigrant inhabited area, there is an open garden called the 'Nepalese market' or *Nepali chowk*. On Fridays, Nepalese migrants gather by the thousands, to not only make phone calls or send remittances but also to buy Nepali and Hindi CDs and DVDs and other items brought over from Nepal. Dozens of men, who have pirated CDs and DVDs during the week, sell them on the black market. Business is thriving in the packed garden, as migrants gather for food and drinks. This temporary place, no trace of which remains once the workers go back to their camp, is the main meeting point for the Nepalese scattered in and around Doha. It definitely is a place of some importance where they can find an atmosphere similar to the one at Ason Tole in Kathmandu during the rush hour.

Photo 1. Nepalese migrants from Bajhang always hang around here, near a furniture wholesaler who hires them. This is part of their territory. Pithoragarh, India, 2002.

Photo 2: In this cheap food outlet migrants can get typical Nepalese food which they eat among fellow compatriots. Pithoragarh, India, 2003.

Photo 3: Nepalese restaurants in Doha are far more ornately decorated and furnished than in India. As a kind of folklorisation trend, Nepalese artefacts are arranged to create a Nepalese atmosphere, whereby workers identify more readily with the restaurant. Doha, Qatar, 2006.

Photo 4: A worker in a labour camp has written his name above his bed to make it more personal. Customising one's private space is also a way of making it more intimate and of establishing landmarks in a foreign land. Doha, Qatar, 2006.

Nepalese migrant territories abroad are almost completely determined by employment. Due to their foreign status, territories built by Nepalese migrants in public spaces are all but exclusive. They are not built around legal institutions, nor do they have to be protected from locals or other intruders. They have a temporary dimension, hence they only exist in the daytime, and leave little trace on the city's landscape. The extreme visibility of migrants (at least in Uttarakhand and in Doha) contrasts with the invisibility of migrant territories once they leave the public space. One has to enter migrants' rooms to discover more arranged, exclusive and shared places.

Private spaces: rooms as refuges

Building private spaces is a way of finding refuge from an environment described as inhospitable and where migrants feel they are forced to live.[15] Private spaces are places where migrants' actions are far more concrete and effective compared to what they are in public places. The second 'shell' (Moles and Rohmer 1988) of their territory, where migrants find a varying degree of intimacy, is their dwelling place. There are striking differences in the way migrants live (*habiter* in French) in India and in Qatar.

In India, and especially in the capital, the strength of social networks, based on caste and village belonging, explains how migrants live together. In towns in Uttarakhand and in New Delhi, they occupy rented rooms, sometimes in former stables, as in Pithoragarh or in Mussoorie, or huts they build themselves like the ones in the Delhi slums. Sometimes they live in lean-tos adjacent to new buildings where Indians live. The difference in lodgings between Indians and Nepalese reflects the subordination of the latter. Although migrants spend year after year in India, their living conditions seldom improve, except for a minority of them who manage to step out of the coolie or watchman niche they usually get stuck into. The investment in rooms is minimal and affective ties to them are almost non-existent: they only have functional attributes. The word *dera* (temporary rented accommodation) is used, never *ghar* (home). Beds are a piece of cardboard or at best they have a wooden frame. Clothes hang on nails or on a rope, and cooking is done in a corner of the room. Only a few migrants, especially established couples (as in Nepalese villages), make an effort to fit out their room, painting it in bright colours and hanging Bollywood-type posters on the walls. To sum up, in India, living and lodging conditions are not so different from the ones in the village where any privacy is almost completely impossible and basic comforts non-existent. The bareness of migrants' rooms in India recalls the bareness of interiors in Nepalese houses. Private places do not really present migrants with a new life.

In Qatar, spatial segregation is at its greatest since all labour camps are confined to areas reserved for unskilled workers, far from Doha city centre (Bruslé 2010a).[16] This relegation confines workers to special areas from where they can only escape on Fridays, if buses are available, which is not always the case. The zones where labour camps are located, such as the main one called the Industrial Area, contrast with the rest of the country. Roads are unpaved, badly maintained and unlit at night-time. No trees are to be seen anywhere. Scores of lorries, trucks and buses pass along the roads there since many businesses, garages and factories share the area with labourers' dwelling places. Dust flies everywhere. Labour camps come in various forms, whether made up of two-three storey buildings or of mobile homes. Depending on the size of the camp and the number of workers, four to twenty men live in each room, sometimes with three-tiered beds. As migrants stay for at least two or three years[17] and as they earn more money than in India, they invest more in their room. Migrants get together to buy TVs, DVD players and hi-fi sets, and sometimes computers. The walls are decorated with posters or pictures from magazines. If the difference with rooms in India is striking, the most astonishing feature of these rooms is the way beds are arranged as the first 'shell' of migrants' territory. In crowded rooms, the only way to build oneself a private space is around one's bed. Sometimes the name of the migrant is written on the wall over their bed or on the bed itself: this represents an active process of appropriating space, a way of marking a frontier between them and the others, that is, to say: "this is my place" (Photograph 4). More commonly, family photographs or any kind of decoration hang on the walls. Those who are more ingenious set up a television and DVD player for their personal use. Customising the bed surrounds makes the bed the ultimate and practically the only place for any intimacy, especially when hidden by a curtain. This is their basic space though its borders can actually be crossed, given that Nepalese migrants' often seek each other's company even when watching a movie or reading a newspaper. The bed is never a closed territory. Although the private space is reduced to a minimum, it is not completely exclusive. It also denotes a quest for intimacy and comfort that are relatively new things, compared to what villagers experience in their own houses where private rooms and beds are not always a regular fixture.

The value of the migration territory depends on achievements

The migration territory develops during the day-to-day routine, at work and in places where migrants live. Although spatial segregation is a major dimension of migrants' conditions, the latter manage to make the foreign space their own, by carving out more or less elaborate spatial niches, depending on the institutions that control their migration and on their will to do so.

The feeling of being exploited and despised, the fact that working in India merely enables most migrants to pay back loans and to satisfy only their family's basic needs attributes little value to the migration experience. It is remarkable that contrary to what is described in Hyderabad (Ali 2007) and in Mexico (Kandel and Massey 2002), where migration is also deep-rooted among the male population and seen as highly positive, Nepalese migrants depreciate the journey to India. Thus, the migration territory is highly devalued compared to the ideal village territory. These mental dimensions of the migration territory reflect the type of working relationship migrants have with their employers, just as places where migrants live and work reflect their social position. Working in India, as in Qatar, means one's status is defined by one's work, no longer by belonging to a caste or by village prestige. This is particularly felt by upper-caste migrants (Brahman and Chetri) for whom, according to Bista (1991), manual work is undignified.

2. Broadening the territories of Nepalese migrants

The links between the places of origin and of migration correspond to the creation of transnational spaces.[18] The territory where migrants spend their lives, and on which the migrants' families rely, transcends the borders of the nation-state and is thus made up of two interrelated parts. The native territory is inherited, carries high values related to family legacy, clan history and belonging. It is the spatial foundation of family life. As far as the migration territory is concerned, it is mainly seen as being useful while conveying values of exploitation and hope, and change (regarding caste-based discrimination for example) and modernity. If we consider that migrants live on two kinds of territory, one despised and the other highly valued, we also have to understand the links that exist between the two of them.

How are new values, or values learnt and enhanced abroad taken into account in the migrants' worldview? Do they prompt a new discourse about the migrants' personal destiny or vision of themselves? How is the migration territory, which represents both relegation and the freedom to discover new spaces, integrated into migrants'living space? How does the fact that migrants live on new territories or in new places hardly similar to the ones at home influence the village territory?

Views and perceptions of the world

In exile, foreign and distant countries no longer only exist in people's imaginations, but migrants are confronted with the materiality of different places, people and occupations. What was merely imagined becomes a real-life experience. The notions of 'home' and foreign land can be re-evaluated.

Migrants to India have a blurred notion of 'abroad' (*bidesh*). According to international law, India is a foreign country but the feeling among Nepalese migrants is that living in India is not really living abroad. The cultural proximity is strong. Modes of behaviour are similar to those in Nepal. Even if for some successful migrants, India can become 'like a home', ambivalent feelings towards India do exist, as demonstrated by a night watchman in Dehradun: "I do not feel that India is outside (*bahira jasto*). To go to a foreign country (*bahira desh*), one needs a passport and visa. But 200 rupees are sufficient to go to Delhi. There is no need for a passport or visa. No one asks us anything. It is not like a foreign country (*bidesh*) here. Yet it is a foreign country, it is not my country (*svadesh*). But I do not have the feeling that India is another country." Even though the status of foreigner governs the lives of migrants in India, feelings of gratefulness towards India also exist. This is particularly the case for two categories of migrant: the ones who reach higher goals, such as running small restaurants or securing a permanent job, while at the bottom end of society, the families who have the impression that being in India enables them to survive, unlike in Nepal where 'only water is available'. In both cases, India becomes home. When links with the village have been severed, when there seems to be no intention of ever going back to their homeland, temporary migrants become permanent migrants and national belonging is bound to change. Nevertheless, the majority of the far western Nepalese in India remain attached to their motherland, even if they are grateful to India.

One way of finding out what *bidesh* represents in children's imagination is to make them draw mental maps. Mental maps of Nepal as seen by young boys in India show that the migrants' world revolves around their district of origin and their place of work. Even if Dhan Raj K. (Fig. 1) knows some Nepalese districts and the names of some Indian states, his spatial knowledge is greatly influenced by temporary migration. There is a continuous flow of information between his native district of Bajura and New Delhi where he stays with his family. On the contrary, Kathmandu is represented as an island in Nepal, surrounded by a blank sea of unknown places. Dhan Raj's map symbolises the remoteness and the periphery of the far western region in Nepal and the strong links it has with India, as portrayed by its inhabitants. The second map, drawn by Tranesh D. from Bajhang, shares some elements with the first one. His representation of Nepal is solid, as shown by the division into numerous administrative entities (Fig 2). However, his knowledge of Nepal is determined by the travelling he does between his home district and India, through the town of Mahendranagar in particular. Both maps show that spatial imagination is the outcome of a national geographical ideology, learnt at school, and of practices, particularly during the migration process.

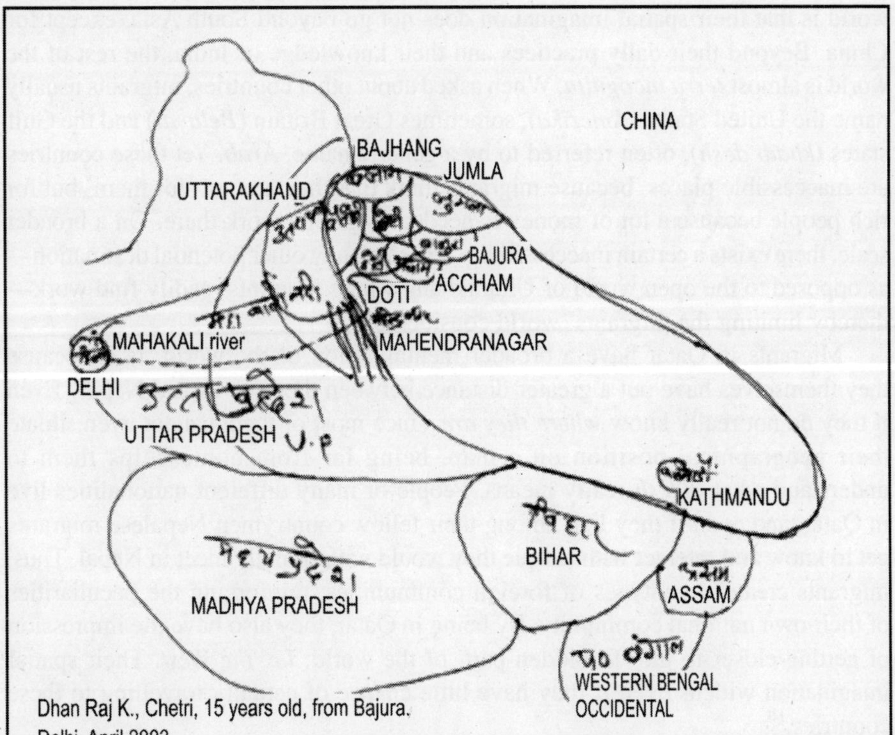

Figure 1. Living space of Dhan Raj K.

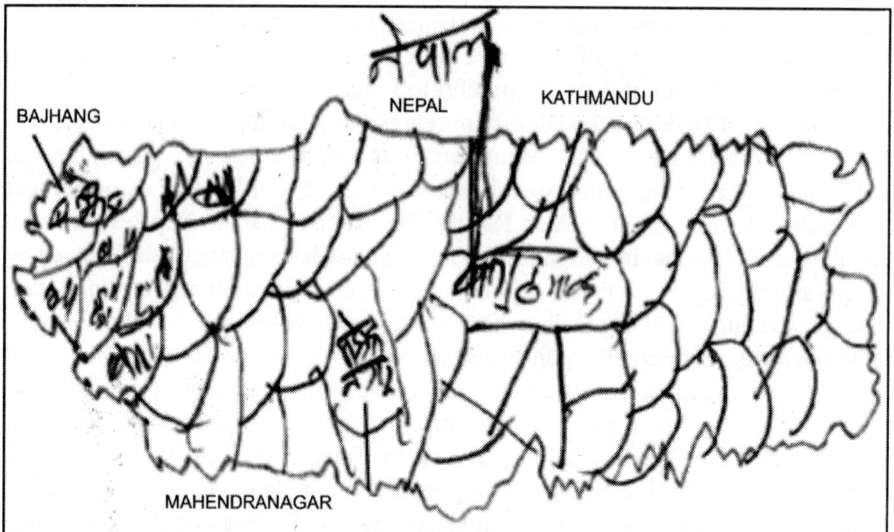

Tranesh Dal J., Brahman, 15 years old, from Bajhang.
Pithoragarh, August 2002.

Figure 2. Nepal as seen from India by Tranesh Dal J.

In India, the striking feature of most far western migrants' perception of the world is that their spatial imagination does not go beyond South Asia, except for China. Beyond their daily practices and their knowledge of India, the rest of the world is almost *terra incognita*. When asked about other countries, migrants usually name the United States (*Amerika*), sometimes Great Britain (*Belayat*) and the Gulf states (*khadi desh*), often referred to by a generic name, *Arab*. Yet these countries are inaccessible places, because migrants think that they are not 'for them' but for rich people because a lot of money is needed to go and work there. On a broader scale, there exists a certain inaccessibility regarding any other potential destination—as opposed to the open world of Uttarakhand where migrants readily find work—thereby limiting the migrants' world considerably.

Migrants in Qatar have a broader mental vision of the world. Is it because they themselves have put a greater distance between themselves and Nepal? Even if they do not really know *where they are*, since most of them cannot even situate their geographical position on a map, being far from home helps them to understand what *bidesh* really means. People of many different nationalities live in Qatar, and even if they live among their fellow countrymen Nepalese migrants get to know and interact with people they would not normally meet in Nepal. Thus, migrants create stereotypes of foreign communities, reinforcing the peculiarities of their own national community. By being in Qatar, they also have the impression of getting closer to the 'forbidden part' of the world, *i.e. the West*. Their spatial imagination widens even if they have little chance of actually travelling to these countries.[19]

Individual conceptions: 'migration changes men'

The migration experience is not lacking in meaning. It alters men's personal conceptions of life and can therefore influence individual projects. Temporary migration to India, and in particular to Uttarakhand, can be called seasonal migration, as the agricultural calendar is the framework from which only a few migrants—those with sufficient labour force at home—can escape. Migrants therefore continue to work in the fields when they return home. On the contrary, working in Delhi as a night watchman or in Qatar provides a permanent break from agricultural work. When boys come home, they may no longer want to plough the land or to get involved in any activity synonymous with hard labour. As in Sainik Basti (Western Nepal), there is a certain disgrace associated with doing certain kinds of manual labour, as "the returnee would feel ashamed to do the same work in Nepal as he does abroad" (Thieme and Wyss 2005: 84). One successful migrant in Dullu (Dailekh district), who was a receptionist at a Valsad hotel (North of Mumbai), managed to buy a house in the village. However, when he came back for a few weeks every year, he was reluctant to plough, to handle the sickle or the hoe. He said he no longer wanted to work as a farmer. He preferred to roam the village in a crisp white shirt. While in Qatar, men have access to a certain form of modern technology, which is not accessible in India or in their village. Brand new cars, roads and buildings create a sense of luxury. In labour camps, there is running water, electricity and air conditioning with none of the associated outages. Migrants from rural areas often discover a modern (*adhunik*) world that fascinates them. A few months after arriving in Doha, the first thing migrants purchase is a cell phone, which is often used to call their family in Nepal. Not unlike teenagers in France for example, young men spend hours talking about their phones, swapping them, trying new applications and so on. As we have seen, a longing for modernity is reflected in the returnees' patterns of consumption. Buying a digital camera and a hi-fi system, even if there is no electricityin their village of origin, is seen as a must. Even if migrants do not benefit from all the advantages a modern country has to offer, they rapidly acclimatise and then tend to consider Nepal as backwards. They appear to be strongly attracted to towns or places with 'modern' amenities, such as shopping malls.

Perhaps it is too soon to advance the hypothesis of an urbanisation of behavioural patterns and aspirations, but distant migration certainly has a role to play in explaining changing practices in Nepal.

Migration, modernity and land

When migrants come home, they bring back "social remittances", that is "ideas, behaviors, identities, and social capital that flow from receiving-to-sending country communities" (Levitt 1998: 926). Migration may lead to a change in the relationship with the place of origin and to a shift in territory for migrants and their families.

For most migrants, being abroad means first and foremost being cut off from their village and from the agricultural work that, among other things, used to shape their relationship with their native place. There is a temporary breach of ties with the earth, when the ex-peasant becomes a worker in a foreign land. If we consider that "topophilic sentiments" are made of "physical intimacy, of material dependence" and "that the land is a repository of memory and sustains hope" (Tuan 1990: 97), farmland is part of the peasant identity. Links are established with the territory through work with the plough and the hoe. A peasant's muscles are called upon for farm work so that the territory and village landscapes may be considered an extension of their body. When the migrant returns home, it is as if everything gets back to normal again. However, as mentioned earlier, it does happen that the temporary break from the land becomes permanent. Migrants have learnt new ways of life, they have been imbued with new values relating to agriculture and rural life, in particular the Nepalese working in the Gulf who have been cut off from their inherited way of life for years. Their relationship with the land has been re-evaluated and very often, the ancestral land is no longer of much value. Migrants are seemingly impregnated by urban values, so much so that they rate villagers as backward. Peasants' self-devaluation makes them seek an activity of greater value, whereby the temporary split with the village territory then becomes a permanent shift in their place of residence.

Among far western migrants, the dominant discourse about land and agriculture underlies the vital links migrants wish to maintain with their land. Buying farmland remains the main objective of migrants, be it in the Tarai or in the hills. To a large part of the migrants interviewed in Qatar, they do wish to pursue their agricultural work, which is the most valued of all. The desire for an urban lifestyle with all its modern amenities does not seem to pervade the far westerners' mentality because they remain intimately attached to their land even though it does not provide them with any security. Elsewhere, the lack of enthusiasm for ploughing, added to the constant contribution of remittances, leads to peasants abandoning land. In Sikles, a village in Kaski district where "more than 48 per cent of the total economically active male population was away from the village" (Khanal and Watanabe 2006: 36), nearly half of the irrigated land (*khet*) and 40 per cent of non-irrigated rainfed terraces (*bari*) are left uncultivated. Although there are various reasons for abandoning land (outmigration, shortage of labour, schooling, poor yields in the agricultural sector, government policies on ownership and land use rights), the migration factor is enough to explain the change in village landscapes.

Secondly, the question of the links between migration and the transformation of village economies and lifestyles needs to be assessed.[20] The scientific debate has shifted from the study of the effects of migration on development to the study of the multilateral links between international migration and local transformation (Taylor 2006). Transformations in villages are not only due to economic remittances but also to the absence of the male migrant, particularly for women who become the head of the household (Kaspar 2005). The question of social mobility in relation to the caste system and the social position in the village also needs be addressed. Unfortunately, given the current stage of research, it is impossible to tackle this question. As regards social remittances, it is difficult to grasp how the values and ideas brought back by migrants actually permeate their family.[21] However, one can appreciate why monetary remittances are used for several major expenditures. Once loans have been paid off, returnees adopt common behaviour patterns seen throughout Nepal, just as in Sainik Basti, where "money is mostly invested in physical capital like a new house or a motorcycle, in another migration, or in the education of a person's children" (Thieme and Wyss 2005: 84). In Raghunathpur VDC (Dhanusa district), where a third of migrants have used remittances to buy land or build a house, migrants working in the Gulf are the first to get a *pakka* house[22] and to fit it out with modern facilities (toilet, tub-well for drinking water) (Karn 2006). Although Karn states that these investments are unproductive, one should consider the multiplying effect of spending in the village and the effects on the community at large (Taylor 1999, 2006).

Finally, international and internal migrations would appear to be interlinked. In a context of a high level of internal settlement migration from the mountains to the hills,[23] the desire to own land in the Tarai is very strong. It is considered a great achievement for migrants to buy a piece of land in the plains. Even though working in India seldom enables migrants to achieve such an aim, this kind of project is every migrant's dream. Night watchmen in Delhi may manage to save enough money to buy a few *kattas* of a paddy-field in the Tarai. Investigations in Qatar among some old-time migrants revealed that this pattern might be different. Along with the desire to quit farming, expressed by many men in Doha, is the desire to leadan urban lifestyle. However, migrants' fascination with cities definitely reveals a generation gap as illustrated by the following story. Bhim Bahadur S. has been a receptionist at a Doha hotel for 13 years. He is also a casual private manpower agent.[24]Thanks to his various jobs, he has managed to save large sums of money he has given to his father to buy a plot of land in Kathmandu. He was also planning to buy buses and to run a transportation company. But his father decided otherwise and bought some paddy-fields near Biratnagar. In the end, Bhim Bahadur went alone to Kathmandu to buy land where he plans to build a house to let.

To fully grasp the actual effects of migration on Nepal's economic growth requires more than just one sketchy life story. The very high level of remittance definitely contributes to the sustainability of rural Nepal but it does not benefit all regions equally (Central Bureau of Statistics 2004). The major risk is a widening gap between regions, and particularly between urban and rural areas, and of course between social classes.

Conclusion: transnational spaces for migrants and their families

The migration territory is not an enclave, it is not landlocked. As part of the overall space where the peasant-migrant lives, it is linked to their original territory, the one that they should never have left. Transnational space is built by the movement of people, by flows of money and information. Daily life spans the place of one's ancestors and income-earning places. Life is sustainable in the former thanks to the latter: complementarity exists between these two spaces with opposing values. As far as migrants in the Gulf are concerned, the way to bridge the gap between these opposing values seems to be a change in their way of life and place of residence. This surely contributes to real social changes that occur in Nepal, particularly among young adults (See Liechty 2002).

The migration territory, made up of particular places, can be described as being the opposite of the village one. In India, all the values attached to it are charged with exploitation, contempt and survival. The migrants' discourse stresses the constraints and a lack of choice: the migration territory is seen as functional, as opposed to the village territory which is associated with emotions and feelings of belonging. No temples or religious places are to be found in migration. The family genealogy is not rooted abroad. Since it is a temporary (*asthayi*) territory, there is minimum investment. Places are appropriated to a minimum degree, simply to make life bearable. As most migrants cannot really improve their standard of living, the migration territory in India is not really a space of hope. It is a territory where migrants are cut off from their land and traditional activity (agriculture), where their control and power over the transformation of the space is reduced to a minimum. The migration territory, contrary to the village territory, is no longer "an extension of the organism, which is marked by visual, vocal and olfactory signs" (Hall 1971). If appropriation does exist, it does not lead to any major transformation. Men have little power over their places of residence.

The migrants' discourse and actions give rise to a paradox. The migration experience and associated places are despised (in Qatar or India), yet time spent away from the village actually tends to drive men away from the village and agriculture. At the same time, being abroad tends to strengthen one's love for one's own country and migrants all say that the best place on earth is their native village (*aphnai thaun ramro lagyo*). In the end, living in a foreign country does not seem to lead to a complete reassessment of the migrants' loyalty to their country and village, which remain the basic places of belonging.

REFERENCES

Ali, S. 2007. "Go West Young Man': The Culture of Migration among Muslims in Hyderabad, India', *Journal of Ethnic and Migration Studies*, 33(1): 37-58.

Bista, D.B. 1991. *Fatalism and Development. Nepal's struggle for modernization*, Calcutta, Orient Longman.

Buslé, T. 2007. 'The World Upside-Down: Nepalese Migrants in Northern India', *European Bulletin of Himalayan Research*, 31:172-183.

– 2010a, 'Living In and Out of the Host Society. Aspects of Nepalese Migrants' Experience of Division in Qatar', *Forum: Qualitative Social Research*, vol.11, no. 2, available at: http://www.qualitative-research.net/index.php/fqs/article/viewArticle/1482/2992

– 2010b, 'Who's in a *labour camp*? A Socio-economic Analysis of Nepalese Migrants in Qatar', *European Bulletin of Himalayan Research*, vol.35-36: 154-170.

Central Bureau of Statistics. 2004. *Nepal Living Standard Survey 2003/04. Statistical report*, Kathmandu, National Planning Commission, HMG Nepal, available on www.cbs.gov.np

Dahal, D.R., A.E. Manzardo and N.K. Rai. 1977. *Land and migration in far-western Nepal*, Kirtipur, INAS.

Debarbieux, B. 1999. 'Le territoire : histoire en deux langues. A bilingual (his-)story of territory', in Chivallon, C., P. Ragouet *et al.* (eds.), *Discours scientifiques et contextes culturels, géographies françaises et britanniques à l'épreuve postmoderne*, Talence, Maison des sciences de l'homme d'Aquitaine,pp.33-46.

Di Méo, G. 1998. *Géographie sociale et territoires*, Paris, Nathan Université.

Foucault, M. 1994 (1984). 'Des espaces autres, hétérotopies', in M. Foucault (ed.), *Dits et Ecrits 1954-1988. Tome IV, 1980-1988*, Paris, Gallimard, pp. 752-762.

Gould, P. and R. White. 1992. *Mental Maps*. London and New York, Routledge.

Hall, E.T. 1971. *La dimension cachée*, Paris, Editions du Seuil.

Kandel, W. and D.S. Massey. 2002. 'The Culture of Mexican Migration: A Theoretical and Empirical Analysis', *Social Forces*, 80(3): 981-1004.

Karn, S. 2006. *Coping with Risks and Managing Livelihoods: Migration in Rural Nepal*, Dhaka, South Asia Migration Resource Network, Refugee and Migratory Movements Research Unit, Occasional Paper 13.

Kaspar, H. 2005. *"I am the household head now." Gender Aspects of Out-Migration for Labour in Nepal*, Kathmandu, Nepal Institute of Development Studies.

Khanal, N.R. and T. Watanabe. 2006. 'Abandonment of Agricultural Land and Its Consequences. A Case Study in the Sikles Area, Gandaki Basin, Nepal Himalaya', *Mountain Research and Development*, 26(1): 32-40.

Levitt, P. 1998. 'Social Remittances: Migration Driven Local-Level Forms of Cultural Diffusion', *International Migration Review*, 32(4): 926-948.

Liechty, M. 2002. '"Out here in Kathmandu": Youth and the contradictions of Modernity in Urban Nepal', in Mines, D.P. and S. Lamb (eds.), *Everyday life in South Asia*, Bloomington and Indianapolis, Indiana University Press, pp.37-47.

Massey, D.S., J. Arango, G. Hugo, A. Kouachi, A. Pellegrino and J.E. Taylor. 1993. 'Theories of International Migration: A Review and Appraisal', *Population and Development Review*, 19(3): 431-466.

Moles, A. and E. Rohmer. 1998 (1972). *Psychosociologie de l'espace. Textes rassemblés, mis en forme et présentés par Victor Schwach*, Paris, L'Harmattan.

Pfaff, J. 1993. 'Migration under Marginality Conditions: the Case of Bajhang', in Interdisciplinary Consulting Group for Natural Resources and Infrastructure Management (INFRAS) and Interdisciplinary Analysts (IDA) (eds.), *Rural-Urban Interlinkages: A Challenge for Swiss Development Cooperation*, Zurich and Kathmandu, INFRAS and IDA, pp.97-107.

Portes, A. 1997. *Globalization from Below: The Rise of Transnational Communities*, WPTC-98-01, Princeton University.

Sharma, J.R. 2007. 'Mobility, Pathology and Livelihoods: An Ethnography of Forms of Human Mobility in/from Nepal', Ph.D. thesis inAnthropology, University of Edinburgh.

Taylor, J. E. 1999. 'The New Economics of Labour Migration and the Role of Remittances in the Migration Process', *International Migration*, 37(1): 63-88.

– 2006. 'International Migration and Economic Development', Turin, available at www.un.org/esa/population/migration/turin/Symposium_Turin_files/P09_SYMP_Taylor.pdf

Thieme, S. and S. Wyss. 2005. 'Migration Patterns and Remittance Transfer in Nepal: A Case Study of Sainik Basti in Western Nepal', *International Migration*, 43(5): 59-98.

Thieme, S. 2006. *Social Networks and Migration. Far West Nepalese Labour Migrants in Delhi*, Muenster, LIT Verlag.

Tuan, Y.-F. 1990 (1974). *Topophilia. A study of environmental perception, attitudes, and values*, New York, Columbia University Press.

Winkler, W.F. 1979. *The Evolution of Caste Organization in a Subregion of Far Western Nepal*, Ph.D. thesis in Anthropology, University of Wisconsin-Madison.

World Bank. 2008. *Migration and Remittances Factbook 2008. Nepal*, available at http://siteresources.worldbank.org/INTPROSPECTS/Resources/334934-1199807908806/Nepal.pdf

Notes

1. The attraction to foreign land is also noticeable regarding education; for example, between September 2006 and September 2007, the number of Nepalese students resident

in Australia rose by 504 per cent, to reach 7,500 (http://www.thaindian.com/newsportal/world-news/504-percent-increase-in-nepal-student-population-in-australia_1005694.html).

2. Private conversation, 2001.
3. The notion of territory has been studied by French geographers such as Debarbieux (1999) and Di Méo (1998).
4. I focused on Nepalese migrants from the far western region of Nepal.
5. One must bear in mind that migrants in the two countries come from different districts of origin.
6. It refers to the way migration "changes values and cultural perceptions in ways that increase the probability of future migration" (Massey, Arango *et al.* 1993: 452). "At the community level, migration becomes deeply ingrained into the repertoire of people's behaviours, and values associated with migration become part of the community's values" (*Ibid.*).
7. Migration as a norm is theoretically formalised by Portes (1997) in a context of cumulative causation of migration trends (Massey *et al.* 1993).
8. India is not included in *bidesh*. According to migrants, *bidesh* encompasses all the countries where a passport is needed. See the second part of this paper.
9. "The low stratum of the local population walks to Kumaon, while more resourceful 'Bangalories' (the name of Bajhangis who work in Bangalore) are able to fly from Bajhang to the Indian border" (Pfaff 1993: 99).
10. Places to rest and sleep with only basic comforts.
11. I hardly met anyone from here in Qatar, whereas there were many men from other parts of Nepal.
12. The question of whether it is different for internal Indian migrants from the working class is not tackled here but is worth considering.
13. Indians, Bangladeshis and Pakistanis have been coming to Qatar since the 1970s. Thus there are many more *ethnic shops* for these migrants than those that target the Nepalese who only began to arrive in large numbers at the end of the 1990s.
14. Their names symbolize their *Nepaliness*: Sagarmatha Restaurant, Himalaya Restaurant, Nepali Bhansa Ghar (Nepalese kitchen), Samsara, Nepali Chowk Restaurant and Mehaman (the Guest).
15. The discourse about the constraints that lead migrants to go abroad is predominant, even though the reasons for becoming an expatriate are far more complex.
16. The denomination 'labour camp' is the one used in the Gulf states. To be more precise, one should talk of 'labourers' camp' or 'residential camp', as workers do not actually work in these camps. The exact figures for the number of workers living in such camps are not known. For a study of a Nepalese labour camp, see Bruslé 2010b.
17. The length of the basic working contract is two years. Many workers renew their contract twice or three times, and even more.
18. The multiple flows between the village and the migration territory include the movement of men, money, information and goods. The use of new information technologies is well developed in Qatar, though almost non-existent in India.
19. Some migrants do reach the West with the help of their sponsor who agrees, in exchange for large sums of money, to take them anywhere. When the destination is reached, the migrant flees from his sponsor who does not try to retain him.
20. These are only research leads, as my fieldwork focused less on these aspects.
21. My fieldwork only concerned the destination places.
22. A house made of brick and cement.
23. The majority of Nepal's population now lives in the Taraï.
24. This man, who had learnt Arabic, acquired financial capital from his amassed social capital.